Measuring ROI in Environment, Health, and Safety

Scrivener Publishing
100 Cummings Center, Suite 541J
Beverly, MA 01915-6106

Publishers at Scrivener
Martin Scrivener (martin@scrivenerpublishing.com)
Phillip Carmical (pcarmical@scrivenerpublishing.com)

Measuring ROI in Environment, Health, and Safety

A Guide to Evaluating EHS Programs, With Case Studies

Jack J. Phillips, Patricia Pulliam Phillips, and Al Pulliam

ROI Institute

WILEY

Co-published by John Wiley & Sons, Inc. Hoboken, New Jersey, and Scrivener Publishing LLC, Salem, Massachusetts.
Published simultaneously in Canada.

For general information on our other products and services or for technical support, please contact our Customer Care Department within the United States at (800) 762-2974, outside the United States at (317) 572-3993 or fax (317) 572-4002.

Wiley also publishes its books in a variety of electronic formats. Some content that appears in print may not be available in electronic formats. For more information about Wiley products, visit our web site at www.wiley.com.

For more information about Scrivener products please visit www.scrivenerpublishing.com.

Cover design by Kris Hackerott

Library of Congress Cataloging-in-Publication Data:

ISBN 978-1-118-63978-8

Printed in the United States of America

10 9 8 7 6 5 4 3 2

Contents

Preface

The Need for this Book: Focus on EHS

If you are part of the workforce, you're directly affected by three important issues: environment, health and safety (EHS). Your employer is concerned about the environment and implements a variety of projects and programs to minimize the organization's impact on it. Leaders are also concerned about your health. The cost of employee healthcare has skyrocketed in the past decade and is greatly impacting the financial health of all types of organizations. And employee safety is no new concern. Maintaining a safe workplace has been a common goal as employers have come to understand that safety is good business.

EHS concerns are often grouped together and managed by one executive. This book is for that executive and his or her entire EHS team, as well as others who support EHS projects, programs and initiatives.

Environmental Issues

Everyday focus on EHS is a relatively recent phenomenon. When global warming first surfaced in the 1980s, it stirred up worldwide debate and concern. However, business leaders resisted the issue, recognizing the extra cost involved in addressing the causes of such a phenomenon. For example, when the United Nations formed the Intergovernmental Panel on Climate Change (IPCC) in 1989, business groups formed their own organizations, the Global Climate Coalition and the Competitive Enterprise Institute, to counter the issues and the debate about the causes of climate change. While most agreed that climate change was occurring, views differed regarding its origins. Finding solutions was not a priority to the business world at large.

Today, most, if not all, businesses recognize that climate change is an issue. The Global Climate Coalition was disbanded in 2002, and most businesses are attempting to solve the problem, resulting in a wave of sustainability, climate change and environmental projects.

From an organizational perspective, addressing environmental issues offers employees, contractors, volunteers and other members the opportunity to influence environmental sustainability through involvement and contribution. The challenge is to create the correct approach to involve these people—one that includes teaching, convincing, communicating, enabling, supporting and encouraging participation in the green process. Mandating change is rarely a recipe for success. Organizations must engage their people and position processes for successful implementation.

The sheer number of environmental projects and initiatives, as well as the fact that not everyone is buying into the issues, brings into focus the need for a systematic way to show the value of those projects. Employees do not always see the need for action because they do not understand the issues or know what they can do to help. Some executives do not understand environmental projects and sustainability efforts. They often feel negatively affected by project outcomes or perceive projects to require unrealistic investments.

Health Issues

Employee health mirrors the health of society. In the U.S. and other countries, citizen health continues to deteriorate. Unhealthy employees contribute to lower productivity, decreased job satisfaction and increased accidents, medical costs, absenteeism and use of sick leave. For example, the annual healthcare cost of obesity in the U.S. is estimated to be as high as $147 billion a year, and healthcare costs related to smoking are approximately $96 billion per year. The effects of smoking are so serious that many organizations refuse to hire employees who smoke. Unhealthy employees put tremendous strain on an organization. The workplace can sometimes make employees sick. For years, employers have implemented different types of programs to protect employees from toxic chemicals, hazardous materials and harmful dusts. Because of this, many organizations are taking initiatives to ensure that employees are healthy through wellness, prevention and health activities.

Safety Issues

Employers have been addressing safety and accident prevention since the beginning of the Industrial Age, when workplace accidents reached near-epidemic proportions. In the early 1900s, much focus was placed on how to change the work environment and procedures in order to avoid incidents and accidents. These efforts continued through the 20th Century. In recent years, much effort has focused on the attitudes and behaviors of employees, including mental alertness and not only an understanding of safety standards but also a willingness to promote them among fellow employees.

Still, accidents and even fatalities happen far too often. The BP explosion and subsequent oil spill in the Gulf of Mexico is an extreme example of a failed EHS management system that cost BP about $50 billion. If an organization does not employ solid and robust safeguards, the consequences can be disastrous; companies are investing heavily because of this reality. Consequently, there is a need and a desire to know whether EHS programs are working, and if they are not, what can be adjusted or changed to make them successful.

This book is for those involved in programs that affect environment, health and safety. The goal of many employers is to protect the environment, keep employees healthy and maintain a safe workplace, but at what cost? Employers need a way to show the value of EHS projects and programs. The following table shows some typical projects undertaken by organizations. The ROI Methodology provides a systematic way to measure success, including ROI for any of these projects.

Examples of EHS Projects	
Environmental Initiatives	Energy saving projects Recycling programs Water conservation projects Waste management solutions Air pollution projects
Health Programs and Initiatives	Industrial hygiene Ergonomics Stress management Health screenings Fitness and exercise Healthy eating and nutrition Smoking cessation programs Obesity programs Sick leave management
Safety Initiatives	Safety management systems Safety leadership programs Safety auditing programs First aid and personal protection Workers compensation Accidental prevention (causation and investigation) Safety incentives System safety Fire prevention and protection Behavior based safety programs Workplace violence Hazardous chemicals and materials

History of the ROI Methodology

The ROI Methodology described in this book was originally developed by Dr. Jack Phillips in the early 1970s. Its application in the training, development, education and human resources fields is unmatched. Dr. Patti Phillips, his wife and partner, used her interest in the application of the ROI Methodology in economic, community and international development to expand its use in government, nonprofit, non-governmental and in private sector organizations. Together, they founded the ROI Institute over 20 years ago and have applied this approach to accountability in more than twenty fields in more than sixty countries. To date, the application of the ROI Methodology includes the following areas of focus:

Broad Applications	
• Human Resources/Human Capital	• Talent Management/Retention
• Training/Learning/Development	• Project Management Solutions
• Leadership/Coaching/Mentoring	• Quality/Six Sigma/Lean Engineering
• Knowledge Management	• Meetings/Events/Conferences
• Organization Consulting/Development	• Marketing/Advertising
• Policies/Procedures/Processes	• Communications/Public Relations
• Recognition/Incentives/Engagement	• Public Policy/Social Programs
• Change Management	• Risk Management
• Technology/Systems/IT	• Ethics/Compliance
• Green Projects/Sustainability Projects	• Healthcare Initiatives
• Safety and Health Programs	• Wellness and Fitness Programs

Al Pulliam has used the ROI Methodology on a variety of environment, health and safety issues at two organizations. He has worked in EHS in both professional and managerial roles for two decades.

All the examples, data and case studies in this book are taken from EHS projects. The ROI Methodology offers common approaches to measuring and evaluating project success so organizations can compare results across functions and channels. More importantly, it allows organizations to develop information that can guide them to improve and reposition projects as they expand across areas of the organization. By addressing EHS projects, executives and managers can drive improvement at the organizational level and improve the environment, employee health and safety.

Motivation to Pursue EHS Projects

Four important forces drive the implementation and maintenance of EHS projects. The first motivation for organizations to implement EHS projects relates to their public image. Organizations want their employees, stakeholders and any other observers to view them as environmentally friendly, health conscious and safety minded. This is a necessity to be considered a "great place to work," the "most admired organization," or the "best organization."

The second motive is to comply with various regulations, laws and standards. To ignore them can be disastrous. For environmental projects, much of the initial effort was driven by regulatory efforts in the 70s. The timing is similar for safety because of major regulations passed by organizations such as OSHA.

The third force is cost savings. Recently, the motivation has been based on cost control and avoidance. When projects and programs are implemented properly, cost savings and cost avoidance are significant. Approximately 80 percent of green projects can produce positive ROIs. Most projects in employee health should deliver positive ROIs if properly implemented. Projects in safety should do the same.

A fourth driver is employee satisfaction, attraction and retention. Employees place a premium on an employer who is protecting the environment, offering a variety of programs to improve the health of employees and providing a safe and comfortable workplace.

These reasons alone make the subject of EHS critical to organizations. Consequently, it is important to ensure that EHS programs are adding value. Following through with the use of the ROI Methodology, this book shows how to measure the success of each of these programs using six types of data.

Audience

This book is designed for those who manage, support, initiate, implement and approve EHS projects and programs. First and foremost, it is for those who create these projects and programs. It provides the framework to plan for success and to measure that success throughout the process to ensure that value is delivered to various stakeholders.

Second, the book is for those who must implement the programs. It shows step-by-step how to collect data early and often in the program, making adjustments along the way. EHS program implementers will find the methodology to be user friendly.

The third audience is those who fund the projects, often the top executives and administrators in an organization. This book shows them that it is possible to measure the impact and financial ROI for investing in any type of program. It often brings in great relief and optimism that there is a proven, credible way to show the value of these important programs.

Next, the book is for individuals who support EHS programs. These are often managers who sometimes see these efforts as overkill or, at times, unnecessary. They question whether a program is working and even when they see safety or health measures improve, they wonder how much of it was related to the project. This book takes the mystery out of these issues and brings this group into an enhanced supporting role.

The last audience is made up of individuals who teach, research and provide consulting for EHS programs. These are professors, researchers, independent consultants and even suppliers to this industry. Individuals who are very critical to the success of EHS programs will find this book a useful tool to show program value in a very credible way.

Collectively, this work provides many tips, tools and techniques that satisfy the diverse needs of these five audience groups.

Flow of the Book

The book first introduces issues in the EHS role, setting the stage for the rest of the book. It explains how the measurement and evaluation processes for EHS have evolved and describes some of the challenges along the way. It shows the progress that has been made and the need for a consistent measurement system. The ROI Methodology is introduced as a systematic, logically proven process for the EHS occupation. Through the next chapters, the ROI Methodology is described, starting with initial alignment and objectives and expands on data collection and analysis. The last chapters focus on reporting results and sustaining the measurement culture as a routine part of the process—not just focusing activities on these areas, but also achieving results. The organization expects a program to not only drive safety and performance, but also to represent a great investment. The table of contents illustrates this flow in more detail.

Acknowledgements

No book is ever the work of the authors alone. Many individuals, groups and organizations have participated in the development of this book. We owe particular thanks to the hundreds of clients we have had the pleasure to work with in the past two decades. They have helped shape, develop, mold and refine this methodology. Their contributions are evident in this book.

We are particularly indebted to Scrivener Publishing for taking on this project. Phil Carmical was very patient with our delivery schedule. Phil and the entire Scrivener team have been very helpful and we are impressed with their commitment to bringing innovative, cutting-edge processes to the environment, health and safety field.

Special thanks go to Rebecca Henderson for the final editing of this book. We look forward to working with her on many more publications.

From Jack:

Thanks to the team and staff at the ROI Institute. With their help, we are pleased to present another important ROI book. I want to thank Patti. The book was her idea. Thanks, Patti, for your diligent editing in the early stages of publication. Patti is an outstanding writer, researcher, teacher, consultant, educator and spouse. My love and my fondness for Patti grows every year. My life, this book, our family and the ROI Institute are much better because of her.

From Patti:

We have two mantras around our house. The first is *all roads lead to ROI*. Jack and I firmly believe that what life (work life and otherwise) holds for us is a balance of benefits versus costs. The second mantra is *we can only do good if we do well.* This mantra describes our mental model most accurately. Limited resources require individuals, organizations and communities to do right by the resources at hand. These resources include people, planet and profit. Reference to this triple bottom line is not new; however, evidence of its acceptance is still waning. Organizations that invest in EHS are clearly interested in people and the planet, but unless their attention is also turned to the optimal allocation of financial resources, their ability to sustain EHS activities is limited, if not impossible. A critical element in measuring the success of optimal resource allocation is ROI. It is through this book that we hope to communicate to EHS professionals and executives a process that helps ensure they implement the right programs, for the right people, at the right time, for the right investment.

Many thanks go to the thousands of people with whom we have had the opportunity to share our process. Their input and encouragement keep us moving forward as we tackle new and interesting issues. Thanks, especially, to Phil Carmical at Scrivener Publishing for giving us an opportunity to describe how our process can help EHS professionals; and particular thanks to our co-author, Al Pulliam, for contributing his expertise in the field of EHS. Having been an executive in the EHS arena for years, he knows the importance of ensuring a balance between EHS investment and the outcomes generated by that investment.

Many thanks go to our team at the ROI Institute. They keep us on track, even when we try hard to veer down a different path.

Finally, and most importantly, thank you, Jack. I've always said Jack gives away more than he gets in return. It is certainly true when it comes to me. For 16 years, I have engrossed myself in a process he developed years ago, finding new and interesting ways to apply and teach it. His continuous encouragement keeps me moving, and he routinely challenges me to do more. I've heard it said that if a couple can complete each other's sentences—or even better, communicate without saying a word—their relationship is a bond not to be broken. We have that; and Jack, while your ROI may not always be positive, I promise to make the intangibles well worth the investment!

From Al:

First I would like to thank Patti and Jack Phillips for inviting me to participate in the writing of this book. Its concepts and methods have an important place in the EHS professional's toolbox. To succeed at advancing EHS programs or projects that compete for scarce resources, the EHS professional has to not only think like a business person but also communicate like one. EHS professionals need to be able to turn what they believe intuitively into a credible business case for their projects, in a language understood by executives. It doesn't take much reading on the EHS professional social media sites to see a question or comment about "management" not approving a project or cutting funding for a program. Managers and professionals who can communicate value win budgets. This book is timely for the profession.

Certainly there is always the person behind the scenes who makes things happen. I would like to thank Nicole Mallory of the ROI Institute for helping keep me on pace. And thank you to my past colleagues from the former Bayou Steel Corporation. During my tenure at Bayou Steel, I had the opportunity to experience most all of the challenges and rewards an EHS career can offer. The lessons learned there have served me well.

List of Authors

Jack J. Phillips, Ph.D.

As a world-renowned expert on accountability, measurement, and evaluation, Dr. Jack J. Phillips provides consulting services for *Fortune* 500 companies and major global organizations. The author or editor of more than 50 books, Phillips conducts workshops and makes conference presentations throughout the world. His expertise in measurement and evaluation is based on more than 27 years of corporate experience in the aerospace, textile, metals, construction materials, and banking industries. This background led Phillips to develop the ROI Methodology—a revolutionary process that provides bottom-line figures and accountability for all types of learning, performance improvement, human resource, technology, and public policy programs. His work has been featured in the *Wall Street Journal, Business Week, Fortune* Magazine, and on CNN. Phillips also serves as president of ISPI 2012-2013. He is chairman of the ROI Institute, Inc., and can be reached at (205) 678-8101, or by email at jack@roiinstitute.net.

Patti Phillips, Ph.D.

President and CEO of the ROI Institute, Inc, Patti earned her doctoral degree in International Development and her Master's Degree in Public and Private Management. Early in her professional career, Patti was a corporate manager who observed performance improvement initiatives from the client perspective and knew that results were imperative. As manager of a market planning and research organization for a large electric utility, she and her team were responsible for the development of electric utility rate programs for residential and commercial customers. In this role, she played an integral part in establishing Marketing University, a learning environment that supported the needs of new sales and marketing representatives. Internationally known as an accountability, measurement, and evaluation expert, Patti facilitates workshops all over the world and consults with U.S. and international organizations – public, private, non-profit, and educational – on implementing the ROI Methodology™. Patti is the author of *ROI Basics* (ASTD 2006) and *The Bottom Line on ROI* (CEP Press 2002), which won the 2003 ISPI Award of Excellence. She is editor or co-author of *ROI at Work: Best-Practice Case Studies from the Real World* (ASTD Press 2005), *Proving the Value of HR: How and Why to Measure ROI* (SHRM 2005), *The Human Resources Scorecard: Measuring the Return on Investment* (Butterworth-Heinemann, 2001), and *Measuring ROI in the Public Sector* (ASTD 2002).

Al Pulliam, MSPH

Al Pulliam, MSPH, is a veteran Environment, Occupational Health and Safety professional with over 20 years experience. This experience includes EHS management in heavy manufacturing, industrial construction, commercial construction and consulting. He holds a Master of Science in Public Health degree from the Tulane University School of Public Health and Tropical Medicine.

1

Environment, Health and Safety is Everywhere

Proliferation of the Field

Abstract

This initial chapter describes the vast scope of topics and issues in the EHS field. After discussing the new role of EHS, it tackles environmental initiatives, such as energy-saving projects, recycling programs, water-conservation projects, waste-management solutions and air pollution projects.

Next, a variety of safety issues are explored, including safety management systems, safety leadership programs, safety auditing programs, first aid and personal protection, workers' compensation, accident prevention, safety incentives, systems safety, fire prevention and protection, behavior-based safety programs, workplace violence, hazardous chemicals and materials and occupational health and safety.

The last portion of the chapter focuses on health issues, including health screenings, healthy eating and nutrition, smoking cessation programs, obesity programs, industrial hygiene, ergonomics and stress management. Finally, the chapter stresses the fact that the changes needed must be managed within the EHS function.

Keywords: EHS topics, energy-saving projects, recycling programs, water-conservation, waste-management, safety, safety leadership, safety auditing, accident prevention, health screenings, ergonomics, stress management, changes within EHS function

Jack Phillips, Patti Phillips, and Al Pulliam, Measuring ROI in Environment, Health, and Safety, (1–20) 2014 © Scrivener Publishing LLC

1.1 The New Role of EHS

No matter what industry, business or occupation, the field of environment health and safety (EHS) is everywhere. In the early 1970s, the United States Environmental Protection Agency (EPA) and the Occupational Safety and Health Administration (OSHA) were established to protect the environment and employees from abuses by the industrial and business communities. In the 1970s and 1980s, the role of the EHS professional who worked in business was primarily one of compliance with new and complex regulatory regimes. Today, to a large degree, there has been a fundamental shift in the way organizations view the EHS effort. While compliance with the vast regulatory burden remains a major function of the EHS professional, organizations have recognized the importance of EHS-related matters to employees, communities and the bottom line. The scope of the EHS profession is vast, and senior EHS professionals are called upon to initiate, evaluate and execute initiatives in a number of areas.

According to a study performed for the Small Business Administration's Office of Advocacy, environmental regulations cost businesses about $281 billion annually. Occupational safety and health regulation costs are estimated at $65 billion annually (Crain, 2010). This is just the estimated annual regulatory burden and does not include proactive initiatives in pollution prevention, health and wellness practiced by many EHS professionals. Below are a few examples of EHS initiatives and programs that the EHS professional faces on a daily basis.

1.2 Environmental Initiatives

Historically, environmental programs and initiatives were compliance based, stemming from a large and complex regulatory framework. The EPA, pursuant to certain milestone legislative laws/acts, promulgated the bulk of these regulations. Some primary laws of concern to the contemporary EHS professional include the Clean Air Act (1970), the Clean Water Act (1972), the Resource Conservation and Recovery Act (1976) (laws addressing hazardous waste) and the Comprehensive Environmental Response and Liabilities Act (Superfund) (1980), plus all of the subsequent amendments for

programs were often limited to the scope of what was required by OSHA. Employees who were part of these programs were managed as such. Now EHS professionals are actively involved with the design of facilities and the implementation of systems to reduce occupational exposure to noise and to keep employees out of hearing conservation programs. With these modifications come the challenges of articulating the value and business impact of the programs as they move farther from compliance-based initiatives. Here are the major safety programs with which the contemporary EHS professional may be involved:

1.3.1 Safety Management Systems

There are two predominant third-party certifiable safety management systems. One is the British System OHSAS 18001. This management system is international in its use and is likely the most recognized safety management system in the world. The second system is OSHA's Voluntary Protection Program (VPP). While any management system can be applied to safety, these are widely recognized standards in the safety area. Certification by either of these safety management systems requires the development of a formal system and a third-party auditor. The decision to implement a formal safety management system may often be made without any consideration to cost or expense. If a company's board of directors is aware that a competitor has recently been certified under OSHA VPP, the decision to implement a system can come without any analysis or thought. Company leadership may not feel the need to formally quantify the value. Albeit intuitive, the company knew that failure to keep up with the competition would have an adverse impact on the brand beyond the cost of implementation.

In cases where there is not a great deal of support for implementing a safety management system, the decision becomes more complex. The joint costs can be significant since there is substantial employee and managerial involvement. The general benefits of an effective safety management system are the promotion of a safe and healthful working environment by establishing a system that identifies and controls safety and health risks, reduces the potential for unwanted events, assists in compliance and positively impacts the bottom line. Quantifying the benefits and comparing them to the existing system is not a typical skill set of the EHS professional, although it should be.

1.3.2 Safety Leadership Programs

Safety leadership programs take many forms. One of the most extensive is General Electric's (GE) Environment, Health & Safety Leadership Program (EHSLP) for entry-level EHS employees. It incorporates a series of rotational assignments across various EHS disciplines in various geographic locations. GE's EHSLP requires a significant time and resource commitment that GE believes returns benefits to the bottom line.

Most safety leadership programs are not as extensive as EHSLP. On some level, however, investments into current and future safety leaders are made in virtually all organizations. Senior EHS professionals are challenged to develop future EHS leaders in ways that bring value to their companies.

1.3.3 Safety Auditing Programs

Safety auditing programs have come a long way from just the "safety man" walking around with a clipboard. Today's comprehensive audit programs include regular formal reviews by front-line supervisors and managers. Facility-level EHS personnel often validate these audits. Corporate internal audits are frequently performed on both management systems and regulatory compliance. In addition to the internal audits, a comprehensive audit system includes third-party certification audits for the management systems and comprehensive compliance by third-party consulting experts. In fact, many of the EHS consulting firms are building business models around program assurance and compliance auditing.

Given the size and scope of a comprehensive audit program, the costs are significant and identifiable. For example, the cost of conducting comprehensive management program assurance and regulatory compliance audits in a 500-person manufacturing operation with two operating facilities can be up to $100,000. The benefits, however, are not clear.

1.3.4 First Aid and Personal Protection

The EHS professional performs a leadership role when it comes to first aid and personal protection. He or she will likely supervise or work closely with the plant nurse or other medical professionals to establish all requirements related to administering first aid.

This could include determining which first aid supplies to keep on hand and choosing the location of eyewash stations and emergency showers.

The OSHA standard now includes regulations about personal protective equipment (PPE). The standard requires that a hazard assessment be performed to determine necessary PPE. The EHS professional must consider all the jobs performed at the facility, the hazards presented by the jobs that cannot be engineered out and the appropriate PPE to be used. The EHS specialist must make decisions regarding what type of protection is required to safeguard hearing, the eyes, face, skull, feet and hands; select protective clothing; and choose appropriate respiratory equipment. In addition to the choice of equipment, systems must be developed for screening employee lung capacity, fit testing and respiratory maintenance.

1.3.5 Workers' Compensation

Depending on the organization, workers' compensation may or may not fall under the EHS expert's job description. In either case, the EHS professional must be familiar with current workers' compensation laws. Workers' compensation claims are a closely watched metric and are considered a prime indicator of the success of the accident prevention program. Simply put, the central question is, "how much is paid out in claims?" The EHS professional must understand how injuries are categorized, who is exempt from workers' compensation, how premiums are calculated and how to conduct proper recordkeeping. While much of the basic workers' compensation duties are straightforward, the EHS professional must be able to know and demonstrate how all of his other programs interact with this important metric.

1.3.6 Accident Prevention (Causation and Investigation)

One of the most important roles of the EHS professional is preventing accidents. An effective investigation will reveal the root cause of the incident. Knowing why the event happened is a threshold matter to providing a corrective action to prevent a re-occurrence.

A variety of accident investigation techniques are available to the investigator. One size does not fit all. A good EHS professional will be able to determine the depth and breadth of the investigation based on the facts as they become apparent. He or she should be skilled in interviewing witnesses, understanding the psychology of why incidents or facts may not be reported and in not letting their personal biases shade the facts.

In addition to good investigative techniques, the EHS professional must know the basic theories of accident causation. These include, but are certainly not limited to, the domino theory of accident causation, the human factors theory of accident causation, systems theory of accident causation and the combination theory of accident causation.

By coupling good investigative techniques with knowledge of why unwanted incidents occur, the EHS professional can make not only determinations about the causes, they can also use this information to determine the business impact of the corrective actions.

1.3.7 Safety Incentives

Many safety incentive programs are effective at rewarding desired behaviors that result in fewer incidents. Although there is debate as to which metric to use as a basis for determining that incentive, they can be successful. Regardless of the metric used or the size and type of the award, the EHS professional must ensure that program objectives are met.

Some programs offer small giveaways that simply promote safety awareness. Others are designed to provide somewhat more meaningful cash awards. Still, other programs may impact significant portions of variable income for supervisors and managers. In either case, the EHS professional must be able to isolate any cause and effect relationships and quantify the business impact of the program to make good business decisions.

1.3.8 System Safety

System safety is a specialized area in the EHS field. It was originally developed in the aerospace industry and gained formal acceptance by the Department of Defense and NASA in the 1960s (System Safety Society, 2002). The basic premise of system safety is that safety can be built into a machine or a system. Not all EHS professionals are system safety experts but all must have a grasp on the concepts of system safety and some of the techniques used. Tools utilized by the system safety professional include preliminary hazard analysis, subsystem hazard analysis, failure modes and effects analysis, technique for human rate error prediction, fault hazard analysis and fault tree analysis.

A system safety technique in which all EHS professionals should be well versed is the job safety analysis (JSA). The JSA basically lists the steps of each particular job, analyzes the potential hazards of each step and addresses the safety precautions and safety equipment necessary to complete them. In organizations that use the JSA or a similar system, the JSA is the first document reviewed in an incident investigation.

Therefore, even though most EHS professionals are not experts in the field of system safety, they should be able to understand, interpret and communicate system safety analyses.

1.3.9 Fire Prevention and Protection

The subject of fire protection and the safety professional could fill volumes. Fire safety includes all the concepts of fire prevention plus fire response. Prevention programs include hot work permit programs, infrared analyses to ensure equipment is operating at optimum temperatures, maintenance programs to ensure clean ventilation systems, combustible dust programs and general housekeeping programs.

Fire protection and response is a significant responsibility for the EHS professional. Programs and systems that EHS professionals manage include evacuation programs, detection and alarm systems, fire brigades, fire extinguisher management and training, standpipe and hose systems, sprinkler systems and dry chemical systems.

Due to the potential loss of life and property because of a fire, considerable resources are devoted to the prevention and control of fires in the workplace. EHS professionals are integral in implementing comprehensive fire prevention and protection programs.

1.3.10 Behavior Based Safety Programs

Behavior based safety (BBS) programs are complementary programs within a safety management system designed to encourage all employees to recognize the daily importance of safety. BBS programs are not designed to replace or supplant other portions of a safety management system. BBS takes a process approach that views safety performance as a long-term development that can be improved.

A variety of ways to approach and implement BBS programs are available, including some off-the-shelf systems on the market. Some EHS managers develop their own systems. For example, if organizations use some form of W. Edwards Deming's plan-do-study-act (PDSA) as part of their quality systems, this system can be easily adapted to a BBS effort. Key to these types of programs is the integration of behavioral management into the safety management program. Safety professionals must understand the concepts of BBS in order to integrate programs throughout the organization and to understand the value proposition the program brings.

1.3.11 Workplace Violence

A relatively new phenomenon in the EHS profession is the prevention of workplace violence. The degree of workplace violence is almost epidemic. There are nearly two million violent victimizations per year in the workplace. Occupational homicide has been a leading cause of death by injury in the workplace for women. One in four U.S. workers will be attacked, threatened or harassed during their working career (Friend, 2010).

Obviously this will become an area of focus for not only the EHS professional but for the country as well. No one is immune to the threat of workplace violence. As of this writing, OSHA has published some guidance on the subject of workplace violence but has not concluded any rulemaking on the subject. The EHS professional will be heavily involved in the initiation of workplace violence abatement programs.

1.3.12 Hazardous Chemicals and Materials

As a result of catastrophic industrial incidents in the 1980s, OSHA implemented its Process Safety Management (PSM) standard. PSM will continue to be a significant program for the EHS professional to facilitate and manage. The standard prescribes a systematic approach to analyzing the safety of systems that store, handle or produce certain hazardous chemicals or materials. In some instances, companies that may not fall under OSHA's PSM standard are developing their own risk based management systems of hazardous materials.

This will continue to be a significant issue in the field of EHS. In addition to analyzing processes, operational procedures must be developed, employees and contractors must be trained and sensitive facilities must protect these systems against acts of terrorism.

1.3.13 Occupational Health Programs and Initiatives

This is the area where the EHS field has experienced the most change over the last few years. Companies are expanding traditional health programs and initiating many new proposals. In addition to focusing on workplace stresses, these programs influence employees in all parts of their lives. This phenomenon is driven by two factors. The first is the cost of employee health care. From 2002 to 2008, the annual average cost of employee health care per employee nearly doubled. In 2002 the cost was approximately $4,336 per employee. In 2008 the annual average cost was $8,331 (Bray, 2009). In 2012 the national cost, according to the global consulting and outsourcing business Aon Corporation, was approximately $10,475 per employee (PR Newswire, 2011).

The second driver for this focus on health is the cost of absences and lost productivity. In fact, a study done by the Wharton School of Business found that the business cost of absenteeism is 28 percent higher than the worker's wage (Bray, 2009).

These cost drivers are causing employers to develop new programs and initiatives to improve the bottom line. These programs are aimed at reducing absences, reducing healthcare costs and improving worker productivity on the job.

1.4 Health and Fitness

There is no doubt that wellness and fitness is positive for health and well being. The question may be, "what is the right type of program for my organization?" This is what the company must evaluate and the choices are endless. Many company campuses are equipped with weight rooms, tennis courts and jogging trails. Some companies offer free gym memberships or encourage and sponsor competitive events. As with many of the newer health and wellness programs, isolating the value of a specific program becomes difficult. With current spending on wellness programs equaling approximately 2 percent of a corporation's total insurance claims, defining that value becomes more important (Society of Human Resource Management, 2010).

1.4.1 Health Screenings

One element of an overall health and wellness program is a health-screening plan. These are designed to provide baseline information about both new and existing employees' physical state. These programs supplement existing programs that ensure the employee is physically capable of performing their job. For instance, to operate an overhead crane the employee has to meet certain correctable vision standards. Employees likely know if they have monocular vision. On the other hand, if employees are long-term smokers and have trouble getting up stairs that is an opportunity to improve their health by encouraging a smoking cessation program, additional medical treatment and a very limited exercise program. Presented in a positive way, employees are encouraged to improve their heath resulting in gains associated with healthy and productive employees.

1.4.2 Healthy Eating and Nutrition

As with exercise, the connection between good health and nutrition is well established. Companies are developing programs to encourage healthy eating both at home and in the workplace. Campuses with in-house dining services are offering healthier products. Some companies are simply buying healthy breakfasts and lunches for their employees. Other options are swapping out the traditional snacks in vending machines with healthier alternatives.

While promoting healthy eating is not an overly expensive proposition, it is still a program that someone in the organization must design, implement and manage.

1.4.3 Smoking Cessation Programs

Counter to fitness and healthy eating being positive for employee health, the connection between smoking and poorer long-term health is also well established. Most experts agree that smoking takes about a decade off a person's life. In addition, the habit is very expensive for employers. In some cases smoking cessation in the work environment is fairly easy to implement. Simply eliminate smoking anywhere in or around the facility. While there may be some push back from a group that has been going to a smoking pit for the last few years, overall, a non-smoking facility is well received by staff and employees. For those with difficulties quitting, at least while at work, there are intervention options.

In addition to eliminating smoking from the workplace, companies are looking to provide financial incentives for employees to completely quit. While there are a few HIPPA hurdles, companies are establishing two-tiered health insurance programs. Smokers pay a substantially heftier premium than non-smokers.

In addition to addressing smoking among existing employees, employers today are refusing to hire people if they are smokers. Being a smoker is not a protected status, and companies that are serious about these issues are not hiring those who currently smoke. According to the CDC, smoking accounts for $96 billion in direct medical costs and another $97 billion in lost productivity and premature death. Business continues to take smoking seriously.

1.4.4 Obesity Programs

Another of the health issues plaguing this country and its workforce is the obesity problem. The health effects of obesity are well known. Obesity can lead to a number of conditions:

- Type 2 diabetes
- Heart related illnesses such as high blood pressure and stroke
- High cholesterol
- Osteoarthritis
- Gall bladder disease
- Liver disease and other illnesses.

Obese workers are less healthy, miss more work and drive up insurance costs. Companies are implementing programs to combat obesity. One place to start is with employee health screening programs. As a part of that program, encouragement to lose weight and to improve health can be initiated in a positive way with the privacy of medical staff. Other obesity program efforts include

- Educational programs
- Activities programs
- Individual treatment
- Moral support programs

Any and all of these programs are being used as obesity reduction initiatives in the business community.

1.4.5 Industrial Hygiene

While the role of the industrial hygienist remains devoted to the anticipation, recognition, evaluation, prevention and control of stressors arising from the workplace, a considerable amount of industrial hygiene work has moved from the shop floor to the main office. Today's industrial hygiene focused EHS professional spends a great deal of time working on programs such as indoor air quality (sick building syndrome, second hand smoke), occupational diseases such as AIDS and other blood borne pathogen programs and cumulative trauma disorders.

In addition to the newer programs, the industrial hygienist continues to develop and initiate programs that include:

- Managing chemical exposures
- Detecting and controlling exposures in the areas of radiation (ionizing and non-ionizing), noise and illumination
- Emergency response and community right to know

More than ever, the programs within the industrial hygiene field impact all employees.

1.4.6 Ergonomics

With the rising cost of health care and the increase in musculoskeletal disorders (MSDs) as a portion of workplace injuries, much focus has been placed on the field of ergonomics in the workplace. In fact, MSDs are the most common form of workplace illness in industrialized nations (Bray, 2009). These MSDs include carpal tunnel syndrome, repetitive strain injuries and cumulative trauma disorders.

Comprehensive ergonomic programs are being developed. These programs include extensive training for employees and management, surveillance of data to spot trends early, case management of all MSD illnesses, job analyses and design to address ergonomic risk factors (force, repetition, awkward postures, static postures, vibration). While these programs address the multidisciplinary sciences addressing the interface between the employee and the work performed, the value of these programs is often unclear. Productivity improvements are hard to isolate. Also MSDs may be caused or aggravated by activities outside the work area. Finally, the treatment of these illnesses is different for each individual. While it is widely believed that a sound ergonomics program brings value to the business, determining that value is a difficult task.

1.4.7 Stress Management

According to the Institute of Stress, employers lose $300 billion annually due to excessive worker stress. This is before the impact of health care costs, which are nearly 50 percent higher for workers reporting high levels of stress (Bray, 2009). Given these numbers, stress management/reduction programs are getting ample attention in the business community. Companies are making various efforts to reduce stress in the workplace. Some of these programs include the following:

- Bringing pets to work
- Assistance with time management
- Classes on financial wellness
- Time off for exercise
- Time for meditation
- Mutual support pairings

While the value of some of these programs may remain questionable, employers have recognized that stress arising from the workplace impacts the bottom line and are taking action.

1.5 Managing Change in EHS

The landscape is covered with all types of EHS initiatives and projects. EHS is everywhere. It touches every part of a business' operations and each employee within the business. Still, these programs must be integrated, managed and properly implemented to reap the greatest rewards. There are several factors that help drive the changes that are taking place in the EHS profession:

1.5.1 Employer Image

As we noted in the introduction, historically, the number one driver for implementing EHS projects in the past was compliance. Now EHS is not only about compliance, it is also about the image it presents of an organization. EHS is in vogue in all types of organizations. Organizations recognize that it is in their best interests for their constituents, consumers, employees, stakeholders and the general public to view them as environmentally friendly, safe and healthy places to work while serving as valuable members of the communities in which they operate.

1.5.2 EHS and The Bottom Line

To convince a group of money-conscious executives to undertake EHS projects, a method must exist to show there is value in these projects that will ensure continued funding and growth.

In our work at the ROI Institute, we have the opportunity to see hundreds of studies each year in a variety of different applications. For those that focus on EHS related subjects, the rate of failure (i.e., a project delivering a negative ROI) is about the same across all functions and it is typically low, usually ranging from 20 to 30 percent. Most of those projects are adjusted to ensure that they deliver value in the future and only about 10 percent are discontinued. A project is discontinued only if it cannot show business value when that is the principal reason for implementation.

The MIT Business of Sustainability survey of more than 1,500 worldwide executives and managers underscores how value is created in environmental projects and initiatives (Berns, 2009). Figure 1.1 shows the different avenues for value creation, all leading to profits, cash flow and total shareholder return.

As we continue to work to manage the integration of EHS into organizations, one of the main challenges is to convince a variety of stakeholders about the value the projects deliver, up to and including the financial ROI.

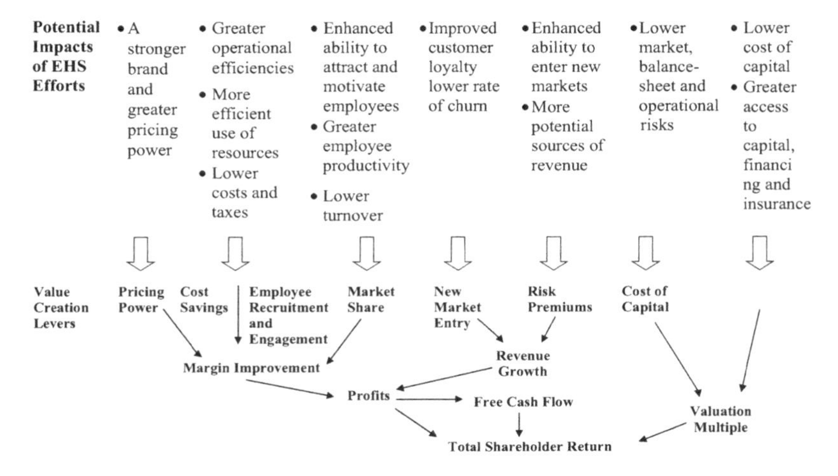

Figure 1.1 How EHS Affects Value Creation. (Adapted from Berns et. al., 2009)

1.6 Final Thoughts

This opening chapter underscores the vast scope and opportunity that has arrived in the EHS profession. EHS projects and programs are being implemented in every aspect of the business model. Although there is great opportunity and promise for these programs, unfortunately, not all of these efforts work as well as intended. The next chapter focuses on some of these issues and problems and it underscores why the ROI Methodology is needed to ensure that EHS projects are successful and bring value to the business.

2

Is It Worth It?

The Value of EHS Initiatives

Abstract

This chapter focuses on the value of EHS initiatives by first exploring the measurement system that is the heart of this book. The levels of evaluation for any EHS project include reaction, learning, application, impact and ROI. The chapter also discusses how value systems have evolved and how they are often misused or misunderstood. The significant focus is on converting the impact data to monetary values and how money has become the necessary value. "Show Me the Money" is a very common request from executives in reaction to proposed EHS projects and solutions. The chapter explores the rationale for focusing on impact and ROI in today's environment. Last, it focuses on criteria for an effective evaluation system, outlining why the ROI Methodology meets these criteria. Additional explanation of the methodology is provided to indicate the need for this comprehensive system in today's EHS function. The chapter ends with the requirements needed from executives to drive value and implement an evaluation system that shows value and makes adjustments and improvements along the way.

Keywords: Reaction, learning, application, impact, ROI, value systems, show me the money, evaluation system, measurement systems, evaluation challenges

Jack Phillips, Patti Phillips, and Al Pulliam, Measuring ROI in Environment, Health, and Safety, (21–48) 2014 © Scrivener Publishing LLC

2.1 EHS Killers

While focus on EHS has been around for decades, efforts to get buy-in for EHS projects beyond the compliance argument have been minimal. The death of potential and existing funding of EHS projects is driven in large part by perceptions based on something other than reality. But, as the old adage goes: perception is reality. Figure 2.1 summarizes the EHS killers.

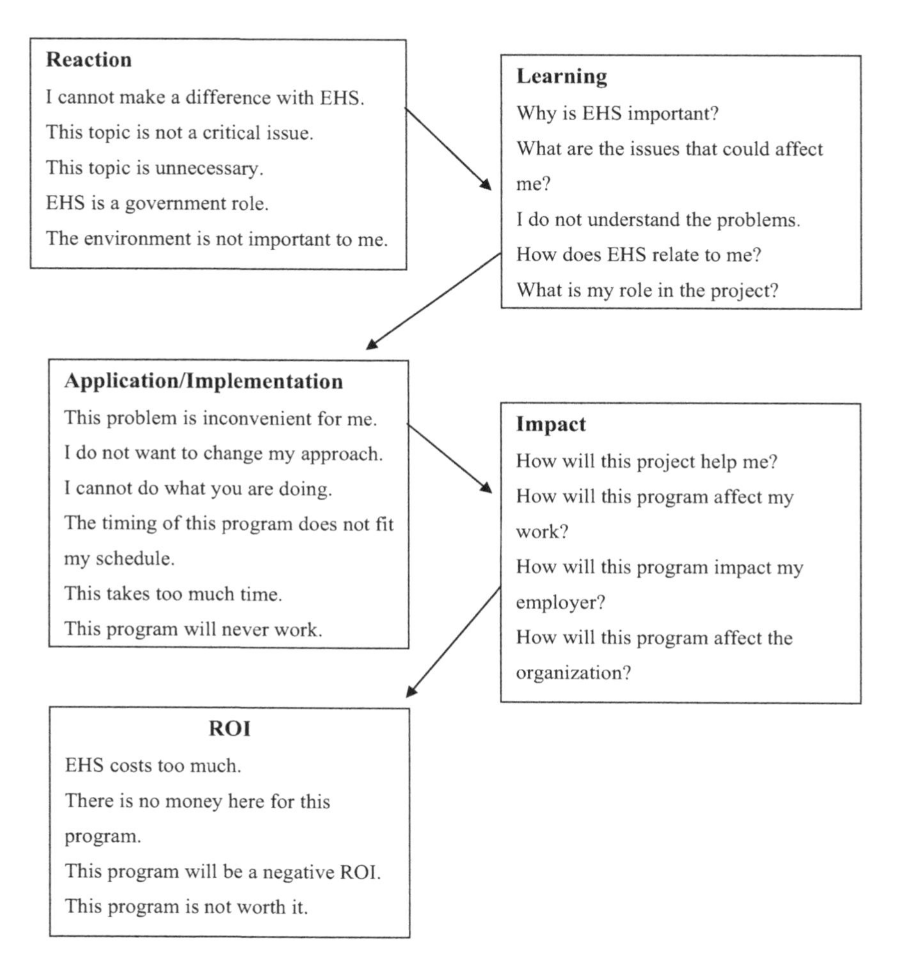

Figure 2.1 EHS Killers.

2.1.1 Reaction

As the "Reaction" section of Figure 2.1 shows, many people are still apathetic or believe their actions can make no real difference. They seem to think that EHS is someone else's problem or that the only way to change it is through regulations and laws. In the book *Green to Gold*, the authors suggest several reasons why many environmental projects fail, including:

1. Visions that see the trees but not the forest
2. Misunderstandings about green issues
3. Expectations that green will always cost more
4. Subtle thinking
5. Claims outpacing actions (Esty & Winston, 2006)

These initial reactions to environmental projects are similar to those of other EHS projects that take the organization beyond compliance. As such, organizations must position these projects so employees and potential participants perceive them as relevant and valuable to themselves and the organization.

2.1.2 Learning

Try this experiment. Take a few elements that are important in an EHS program, such as industrial hygiene, and ask your friends and colleagues who are not in the EHS profession if they understand their meaning. Chances are they may have a rudimentary grasp of the subject because they heard the phrase. They cannot, however, really describe what these elements encompass. EHS issues are now a part of academic course curricula, but for the majority employees in an organization—not to mention the world's population—there is only a limited understanding.

A simple environmental greenhouse gas (GHG) issue such as changing to an alternative type of light bulb can have a big impact. For example, a Swedish power utility developed a chart describing what methods would be useful for cutting GHG. Some of these methods are more expensive than others and some of them actually have a double advantage (i.e., by following these methods, people in companies could both cut emissions and save money).

One method with a double advantage is to change traditional light bulbs to more energy-efficient ones. In the United States, for example, lighting represents about 20 percent of all electricity usage. A standard incandescent light bulb costs around two dollars and uses about $20 of electricity per year. In contrast, a low-energy bulb cost about eight to ten dollars but only uses about four dollars of electricity per year. With a cost reduction of $16 per year and an investment of $10, the benefits clearly exceed the costs.

Yet, even with a notable financial impact, getting people to actually change the light bulb requires that they recognize and embrace the importance of doing so. This takes education, information and persuasion. People need to know what to do, how to do and why do it before they will actually do it.

2.1.3 Application and Implementation

Perhaps the greatest problem with EHS projects is the inability to change existing behaviors and habits. Much of the time there are only half-hearted attempts at applying or implementing processes focusing on desired outcomes. Sometimes people do not do what is needed because it is inconvenient, it requires a change of habit or they think they cannot do it. Others think that it may take too much time and see too many barriers before the actual success of the initiative. Still, others see that they can do the necessary tasks but need support.

For example, a behavior based safety program may be resisted because employers may not see the connection between behavior and safety. Also, employees may not use the wellness and fitness center because they do not see a connection between using the center and improving their health. A GHG reduction initiative to change to energy-efficient lights highlights the difficulty of getting a person to change current practice (Schendler, 2009). Many people prefer standard bulbs and do not see the value or care about the value of making the switch. They often see only that the energy-efficient bulbs cost more.

After much political wrangling, you manage to install energy-efficient lighting in a high-end hotel restaurant. The project will save thousands of dollars in electricity costs while preventing tons of carbon emissions from entering the atmosphere. It is the "rubber meets the road" of the sustainability movement, the blue-collar work of the climate battle. The restaurant opens and the manager is put off by the sight of compact fluorescent bulbs. He removes the bulbs, throws them out and replaces them with inefficient halogens. Not because he is ignorant or because he does not care but because he has a business to run and he is doing it the best way he knows how. His perspective is, "you do not put energy-efficient fluorescent bulbs in a fancy restaurant any more than you would put Cool Whip on an éclair."

Nonetheless, this is what your sustainability efforts have brought you: a wasted design and installation fee, inefficient lighting, the manager's loss of faith in green technology, hundreds of expensive compact fluorescent bulbs that instead of being reused (at the very least) are now leeching costs for new bulbs and installation. This true story happened a decade ago at Aspen Skiing Company. And there has been no improvement in that restaurant's lighting since. (Schendler, 2009).

2.1.4 Impact

While compliance is a major aim for EHS projects, many sponsors of projects want to know the impact of a project and the specific measures it will drive. Those involved want to understand how it will affect them personally and professionally. Others want to know the impact on a community group or city where they live. Some are interested in what affect it has on the environment or how it helps the safety performance. Unfortunately, this evidence is often needed before the decision to invest in the project is made. If the project involves savings in workers' compensation costs, for example, some want to know *before investing* how much will be saved. While there may be enough credible data to make a reliable forecast, the evidence needs to be strong.

A colleague of ours tells a story about proposing that a company should replace its light bulbs with energy-efficient ones. The change would be expensive but the savings would quickly pay for the costs. When he presented the idea to the CEO, the response was, "I don't believe the lights will save money." After much discussion, he decided to demonstrate the value of changing the bulbs. At the next meeting of the company's senior executives, our friend took a wattmeter and showed the energy usage of the different light bulbs. When he tested incandescent light bulbs, the meter moved rapidly.

When he tested the compact fluorescent bulbs, the meter slowed to a near standstill. Still, the executive was reluctant to spend the money. Only after much additional discussion, analysis and comparison did the executive agree to invest. When this level of difficulty occurs with every project, some change proponents may give up, finding it easier to avoid the issue altogether.

From our work at the ROI Institute, we recognize that impact data are the most critical data executives want to see. Yes, the ROI is important but executives are often willing to invest in EHS projects if there are demonstrable intangible benefits that do not figure into the ROI calculation. Therefore, impact data are critical and must be developed for many executives to invest.

2.1.5 ROI

Some people believe that EHS projects beyond basic compliance will result in a negative payoff. In fact, there is often an impression among the financial groups in organizations that programs beyond compliance come at a premium. Perhaps this is based on history when, years ago, health care costs were cheaper, workers' compensation settlements were more company favorable and the more subtle health impacts arising from the workplace were less well known. In reality, most well designed EHS programs can actually save money in the long term, but the intuitive perception still exists that the costs outweigh the benefits, resulting in a negative ROI. As such, people begin with the perception that an EHS project is an expense that outweighs any marginal added value.

2.1.6 EHS Facades

In addition to the perceptions that kill forward motion with EHS initiatives, some organizations resort to putting on an EHS facade, which includes a bit of trickery and deception, much of which is purposeful. Wellness and fitness consultants make claims that cannot be verified. External safety trainers boast of huge safety improvements caused by their training. Safety equipment manufacturers suggest improvements that cannot be substantiated. The environmental field has greenwashing. According to Lori Lake, President of GreenTV.com, greenwashing is the act of misleading consumers and other stakeholders regarding the environmental practices of a company or the environmental benefits of a product or service. This "greenwashing" applies to EHS projects in general:

1. *Sin of the hidden trade-off.* A claim suggesting that a company is environmentally friendly or safety conscious based on a narrow set of attributes without attention to other important EHS issues. Having an award for safety is not necessarily evidence of a safe workplace just because it comes from a respected source.
2. *Sin of no proof.* An EHS claim that cannot be substantiated by easily accessible supporting information or by a reliable third-party certification. Common examples are that zero injuries are our goal.
3. *Sin of vagueness.* A claim that is so poorly defined that its real meaning is likely to be misunderstood by the consumer. The term *biodegradable* is an example. While on many fronts a biodegradable waste stream may be preferable to wastes that are non-biodegradable and environmentally persistent, biodegradable wastes are the ones that cause excessive biochemical oxygen demand in natural water systems. Just because something is biodegradable doesn't make it inherently good.
4. *Sin of worshipping false labels.* A process that, through either words or images, gives the impression of third-party endorsement where no such endorsement exists. For instance, a manufacturing facility can claim that it is approved by the state safety agency, when in fact, the state does not issue an approval rating of the process.
5. *Sin of irrelevance.* A claim may be truthful but unimportant. A company may claim that their employees are healthier than their industry average. If the industry as a whole is operating in the darker ages in terms of its employee health programs, the fact is irrelevant.
6. *Sin of lesser of two evils.* Claims that may be true within the process category, but with risks that distract the stakeholders from the greater environmental impacts of the category as a whole. An example would be the assertion that hazardous wastes are recycled opposed to eliminating their generation.
7. *Sin of fibbing.* EHS claims that are simply false. A common example is manipulating OSHA recordkeeping incident rate information.

Creating an EHS facade of going beyond compliance by committing one or more of these seven sins represents a lack of buy-in and awareness. These gaps lead to inaction or inappropriate action affecting outcomes associated with viable EHS projects. While it is disappointing that attitudes, perceptions and lack of interest and purposeful (or accidental) distortion of an organization's EHS reality seem to get in the way of making progress with new EHS efforts, these barriers exist with almost any type of change. For change to occur, even when it comes to EHS, people must have a favorable reaction, understand the issues and take appropriate action. The good news is that most people want to do the right things when it comes to the environment and their employee's safety and health.

2.2 Value Redefined

Everyone has an opinion about value and these opinions vary considerably. A person's reaction to involvement in an EHS project is representative of that person's perceived value. People may see it as important, necessary and valuable, or they may see it as unnecessary, irrelevant and a waste of time. As they become involved, they gain awareness of issues, principles, trends, terms and concerns about environment, health and safety issues. This new awareness also represents value.

"Show me the money" represents a value statement toward EHS projects. Many initiatives are measured by activity, such as the number of people involved, the number of actions taken and the time to complete a particular project or initiative. Less consideration is given to the benefits derived from these activities. Today, the value definition has shifted. Value is defined by results more than activity. More frequently, value is defined as a monetary benefit when compared to cost.

The ROI Methodology described in this book can "show the money" in a credible way. The process had its beginnings in the 1970s and has since expanded in recent years to become the most comprehensive and broad-reaching approach to demonstrating the value of investing in projects of all types.

2.2.1 Types of Values

Value is determined by stakeholders' perspectives, which may include organizational, spiritual, personal and social values. Value is defined by consumers, employees, taxpayers and shareholders. Capitalism defines value as the economic contribution to shareholders. The Global Reporting Initiative (GRI), established in 1997, defines value from three perspectives: societal, environmental and economic. To put it another way, value is defined in terms of people, planet and profits.

Even as projects, processes and programs are implemented to improve the social, environmental and economic climates, monetary value is often sought to ensure that resources are allocated appropriately and that investments reap a return. No longer is it enough to report the number of projects initiated, the number of participants or volunteers involved or the awareness generated through an EHS project. Stakeholders at all levels—including executives, shareholders, managers and supervisors, project designers and employees—are looking for outcomes, and in many cases, the monetary values of those outcomes.

2.2.2 The Importance of Monetary Values

Some critics assert that too much focus is placed on economic value. But money is the ultimate normalizer. It is the monetary resources that allow organizations and individuals to contribute to the greater good. But, as with any other resource, monetary resources are limited and they can be put to best use, underused or overused.

Economics 101 describes the utilization of resources in terms of a simple production possibility curve as shown in Figure 2.2. The curve represents the limitations on resources for any two given options (A-axis and B-axis). The point along the curve represents efficient use of resources. This is known as Pareto Efficiency or Pareto Optimality. The point inside the curve represents inefficient use of resources, known as Pareto Inefficiency. If resources are to be utilized optimally, then a tool must be used that equally compares the benefits of funding project A versus project B. This is where ROI comes in. By normalizing the benefits of both projects using money (the same type of indicator as the investment), decisions about whether to allocate resources to project A versus project B are easier and more equitable.

Organizations and individuals have choices about where they invest these resources. To ensure that monetary resources are put to best use, they must be allocated to programs, processes and projects that yield the greatest return. The good news is that most EHS projects can create a positive return for their investors.

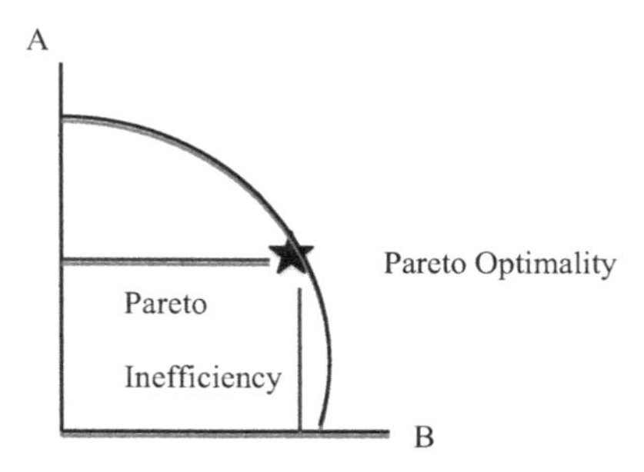

Figure 2.2 Production Possibilities Curve.

2.2.3 The "Show Me" Generation

Figure 2.3 illustrates the requirements of the "show me" generation. "Show me" implies that stakeholders want to see actual data (i.e., numbers and measures). This request is the initial attempt to see value in projects. This request has evolved into "show me the money," a direct call for financial results. But this alone does not provide the monetary evidence needed to ensure that projects add value. The assumed connection between projects and value must give way to the need to show the *amount* of connection. Hence, "show me the real money" is an attempt at establishing credibility. This phase, though critical, still leaves stakeholders with an unanswered question: "Do the monetary benefits linked to the project outweigh the costs?" This final question is the mantra for the new "show me" generation. "Show me the real money and make me believe it is a good investment." The new generation of EHS project sponsors also recognizes that value is more than just a single number, value is what makes the entire organization system tick, hence the need to report value based on people's various definitions.

2.2.4 The New Definition of Value

The changing perspectives of value and the shifts that are occurring in organizations have all led to a new definition of value. Value is not defined as a single number. Rather, its definition is composed of a variety of types of data. Value must be balanced with quantitative and qualitative data as well as financial and nonfinancial perspectives. The data sometimes reflect tactical issues, such as activity as well as strategic issues, such as ROI. Value must be derived at

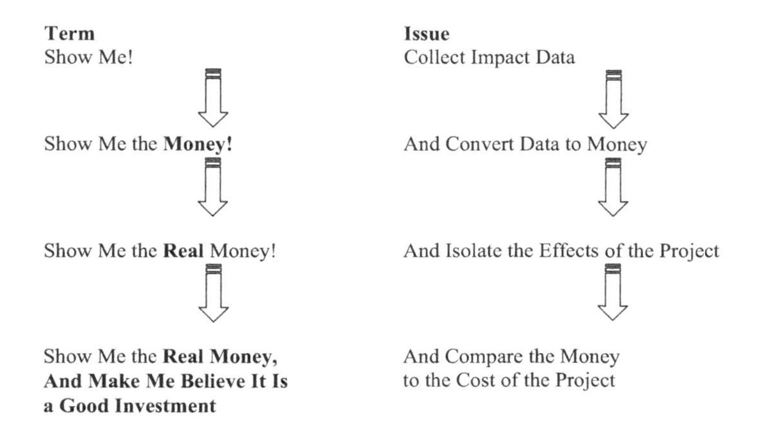

Figure 2.3 The "Show Me" Evolution.

different time frames and it does not necessarily represent a single point in time. It must reflect the value systems that are important to all stakeholders. Data must answer questions such as: So what? How much? How do you know? What is it worth? How can we improve?

Data composing value must be collected from credible sources, using cost-effective methods and value must be action oriented, compelling individuals to change habits and make adjustments in their processes. And yes, value may be intangible, not converted to money.

The processes used to calculate value must be consistent from one project to another. Standards must be in place so that results can be compared. These standards must support conservative outcomes, leaving assumptions to decision makers. The ROI Methodology presented in this book meets all these criteria. It captures six types of data that reflect the issues contained in the new definition of value: reaction, learning, application and implementation, impact, return on investment and intangible benefits, those impact measures not converted to money but that still represent important outcomes.

2.3 Why Now?

In the past decade, a variety of forces have driven additional focus on measuring the impact of EHS projects, including the financial contribution and ROI. These forces have challenged the ways of defining EHS success.

2.3.1 Project Failures

Almost every organization encounters unsuccessful projects, including EHS projects that go astray, costing far too much and failing to deliver on promises. Project disasters occur in business organizations as well as in government and nonprofit organizations. Some project disasters are legendary. Some are swept into closets and covered up but they exist and the numbers are far too large to tolerate. They are rarely reported in the literature or media. These failures have generated increased concern about measuring project and program success—before, during and after implementation. Many critics of these projects suggest that failure can be avoided if:

1. The project is based on a legitimate need from the beginning.
2. Adequate planning is in place at the outset.
3. Data are collected throughout the project to confirm that the implementation is on track and adjustments are made accordingly.
4. An impact study is conducted to detail the project's contribution.

Unfortunately, these steps are sometimes unintentionally omitted, not fully understood or purposely ignored; hence, greater emphasis is being placed on the processes of accountability.

2.3.2 Project Costs

The costs of projects and programs continue to grow. As costs rise, the budgets for these projects become targets for others who would like to have the money for their own projects. What was once considered a mere cost of doing business is now considered an investment to be wisely allocated. These days, the annual direct costs of EHS initiatives are hundreds of billions of dollars in the United States. Some large organizations spend well over one billion dollars every year on their combined EHS efforts. With numbers like these, the huge field of environment, health and safety is no longer considered a compliance expense (as it was twenty years ago), rather, it is regarded as an investment and many executives expect a return, in both EHS terms and economically.

2.3.3 Accountability Trend

A consistent and persistent trend in accountability is evident in organizations across the globe. Today almost every function, process, project or initiative is judged based on higher standards than in the past. In organizations, various teams are attempting to show their worth by capturing and demonstrating the value they add. They compete for funds; therefore, they have to show value. For example, the research and development role must show its value in monetary terms to compete with mainstream processes such as sales and production, which for more than a century have shown their direct monetary value. EHS is not exempt from this trend. While there have been a few studies describing the cost-benefit comparison of EHS projects, the overall cost and elusiveness of EHS calls for a deeper look into accountability.

2.3.4 Business Focus of EHS Managers

In the past, managers of support functions (such as EHS) in government, nonprofit and private sector organizations had more technical than business experience. Today things have changed. While many of these managers still come from engineering and science backgrounds, the most successful of these mangers will supplement their technical backgrounds with formal business education or, at a minimum, thoroughly master specific business concepts. These enlightened managers are more aware of bottom-line issues in the organization and are more knowledgeable about operational and financial concerns. They have studied the use of ROI in their academic preparation, where the ROI Methodology was used to evaluate purchasing equipment, building new facilities or buying a new company. Consequently, they understand and appreciate ROI and are eager to apply it in the area of EHS management.

2.3.5 The Growth of Project Management

EHS initiatives and efforts represent projects in organizations—projects that must be managed with a schedule and budget. Few processes in organizations have grown as much as the use of project management. Just two decades ago it was considered a lone process, attempting to bring organizational and management structure to project implementation. In contrast, today, the Project Management

Institute (PMI), which offers three levels of certification for professional project managers, has more than two hundred thousand members in 125 countries. Jobs are being restructured and designed to focus on projects. The growing use of project management solutions, tools and processes requires a heavy investment for organizations. With this growth and visibility come requests for accountability. Critics want to know if they are producing results—and this includes the management of EHS projects.

2.3.6 Evidence-Based or Fact-Based Management

Recently, an important trend in the use of fact-based or evidence-based management has worked its way into EHS projects. Although many key decisions have been made using instinctive input and gut feeling, more managers are now using sophisticated and detailed processes to show value. Important decisions must be based on more than gut feeling or the blink of an eye. With a comprehensive set of measures, including financial ROI, better organizational decisions regarding people, planet and profit are possible.

When taken seriously, evidence-based management can change how every manager thinks and acts. It is a way of seeing the world and thinking about the craft of management. Evidence-based management proceeds from the premise that using better, deeper logic and facts, to the extent possible, helps leaders do their jobs better. It is based on the belief that facing the hard facts about what works and what doesn't and understanding and rejecting the total nonsense that often passes for sound advice will help organizations perform better. This move to fact-based management sometimes expands measurement to include ROI.

2.3.7 Benchmarking Limitations

For years, executives have been obsessed with benchmarking, using benchmarking studies to compare every type of process, function and activity. Unfortunately, benchmarking has its limitations. First, the concept of best practices is sometimes an elusive issue, particularly with EHS projects. Results of all benchmarking projects do not necessarily represent best practices. In fact, they may represent just the opposite. Benchmarking studies are developed from organizations willing to pay to participate. So, while these organizations may be doing well, they do not represent the many others who are investing their resources in getting the job done. Second, what one organization needs is not always the same as that of another. A young organization will have needs (safety, environmental and otherwise) that are different from an experienced organization with a twenty-year history of EHS efforts. A specific benchmarked measure or process may be limited in its use by some. Finally, the benchmarking data are often lacking measurement data, reflecting few if any measures of success and financial contributions with ROI values. Because of this, executives have asked for more specific internal processes that can show these important measures.

2.3.8 The Executive Appetite for Monetary Value

For years, managers and department heads in public relations, public affairs, corporate communications and human resources convinced executives that their processes could not be measured and that the value of their activities should be taken on faith. Unfortunately, many EHS projects were also implemented as "faith-based" initiatives. Today, executives no longer buy that argument; they demand the same accountability from these functions as they do from the sales and production areas of the organization. "Show me the money" is more than a line from a Tom Cruise movie. Top executives are requiring organizations to shift their measurement and evaluation processes to include the financial impact and ROI.

2.4 EHS Chain of Impact

Sometimes it is helpful to think about the success of an EHS project in terms of a chain of impact that must occur if the project is going to be successful in terms of business contribution. After all, if there is no business contribution, it is unlikely the project will be implemented. The chain of impact includes the five categories, or levels of outcomes, discussed in general terms previously. These five levels of outcomes range from reaction to ROI, but the chain of impact begins with inputs to the process. Inputs, referred to as Level 0, define the people involved in the project, how long it will take it to work, the cost, resources and the activities involved. Obviously, these data are essential to move forward with a project, but they do not speak to the success of the project. It is through reaction, learning and application of knowledge, skill and information that a positive impact on business measures is realized. Stakeholders recognize how much impact is due to the project because a step occurs that will isolate the effects of the project from other influences. Impact measures are converted to monetary terms and compared to the cost to determine the ROI. In addition to these outcomes, intangible benefits are reported. Though they are not a new level of data, intangibles are representative of impact measures and are always reported in addition to the monetary contribution of a project. Figure 2.4 represents this chain of impact that occurs through the implementation of EHS projects.

Reaction is the first level of outcome. Project participants must see the value in EHS initiatives. They must perceive the project is important, necessary, useful and practical. If the reaction is adverse, the project is dead.

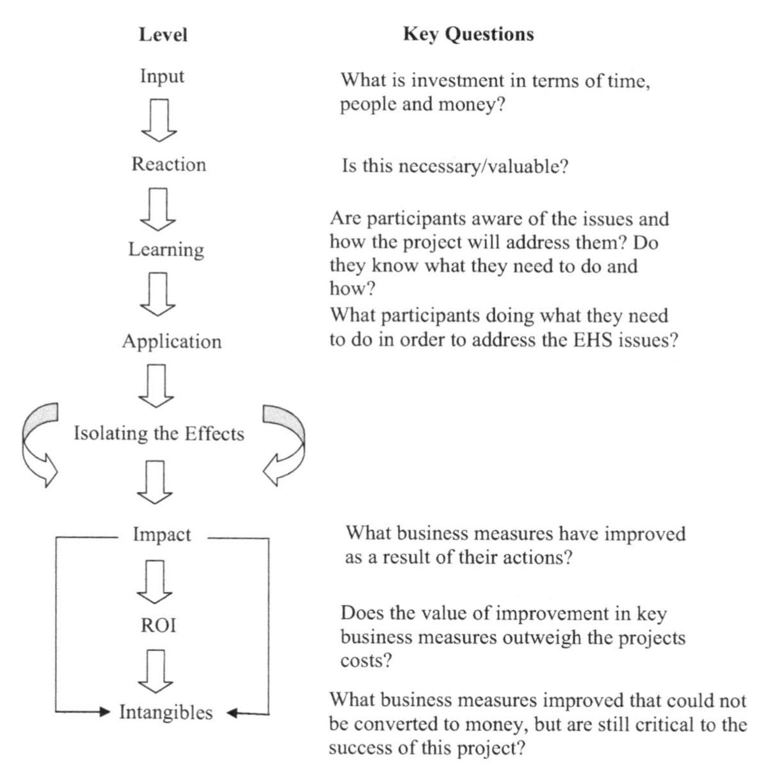

Figure 2.4 EHS Chain of Impact.

Next, there must be learning. With EHS projects, there is a tremendous amount of knowledge acquisition. Learning involves several issues. Participants involved must understand the issue itself, what they need to do, what is involved in the project and be aware of the project's potential success. If participants don't know what to do, how to do it, or why they should do it, the project is dead.

The next level is application, which represents action on the part of the participants involved in the project. They must change their habits, take action and new steps, change procedures and influence others to do the same. This is the activity (i.e., action, behavior) that is essential for the project to be successful. Unfortunately, it is at this level that projects usually break down. Barriers to implementation occur for a variety of reasons. These barriers must be identified and either removed, minimized or circumvented. In addition to avoiding or reducing barriers, enablers must exist to support successful implementation. These must be identified and enhanced to allow people to fully engage in projects and have success with them.

Impact is the next level of outcome. Outcomes at this level represent a consequence or improvement in key performance indicators resulting from application. Impact data answer the questions, So what? What will be the impact? Will there be a cost savings? Will there be a reduction in waste? Will there be improvement? As project implementation succeeds, efficiencies result, meaning we use less electricity or fuel, we reduce lost-time accidents, injuries and safety violations and we increase productivity, reduce health-care costs and improve the engagement of our employees. These impact measures, which are usually available in organizational record systems, represent the most important data set for top executives, leaders and administrators.

The next level of outcome is ROI. This, for some, is the ultimate evaluation comparing the costs versus the benefits of projects. If the monetary benefits exceed the costs of the project it is perceived as a good investment economically. Positive economic returns are critical to the continuation of any initiative.

The sixth important data set concerns intangible benefits. Again, this is not a new level of outcome measures. Intangibles are the impact measures that cannot be converted to money credibly with minimum resources. These are important and often provide the rationale for many of the early projects an organization selects. Image, reputation and brand are powerful measures, even without the monetary value tied to them. These measures are often reported as intangible. Positive results at each level of the chain of impact are critical to achieving a positive ROI.

2.5 Shortcomings of Current Measurement and Evaluation Systems

For the most part, the current systems of measuring and evaluating EHS projects fall short of providing the proper system for accountability, process improvement and results generation. As we examine the ways in which projects are evaluated, we see ten areas for improvement. Table 2-1 lists each problem or issue and presents what is needed for improvement. It also shows how the ROI Methodology meets all ten of these areas.

2.5.1 Focus of Use

To some people, evaluation is analogous to auditing. Someone checks to see if the project is working as planned and a report is generated to indicate if there is a problem. Many capital expenditures, for example, are often implemented this way. The board approves and project and after it is completed internal auditors produce a board-mandated, follow-up report. This report points out how things are working and/or not working, often at a point when it is too late to make any changes. Even in government, social sciences and education, the evaluations are often structured in a similar way. For example, our friends in the British government tell us that when new projects are approved and implemented, there is always a funding pool set aside for evaluation. When the project is completed, an evaluation is conducted and a detailed report is sent to appropriate government authorities. Unfortunately, these reports reveal that many of the programs are not working and it is too late to do anything about it. Even worse, the people who implemented the project are either no longer there or no longer care. Accountability calls for evaluation reports. But all too often these reports result in punitive consequences, blaming the usual suspects, or they serve as the basis for performance review of those involved.

It is no surprise that auditing with a punitive twist does not work with EHS projects. These project evaluations must be viewed

Table 2.1 Problems and Opportunities with Current Measurement Systems.

Topic	Problem or Issue	What Is Needed	ROI Methodology
Focus of use	Audit focus—punitive slant	Process improvement focus	This is the number one use for the ROI Methodology
Standards	Limited standards exist	Standards needed for consistency and comparison	Twelve standards accepted by users
Types of data	Only one of two data types	Need a balanced set of data	Six types of data representing quantitative, qualitative, financial and non-financial data
Dynamic adjustments	Not dynamic—does not allow for adjustments early in the project cycle	A dynamic process is needed so that adjustments are made — early and often	Adjusts for improvement at four levels and at different time frames
Connectivity	Not respectful of the chain of impact that must exist to achieve a positive impact	Data must be collected at each stage of the chain	Every stage has data collection and a method to isolate the project's contribution
Approach	Activity based	Results based	Twelve areas for results-based processes
Conservative nature	Analysis not very conservative	A conservative approach is needed for buy in	Very conservative: CFO and CEO friendly
Simplicity	Not user friendly— too complex	User friendly, simple steps	Ten logical steps
Theoretical foundation	Not based on sound principles	Should be based on theoretical framework	Endorsed by hundreds of professors and researchers; grounded in research and practice
Acceptance	Not adopted by many organizations	Should be used by many	More than 7,000 organizations using the ROI Methodology

as process improvement opportunities. If a project is not working, then changes must take place for it to be successful in the future. Process improvement is the focus of the ROI Methodology.

2.5.2 Standards

Unfortunately, many of the approaches to evaluate EHS projects lack standards unless the project is a capital expenditure. If it is a capital expenditure, the evaluation process is covered by generally accepted accounting principles (GAAP). However, many EHS projects are not capital expenditures. In these instances, standards must be employed to ensure consistent application and reliable results.

Overall, the standards provided by the ROI Methodology as described in Chapter 4 provide consistency, conservatism and cost savings as the project is implemented. This enables the results of one project to be compared to those of another and the project results to be perceived as credible.

2.5.3 Types of Data

The types of data that must be collected vary. Unfortunately, many projects focus on impact measures alone, showing cost savings, less waste, improved productivity or less energy consumption. These are the measures that will change if a project is implemented.

What is needed is a balanced set of data that contains financial and non-financial measures as well as qualitative and quantitative data. Multiple types of data not only show the results of investing in EHS projects, but they also help to explain how the results evolved and how to improve them over time. As shown in Figure 2.4, the ROI Methodology develops six types of data including reaction, learning, application, impact, ROI and intangible benefits.

2.5.4 Dynamic Adjustments

As mentioned earlier, a comprehensive measurement system must allow opportunities to collect data throughout the project implementation rather than waiting until it has been implemented (perhaps only to find out it never worked from the beginning). Reaction and learning data must be captured early. Application data are captured when project participants are applying knowledge skills and information routinely. All of these data are used to make adjustments in the project to ensure success, not just to report post-program outcomes at a point that is too late to make a difference. Impact data are collected after routine application has occurred and represent the consequences of implementation. These data may not be connected to the project and must be monitored and reviewed in conjunction with the other levels of data. Once the connection is made between impact and the project, a credible ROI is calculated. The ROI Methodology allows data to be collected early and often, using the information gathered to make adjustments throughout project implementation.

2.5.5 Connectivity

For many measurement schemes, it is difficult to see the connection between the project and the results. It is often a mystery as to how much of the reported improvement is connected to the project or if there even is a connection.

Data need to be collected through the process so that the chain of impact is validated. In addition, when the business measure improves, there must be a method to isolate the effects of the project on the data to validate the connection to the measure. The ROI Methodology described in this book provides both methods.

2.5.6 Approach

Too often, the measurement schemes are focused on activities. People are busy. They are involved. Things are happening. Activity is everywhere. However, activities sometimes are not connected to impact. The ROI Methodology is based on results. Not only does the process track monetary results, but it monitors success of a project from rollout to completion, allowing projects owners to make incremental adjustments when necessary. By having a measurement process in place, the likelihood of positive results increases. A complete focus on results versus activity improves the chances that people will react positively, change their attitude and apply necessary actions, which leads to a positive impact on immediate and long-term outcomes.

2.5.7 Conservative Nature

Many assumptions must be made in the collection and analysis of data. If these assumptions are not conservative then the numbers are overstated and unbelievable. This decreases the likelihood of accuracy and buy-in. The ROI Methodology has conservative assumptions known as guiding principles that are CFO and CEO friendly.

2.5.8 Simplicity

Too often, measurement systems are complex and confusing for practical use, which leaves users skeptical and reluctant to embrace them. The process must be user-friendly, with simple, logical and sequential steps. It must be void of sophisticated statistical analysis and complicated financial information, at least for the projects that involve participants who lack statistical expertise. The ROI Methodology is a step-by-step, logical process that is user-friendly, even to those who do not have statistical or financial backgrounds.

2.5.9 Theoretical Foundation

Sometimes measurement systems are not based on sound principles. They use catchy terms and inconvenient processes that make certain researchers and professors skeptical. A measurement system must be based on sound principles and theoretical frameworks. Ideally, it must use accepted processes as it is implemented. The ROI Methodology has been endorsed by hundreds of professors and researchers who have participated in our certification process with a goal of making it better. Over 100 universities have adopted books authored by top executives of the ROI Institute.

2.5.10 Acceptance

Practitioners in all types of organizations must use the adopted measurement system. Too often, the measurement scheme is presented as theoretical but lacks evidence of widespread use. The ROI Methodology, first described in publications in the 1970s and 1980s (with an entire book devoted to it in 1997) now enjoys 5,000 users (Phillips, 1997). It is used in all types of projects and programs from technology, quality, marketing and human resources. In recent years it has been adopted for EHS projects and efforts.

The success of the ROI Methodology will be underscored in detail throughout this book with examples of applications. It is a comprehensive process that meets the important needs and challenges of those striving for successful green projects.

2.6 EHS Leadership: A Requirement for Success

Strong leadership is necessary for projects to work. Leaders must ensure that EHS projects are designed to achieve results rather than just to improve image. These projects and efforts must deliver the value that is needed by all stakeholders. Table 2-2 shows the twelve actions that must be taken to provide effective, results-based EHS leadership, which is critical to delivering results at the ultimate

Table 2.2 EHS Leadership for Results.

1. Allocate appropriate resources for EHS projects and efforts.
2. Assign responsibilities for EHS projects and programs.
3. Link EHS projects and programs to specific business needs.
4. Address performance issues involving the key stakeholders in the project, identifying the behavior that must change.
5. Understand what individuals must know in order to make projects successful, addressing the specific learning needs.
6. Develop objectives for the projects at multiple levels, including reaction, learning, application, impact and yes, ROI.
7. Create expectations for the projects' success with all stakeholders involved, detailing their role and responsibilities.
8. Address the barriers to the successful project early on so that the barriers can be removed, minimized, or circumvented.
9. Establish the level of evaluation needed for each project at the beginning so that participants will understand the focus.
10. Develop partnerships with key administrators, managers and other principle participants who can make the project successful.
11. Ensure that measures are taken and the evaluation is complete with collection and analysis of various types of data.
12. Communicate project results to the appropriate stakeholders as often as necessary to focus on process improvement.

level, ROI. However, only one of the items involves data collection and evaluation - number 11. The remaining leadership areas represent steps and processes that must be addressed throughout an EHS project's cycle. We developed these actions after observing, studying, conducting and reviewing thousands of ROI studies. At the ROI Institute, we know what keeps projects working and what makes them successful. Following these twelve leadership roles can ensure EHS success.

2.7 Challenges Along the Way

The journey to increased accountability and the quest to show monetary value for benefits, including ROI, are not going unchallenged. This movement represents a cultural shift for many individuals, a systemic change in processes and often a complete rethinking of the initiation, delivery and maintenance of green projects and sustainability processes in organizations.

2.7.1 The Commitment Dilemma

Commitment is key to the successful implementation of the ROI Methodology. Some users hope to obtain an immediate ROI value using the ROI Methodology, but there is more to it than a simple calculation. To achieve success, a commitment is needed to make changes when the data reveal that the change is imperative. Also, commitment is needed to properly use the information provided by the process. Finally, there must be discipline and determination to make it work and to implement it successfully.

2.7.2 Lack of Preparation and Skills

Although the interest in proving the value of EHS projects and measuring ROI is now heightened and much progress has been made, these are still issues that challenge even the most sophisticated and progressive functions. One problem is the lack of preparation and skills needed to conduct these types of analyses. The preparation for EHS project leaders often omits necessary skill building for measurement and evaluation. Rarely do the curricula in degree programs or professional development courses include processes and techniques to show accountability at this level. Consequently,

3

Investing in Environment, Safety and Health Initiatives

How Companies Approach the Investment

Abstract
This chapter revisits the rationale for the investment in EHS within an organization. For most organizations, particularly those in the manufacturing and service industries, this investment is huge and growing. The specific amount of investment varies, even within the same field. Ultimately, top executives set the investment level for EHS initiatives. Some rely only on benchmarking, while others adopt more well-defined strategies. Still others opt to avoid the investment altogether. The chapter presents five often-used strategies for setting these investment levels.

Keywords: EHS investment, ROI, minimum investment, overinvesting, benchmarking, avoiding investment, payoff

3.1 Overview

This chapter analyzes five investment options in some detail and offers specific recommendations at the end of each section and at the end of the chapter. Although an investment level may be set initially by using one or more of the these approaches, or some other process, the level of spending on EHS initiatives is an issue that should be reviewed periodically.

The five strategies often used, either intentionally or unintentionally, are:

1. Avoid the investment
2. Invest the minimum
3. Invest with the rest

Jack Phillips, Patti Phillips, and Al Pulliam, Measuring ROI in Environment, Health, and Safety, (49–72) 2014 © Scrivener Publishing LLC

4. Invest until it hurts
5. Invest when there is a payoff

3.2 Strategy 1: Avoid the Investment

Some executives prefer to take a passive role with EHS efforts, attempting to avoid the investment altogether. After all, no one is making them do it. While appearing somewhat dysfunctional, this approach has proven effective for some organizations, depending on the strategic focus, resource accountability and perception of the importance of EHS. For example, many companies outsource manufacturing to India, China, Bangladesh and other countries to lower costs. The primary cost item for this decision is labor cost. EHS is a significant part of labor costs. There are a variety of factors driving this strategy, some of which may be familiar to you.

3.2.1 Forces Driving This Strategy

Several factors motivate an executive to approach investing in EHS initiatives using this strategy. First, when the operating unit has little visibility with the public, the situation is "out of sight, out of mind." Organizational leaders may feel no compelling reason to require investing in EHS when the plant or call center is not in the community. Second, when the organization is small, some small businesses feel they cannot make a difference; they view themselves as only a small part of the picture and investing in the environment movement is not a concern. The third driving force is when organizations have chosen to outsource most of what they do: production, marketing, advertising, distribution, logistics and sales. With only a few employees, EHS is not an issue for them. The fourth factor is when survival of the business is of utmost concern. Some industries and organizations are struggling to survive and when this is the case, spending money to protect the environment and the well being of employees is not at the top of the list. If you have not made a profit in two years, would you consider investing in a wellness and fitness program?

3.2.1.1 Out of Sight, Out of Mind

Many organizations do not need to deal directly with the public. They do not have direct consumers. They do not make products that are sold directly to consumers or provide services that are delivered directly to them. Instead, they sell to other business organizations (B2B). With no consumer contact, there is little pressure from consumers to go green.

Others produce obscure products that are used only as a small part of another product. For the most part, these are small, privately held organizations that serve a specific niche. For example, Buntrock Industries, based in Williamsburg, Virginia, serves only the investment casting industry. Buntrock, a small business, serves others who apply their equipment and materials to the process of investment casting. They do not necessarily invest in EHS initiatives.

Another example, a small manufacturing company, makes plastic mugs for the specialty advertising industry. These mugs always feature someone else's brand or logo, making the company "out of mind" in terms of the ultimate consumer and the public. The plant, which employs about 200 people, is privately held and the owners are naturally concerned about making a profit and employee health. This type of business has a huge impact on the environment, yet they invest essentially nothing in the green movement. The strategy is to avoid it, except when it is legally required. The owners of the company are convinced that investing in EHS projects is not necessarily a wise use of their money, because there is no immediate payoff. They are not convinced that cost savings are forthcoming to this sort of change and they have not taken the time to explore and understand the green issue. Their only requirements are meeting health and safety standards, adhering to requirements for hazardous chemicals and following environmental regulations for disposing of the materials and waste for the plant.

3.2.1.2 Too Small for EHS

While large organizations attract the most attention, most of the businesses in the United States are small. Small businesses represent both a challenge and an opportunity when it comes to EHS issues. Because of the total numbers of small businesses, getting them to embrace the EHS movement will make a significant impact in many different areas. However, the challenge is convincing them to do it. Because they are small (and inconsequential from their perspective), they see no reason to invest in EHS; they see no value.

Some small companies are more willing to change because they sell directly to the customer. For example, patrons of a small dry cleaning business would not necessarily require their dry cleaners to be involved in the green movement or the safety and wellness of employees. However, because the cleaning materials used may include harmful chemicals, consumers might be interested to know that the company is also taking care of the environment. Also, the customers may be concerned about employee safety. In another example, customers of a small print and copy shop might be interested in knowing that the print shop is environmentally sensitive, using recycled paper and sustainable products

For small companies such as these, if customer interest in EHS is strong enough, owners may be compelled to rethink their investment strategy.

3.2.1.3 *Outsourced Production and Services*

Some organizations outsource most of what they do, which leaves only a small central group of employees. They outsource production, marketing and distribution. Because other people are doing most of the work and driving the business, they are less likely to focus on green initiatives. For example, one of our publishers outsources printing, editing, design, marketing, book inventory and distribution. The books are sold in bookstores and there is no sales force. Because the publisher has a small staff, why should they be concerned about EHS? Customers and authors do not visit their offices. There are no external shareholders. There is no pressure for them to pursue EHS projects.

Companies outsourcing manufacturing still find criticism for lack of EHS emphasis. The gruesome fire that killed 112 garment workers in Bangladesh on November 25, 2012 underscored a stubborn problem that has dogged retailers for years. The International Labor Rights Forum says that an estimated 600 garment workers in the country have died in fires since 2005. These persistent safety lapses have fixed a spotlight on U.S. retailers, prompting criticism from labor groups that say companies do not do enough to protect the people who make the clothes they sell. "Workers' lives are on the line while they're making clothing for export to the U.S. market," said Liana Foxvog, spokeswoman for the labor rights group. Stores that sell clothes in the United States are on the end of a supply chain that starts with cheap labor at factories in South Asia. Minneapolis-based Target Corp. is no exception. The November 25 fire happened at a factory called Tazreen Fashions, whose parent company was a Target supplier until 2008. Tragedy struck the Minneapolis company's supply chain directly two years before when a December 2010 fire killed 29 at a factory that supplied clothes to Target, J.C. Penney, Kohl's and Abercrombie & Fitch.

3.2.1.4 *Survival Is an Issue*

Some organizations, particularly those in declining industries and those that must deal with a particular economic crisis because of industry issues or external environment, face the prospect that they may not survive much longer. They may characterize investing in EHS as similar to rearranging the chairs on the Titanic. It is not a priority issue for them, so they avoid it altogether. For example, over a ten-year period, a well-known publishing and self-help company that was traded on the New York Stock Exchange was struggling to survive but making no profit. The stock fell from thirty-five dollars a share to fifty-five cents. During this period, there was almost no focus on EHS projects or sustainability initiatives. Although they were in the public eye, survival was the issue. When will we make a profit? Can we make it? Which strategy is going to work? These are the critical issues. In these circumstances, green initiatives just do not make the top of the list.

Some entire industries are facing these kinds of problems. For example, although newspapers have a tremendous impact on the environment and employee safety, many are going out of business. This is not because of the economic decline, but because of the growth of and demand for digital media. Expectations are that print news will be extinct in the future. It is unlikely that these organizations are focused on going green.

3.2.2 Techniques to Persuade a Change in Investment Strategy

Changing the attitude and behavior of this "no investment" group is problematic, but it represents a great opportunity because of the number of people in this category. Essentially, the goal is for them to move from this strategy of deliberately avoiding the investment to investing more. This fundamental switch requires several actions.

3.2.2.1 *Show the Value*

Educational and awareness sessions are needed to show this particular group that by investing in EHS they can make a difference and reduce costs or increase profit as a result. One person can make a little difference and many individuals can make a huge change. They must see that there is not only a reason to invest in EHS projects, but that there is some value in doing so. They must realize that most EHS projects will result in cost savings, but only if they are implemented properly, if the projects are measured systematically and if the results are used to drive improvement.

3.2.2.2 *Engage Industry Associations*

Almost every type of organization in this category is part of an industry or trade association. By informing their members about environment, health and safety issues, emphasizing what is available, what is possible and what is essential, industry associations can influence these organizations. They can demonstrate the economic opportunity to invest and how they can achieve a positive return on that investment. Educational sessions, workshops, webinars, conferences and newsletters can help. Producing case studies, booklets, pamphlets and other materials will let the members know about EHS issues and what they should do about them. In industries that have a tremendous impact on the environment, this is absolutely critical. Otherwise they will be forced into actions through legislation, which is never the desired position.

3.2.2.3 *Engage the Business Groups*

Just as a trade association may help to persuade reluctant organizations to embrace the green movement, business groups such as the Chamber of Commerce, National Association of Manufacturers and others can push the EHS agenda. These support organizations can provide information, educate, inform and show value in the same way as trade associations.

3.2.2.4 *Engage Cities and Communities*

Local communities can reach reluctant organizations through citizen communications, community groups and networks that bring the EHS message to all of the constituents and business owners. Involving, educating, showing, helping and enabling reluctant businesses are all possible approaches. Cities and communities must set the example. Ideally, a city will be environmentally friendly and promote green projects. The city will use safe work procedures, create safe and healthy work places and promote and implement wellness and fitness programs.

go deeper than that, because it will affect profits too much. Sometimes this is the case with retail stores and restaurants, particularly fast-food restaurants. They are unable to invest heavily because they have such low profits or sometimes no profits. However, there are many good examples in this area where leaders have stepped up to this challenge. Wal-Mart, the world's largest retailer, is obviously a low-margin industry. Yet Wal-Mart is probably having a greater impact on the environment than any other organization, because they are forcing their suppliers to go green, eliminating or reducing packaging, saving paper, (which saves trees) and reducing transportation costs (thereby reducing carbon emissions). It is having less success in promoting the health and well being of its employees.

Wal-Mart's size ensures that it has a lot of clout, but more actions can be taken in smaller retail organizations as well. While many fast-food, quick-stop restaurants ignore this issue, investing only the minimum for obvious image reasons, Starbucks has invested heavily in the environment beyond the noticeable signs in the store.

3.3.2.3 Value

Many organizations invest the minimum because they are not convinced that there is value in EHS projects. For example, they perceive green as a cost—not as an opportunity. They are unaware of what investing in green can achieve for their bottom line as well as the environment. Yes, it is true that not every green project will have a quick monetary return, but the right ones will. Even the projects that have a negative ROI will have a positive payoff in intangibles. The projects that do have a positive payoff can help fund those that do not. They perceive investments in employee health as risky in terms of payoff. They assume that the benefits just do not outweigh the costs. Investments in safety, beyond simple accident prevention, seem to be unnecessary since the payoff is unclear. Fortunately, the opposite is the case. These investments can have a high payoff.

3.3.2.4 Survival

As in the case of investment strategy number one, some organizations invest the minimum because of their need to focus on survival. While they may value the idea of investing more in EHS, their bottom line takes priority over their desire to contribute to a greater good. Survival is the key, so they invest only the minimum in green and remain focused on their core business.

3.3.3 Techniques to Persuade These Organizations to Change Their Strategy

Attempting to influence leaders to change their strategy and invest more than the minimum requires several coordinated efforts. Changes in this category are similar to the recommendations for addressing the previous strategy of avoiding the investment altogether. The difference is that this second group is already investing; they see a need to invest, but they can do much more.

This group needs examples of what others are doing. They need to see the payoff. Positive examples in their type of business or industry can help. Fortuitously, there are hundreds of positive examples, such as Alcoa, Nestle, BASF, Dow, SC Johnson, Shell, DuPont, Cargill and Adidas. They must see and understand the value of investing more in this area.

As with the previous strategy, trade associations can be helpful. These groups must encourage organizations to invest more by showing benchmarking data, presenting real case studies and by showing the economic value of investing more in these projects in a convincing, credible way.

3.4 Strategy 3: Invest with the Rest

Many executives prefer to invest in EHS efforts at the same level that others invest. This approach involves collecting data from a variety of comparable organizations, often perceived as implementing best practices, to determine the extent to which those organizations invest in EHS. The benchmarking data are used to drive improvement or changes, if necessary, to achieve the benchmark level.

3.4.1 Forces Driving the Strategy

There has been phenomenal growth in benchmarking in the last two decades. Virtually every function in an organization has been involved in some type of benchmarking to evaluate activities, practices, structure and results. Because of its popularity and effectiveness, many environmental, safety and health executives use benchmarking to show the value of and investment level for these initiatives. In some situations, the benchmarking process develops standards of excellence from "best practice" organizations. The cost of connecting to existing benchmarking projects is often low, especially when considering the available data. However, when a customized benchmarking project is needed, the costs are higher. Organizations such as the Triple Bottom Line Alliance have benchmarking data. Other helpful benchmarks come from the U.S. Green Building Council, National Safety Council, The Carbon Consultancy, Ltd. and the World Business Summit on Climate Change. These sources provide opportunities to understand and validate investments in green initiatives.

An important force driving the invest-with-the-rest strategy is that it is a safe approach. Benchmarking has been accepted as a standard management tool, often required and suggested by top executives. It is a low-risk strategy. The decisions made as a result of benchmarking, when proven to be ineffective, can easily be blamed on the faulty sources or faulty processes, not the individuals who initiated or secured the data.

Benchmarking is a strategy that should be used in conjunction with other approaches. With its low-cost approach, benchmarking provides another view of the EHS function and the investment it requires.

3.4.2 Benchmark Measures

Investment benchmarks are captured in a variety of benchmarking studies focused on a few measures. For example, it may be helpful to understand how much is invested annually or quarterly for a particular employee, category or group with data showing the investment per employee. This is particularly helpful when EHS projects involve many employees. Similar data can be captured for customers or suppliers. How much should a company invest to educate and assist customers with EHS issues? How much should be invested in suppliers, encouraging, assisting or requiring them to practice EHS? Benchmark data may provide insight.

Another potential measure is the investment in EHS projects as a percent of total employee payroll. The numbers typically range from 2.0 percent to 5.0 percent. Best practice is on the high side. A similar measure is the total investment in EHS as a percent of revenue. This is particularly helpful in the construction, manufacturing and process industries, where the cost of going EHS is very expensive and a portion of the revenue is allocated to this issue. Another measure that is helpful in most situations is to consider EHS initiatives as a percent of operating costs. This measure recognizes that most EHS projects come from and support, the operational issues in an organization and it is helpful to compare the costs with other operational expenditures.

The total investment in EHS can be divided into different job groups, categorized by department, division, regions or units. Still, other ways to analyze the investment is by functional categories in the life cycle of an EHS project, such as analysis, design and development, implementation and delivery, coordination and management and measurement and evaluation. This type of breakdown can be helpful. The investments in analysis and measurement and evaluation are usually too low.

For cities and community groups, similar sets of benchmark measures are sometimes available. The investment per citizen, the investment as a percent of budget and the investment as a per-cost of tax revenue are examples.

In short, the investment number must represent a meaningful value for the organization, particularly when it comes from organizations representing best practices.

3.4.3 Concerns with This Strategy

Several issues that often inhibit the benchmarking process should be addressed. Benchmarking involves three challenges. The first challenge is to understand the sources that currently exist for benchmarking studies. Respected organizations are needed for benchmarking studies because having credible data is important. It is even more difficult to benchmark at the international level. A replication process is necessary for benchmarking in each country.

The second concern is the organizations participating in the benchmarking study. Data must come from organizations regarded as best practice leaders that are similar to other organizations. Not all benchmarking sources represent best practices. Often, they include any organization willing to pay the price to participate.

A third concern is the benchmarking measurements. The metrics must be meaningful, respected and comparable. Some benchmark reports contain data that are not easily replicated or easily obtained in organizations. Businesses use "competitive intelligence" to drive business decisions based on comparative and competitive data. It is important for EHS executives to determine the right measures for their own organization and management. Once shared, the measure should be used routinely in order to make comparisons. Measures should be replicable and easily obtained.

3.4.4 Customized Benchmarking

Concerns about benchmarking may leave executives with little choice but to develop a customized benchmarking project to address their organization's interests and needs. If more organizations developed their own benchmarking studies, there would be more available data from the various partners. Figure 3.1 shows a seven-phase benchmarking process that can be used to develop the custom-designed benchmarking project.

3.4.5 Advantages and Disadvantages of This Strategy

Benchmarking satisfies a variety of needs and is used in several important applications. It is helpful in strategic planning for the EHS function and in decision-making processes used to determine the desired investment level. Information and measures derived from the benchmarking process can enable executives to meet

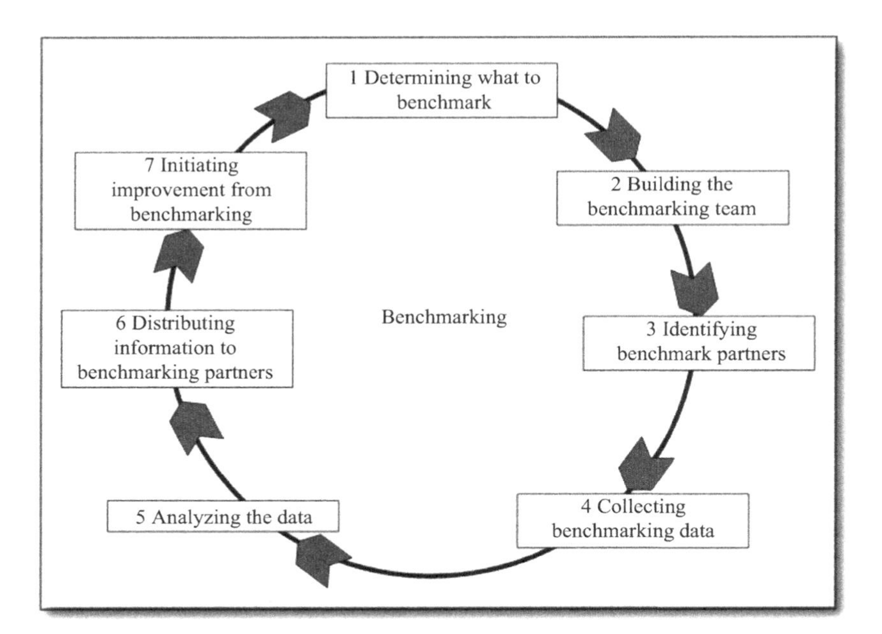

Figure 3.1 Phases of the Benchmarking Process.

strategic objectives. Benchmarking is also useful in identifying trends and critical issues for EHS project management. Measures from benchmarking can become the standard of excellence for an activity, function, system, practice, program or specific initiative. It has become an important measurement tool for senior executives.

The benchmarking process is not without its share of problems, consequences and issues, however. Benchmarking must be viewed as a learning process, not necessarily a process to replicate what others have accomplished. Each organization is different; what is needed in one organization may not be the same as in another. Also, developing a custom-designed benchmarking project is time consuming. It requires discipline to keep the project on schedule, within the budget and on track to drive continuous process improvement. Determining what the best practices are is an elusive goal. Benchmarking can create the illusion that average data, taken from a group willing to participate in a study, represents best practices. National and international data are a difficult issue that often limits benchmarking as a global tool. Finally, benchmarking is not a quick fix; it is a long-term improvement process and one that needs to be continually replicated to be valuable and accurate.

3.5 Strategy 4: Invest Until It Hurts

While some organizations invest at the same level of other organizations, many operate under the premise that more is better. They overinvest in EHS. The results of this approach can be both disappointing and disastrous. A few executives do this intentionally; most do it unknowingly. Either way, this is a strategy that deserves serious attention, because the investment in EHS is beyond what is needed to meet the goals and mission of the organization. Executives approve or implement almost every EHS project they see or discover and explore every idea that comes over the horizon.

3.5.1 Rationale for the Strategy

Some advocates suggest that overinvesting in EHS initiatives is not an important issue—after all, they think, the more you invest, the more success with green projects. The environment is protected, employees are safer and the workplace and staff are healthier. Also, they suggest that this issue is too important for overinvestment to even be possible. You cannot spend too much on protecting the environment and employees. However, others will argue that overinvesting occurs regularly and is unnecessarily burdening organizations with excessive operating costs. Overinvesting puts pressure on others to follow suit, thus creating an artificial new benchmark. As noted, this is often not a deliberate strategy. Rather, executives are usually unaware that the increased spending is not adding value.

3.5.2 Signs of Overinvesting

Many signs indicate that companies are overinvesting in EHS projects. For example, consider the comments of the CEO of a major retail company, who had to announce a disappointing financial performance. In an interview, the CEO indicated that the company's poor performance was due, in part, to the excessive amount of sustainability investment. In this case, employees enjoyed participating in green projects and store managers supported these efforts to the extreme. The result was that employees were away from work and there was not enough staff to serve the customers, causing customer dissatisfaction and ultimately loss in revenue. There were no data to show the value of these green projects.

In another example, an automotive supplier, once the shining star for EHS work, had a commitment that 25 percent of operating expenses would be invested each year in EHS projects. Manager bonuses were attached to this goal and were trimmed significantly if targets were not met. As expected, all types of EHS projects were initiated. Some employees and managers complained that they were taking on too many projects, often unrelated to their work, simply to meet this goal. What was once designed to show a commitment to EHS turned into an expensive practice and, in some cases, a major turnoff in the eyes of employees. Some employees suggested that the money gained from these projects should be distributed to them as a special bonus. The company had little data about the success of these projects. When the economy dipped, the funding stopped. No data were available to show the results of the projects.

There has to be a balance for spending enough to make employees safe and healthy and overspending "just to be safe". We all have seen workplaces that take safety to an extreme, leaving us wondering about the need for all the effort. For example, aircraft designers have said for years that an aircraft could be designed so that it would not crash. Back up and redundant systems will always keep it going. The only problem is that there would be no room for passengers or cargo. While this is an extreme example, the point is well taken. Avoid investing too much.

3.5.3 Forces Driving This Strategy

Several forces cause excessive spending. Some of these are realistic challenges but others are mythical. Either way, they cause firms to overinvest. In the last two decades, EHS has been a battle cry of many organizational leaders. A few executives were willing to do almost anything to focus on the issue. This often led to investing excessively in EHS, well beyond what would be necessary or acceptable in many situations. The conventional wisdom was that offering all types of EHS initiatives would fix the problem and was necessary for business survival. However, many organizations, even industries, were able to do well without having to resort to this strategy.

Some executives spend excessively to remain competitive in the market. They must attract and maintain highly capable employees and have a flawless image with consumers. Consequently, they are willing to invest heavily in EHS projects. They sometimes offer projects that can cost the company in operating profits. They want certain capabilities and are willing to invest to keep their talent. The healthy, safe, environmentally sensitive image becomes an important competitive recruiting strategy.

Some executives have an appetite for new fads. They have never met one they did not like, so they adopt new fads at every turn, adding additional costs. The landscape is littered with EHS projects and dozens of solutions. Once a fad is in place, it is hard to remove; this adds layers of projects and goes beyond what is necessary or economically viable. Some of these executives become the leaders of this movement, garnering much recognition and praise and consequently creating more projects.

Spending too much on EHS initiatives can occur because executives are unwilling or are unable to conduct the proper initial analysis to see if the project is needed. A proper analysis will indicate if the specific EHS project is the right solution to a particular problem or concern. Without the proper analysis, EHS projects are implemented when they are not needed, possibly wasting money.

Some executives spend an excessive amount on EHS because they can afford to do so. Their organizations are profitable, enjoying high margins and ample growth and the executives want to share the wealth to protect employees and save the planet. Many high-tech companies have made tremendous amounts of money and their CEOs spend excessively on environmental, health and safety issues because they feel it is a necessity. However, there are little data to show the value of these expenditures. Furthermore, when the economy turns sour, a new CEO is appointed or the company is sold, the company may not be able to sustain these projects.

3.5.4 Concerns with This Strategy

Obviously, spending excessively is not a recommended strategy. There are many potential problems with this approach, not only for the company but also for the industry. The most significant disadvantage of overinvestment is less-than-optimal financial performance. By definition, this strategy involves investing more than necessary to meet the objectives of the organization. While some increases in EHS investment yield additional financial results, for many EHS projects there is evidence of a point of diminishing return, where the added benefits peak and then drop as investments continue. This relationship between performance and investment in EHS initiatives is depicted in Figure 3.2. Excessive investing can eventually deteriorate performance in the organization, particularly in industries where the EHS expense is an extraordinarily high percentage of the total operating cost.

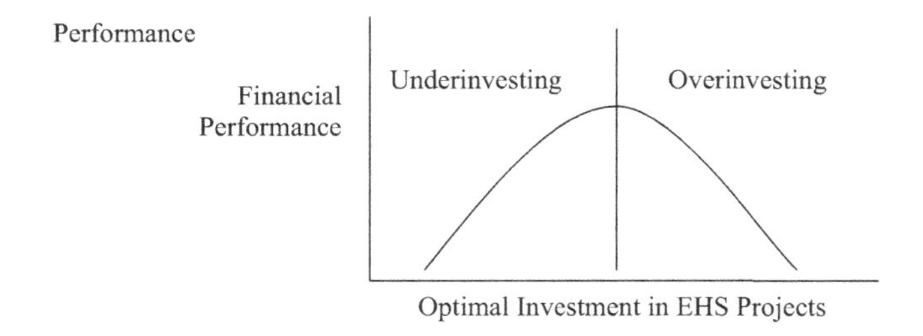

Figure 3.2 The Relationship Between Investing in Green Projects and Financial.

3.6 Strategy 5: Invest as Long as there is Payoff

Some executives prefer to invest in EHS projects when there is evidence that the investment is providing benefits. They often compare monetary benefits of EHS projects with the costs of the projects. This strategy is becoming more popular following the increased interest in accountability, particularly with the use of return on investment (ROI) as a project evaluation tool. With this strategy, all EHS initiatives are evaluated and a few key projects are measured at the ROI level. The ROI in these projects is calculated the same way as the ROI is calculated for investments in buildings or equipment.

3.6.1 The Strategy

This strategy focuses on implementing a comprehensive measurement and evaluation process for expenditures in an organization. This involves the possibility of capturing different types of data in the chain of impact. These data include reaction, learning, application, impact, ROI and intangible benefits. Using this philosophy, only a small number of projects are evaluated at the ROI level, whereas every project is evaluated at the reaction level. Also, when business impact and ROI are developed for EHS projects, one or more techniques are used to isolate the impact of the projects on the business data.

3.6.2 Forces Driving Change

Although the trend toward additional accountability has been increasing over the last decade, there are several reasons why this strategy is critical at this time. In the last few years, the demonstration of the value of EHS projects to the organization has been at the forefront of executive agendas. With this mandate, EHS team members have had to develop the skill to communicate with other managers the contribution to the financial bottom line in the language of business. In a world where financial results are measured, a failure to measure the success of green initiatives destines the project team to risk oversight, neglect and potential failure. The leader responsible for EHS initiatives must be able to evaluate, in financial terms, the costs and benefits of selected projects and practices.

The increasing cost of EHS is another driving force. As we discuss throughout this book, investment in EHS is quite large and growing. As these budgets continue to grow—often at a growth rate outpacing other functions of the organization—the costs alone are requiring some executives to question the value of the investment. Some executives are requesting that the impact of EHS projects be forecast before they are implemented. In some cases, the ROI is required at budget review time.

Let's say that a production manager proposes investing in new technology to help speed up the production and decrease unit production cost. To ensure the benefits of this investment are clear, the production manager compares the resulting cost savings to the cost of implementing the new technology, providing a forecast ROI. Because of this effort, decision makers can clearly see how investment in the technology compares to other proposed investments. EHS professionals must compete for scarce organizational resources in this same vein. While requesting funds on the basis of environmental impact can be effective, it is more effective when environmental impact or morale is associated with organizational cost savings (or profit increases) and how they match up to amount of the requested funds.

Another driving force is the role of the Chief Financial Officer (CFO). *The Economist* magazine labeled the CFO the most important executive. The CFO views EHS as an unavoidable cost of business. Nevertheless, when considered as a collection of smaller investments, there are clearly choices to be made. Which EHS projects are worth the investment? If managers can gain some sense of return on these different options then they can ensure that money is being put to the best use. This may not always mean putting a monetary value on the different choices but perhaps understanding their effect on key nonfinancial indicators, such as customer retention.

A final driving force is that more EHS leaders are managing the function as a business. These executives have operational experience and, in some cases, financial experience. They recognize that EHS projects should add value to the organization and, consequently, these executives are implementing a variety of measurement tools, even in the hard-to-measure areas. These tools have gradually become more quantitative and less qualitative. ROI is now being applied in EHS initiatives just as it is in technology, quality and product development.

3.6.3 The ROI Methodology

In keeping with the strategy to invest as long as there is a payoff, many organizations are employing comprehensive accountability processes including ROI. The ROI Methodology, the process described in this book, is one of the most comprehensive and credible approaches to account for and improve upon investments in EHS projects. This process helps decision makers understand which projects are paying off in terms important to all stakeholders. To develop a credible approach for calculating the ROI in EHS projects, several components must be developed and integrated. This strategy comprises five important elements, which form the basis for this book:

1. An evaluation framework is needed to define the various levels of evaluation and types of data, as well as to determine how data are captured. This framework includes how to connect the project to business needs.
2. A process model must be created to provide a step-by-step procedure for developing the ROI calculation. Part of this process is the isolation of the effects of a project from other factors in order to show its monetary payoff.
3. A set of operating standards with a conservative philosophy is required. These "guiding principles" keep the process on track to ensure successful replication. The operating standards also build credibility with key stakeholders in the organization.
4. Successful applications are critical to show examples of how ROI works with different types of EHS projects and initiatives.
5. Resources should be devoted to implementation to ensure that the ROI Methodology becomes operational and routine in the organization. Implementation addresses issues such as responsibilities, policies, procedures, guidelines, goals and internal skill building.

Together, these elements are necessary to develop a comprehensive evaluation system that contains a balanced set of measures, has credibility with the various stakeholders involved and can be easily replicated.

3.6.4 Advantages and Disadvantages of This Strategy

Employing the ROI Methodology to help manage this last investment strategy has several important advantages. With it, the sustainability team and the sponsor who requests and authorizes an EHS project will know the specific contribution of a project in a language sponsor that all stakeholders understand. Measuring the ROI is one of the most convincing ways to earn the respect and support of the senior management team—not only for a particular project, but for the EHS function as well.

Because a variety of feedback data are collected during project implementation, the comprehensive analysis provides data to drive changes in processes and make adjustments during implementation. Throughout the cycle of project design, development and implementation, the entire team of stakeholders focuses on results. If a project is not effective and the results are not materializing, the ROI Methodology will prompt modifications. On rare occasions, the project may have to be halted if it is not adding the appropriate business value, but only if it was specifically designed to add business values.

Note that this methodology has some important barriers to success. It is not suitable for every organization and certainly not for every EHS project. For one thing, the ROI methodology adds additional costs and time to the green budget, although not a significant amount—typically no more than 3 to 5 percent of the total direct EHS budget. The additional investment in ROI should be offset by the results achieved from implementation. This cost barrier often stops many ROI implementations early in the process.

Moreover, many sustainability staff members may not have the basic skills necessary to apply the ROI Methodology within their scope of responsibilities. An EHS project may not focus on results, but on qualitative feedback data and occasional cost-savings data. It is necessary to move EHS projects from an activity-based practice to a results-based practice, which often requires skill-development in the ROI Methodology for the team. In some cases, staff members do not pursue ROI because they perceive an ROI evaluation as an individual performance evaluation instead of a process improvement tool. This misconception can be remedied through education and skill building.

3.7 Final Thoughts

As this chapter shows, some investment strategies work better than others. Avoiding the investment is a disappointing and perhaps disastrous approach. Investing the minimum is appreciative but not enough, more still needs to be done. Investing with the rest is a moderate attempt at making investment decisions, although, following the crowd may not be best for a specific organization and their stakeholders. Overinvesting has its own unique problems and while more is being accomplished, it can have a negative effect that may ultimately reduce investments. This leads to the fifth strategy, the desired approach. Invest when there is a payoff.

Figure 3.3 shows the strategies for shifting the investments toward the desired results. For example, the leaders who are avoiding the investment need to at least move to the minimum. However, it may be difficult in any reasonable time frame to move this group beyond investing the minimum since they will invest more if they see a reason to do it.

Leaders in the second strategy are already investing; they just need to invest more. While they should invest at least to benchmark levels, ideally they should look beyond that and invest when there

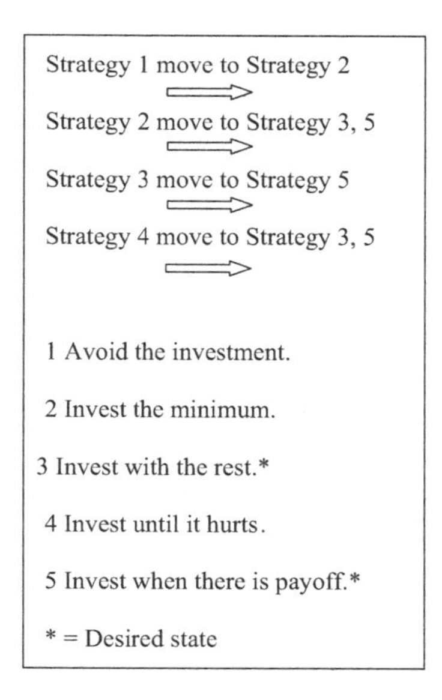

Figure 3.3 Shifting the Investment Strategy Toward the Desired State.

is a payoff. Thus, the third and fifth strategies would be the goals for those stuck at investing the minimum.

Leaders who are investing with the rest need to continue to work to make sure that there is a payoff, capturing data to ensure that their investments are working as desired. This may mean that there should be more or less investment but the crucial element is to show the value in some organized, credible way.

Investors need to ensure that new projects are working. While they may be overinvesting intentionally or unintentionally, an accountability system requiring the evaluation of every project and even some to ROI is a sensible, rational approach to resource allocation that stakeholders will accept and appreciate.

Finally, investing when there is a payoff is an emerging strategy to show the organization a return on their sustainability investment. The remainder of the book focuses on how to accomplish this in a credible, feasible way.

4

The ROI Methodology: A Tool to Measure and Improve

Abstract

In this chapter we introduce the ROI Methodology in some detail, showing how it has evolved into five different components. The first component is the results framework, which represents the types of data. These levels of data are important in the evaluation. As described earlier, results are measured at five levels (reaction, learning, application, impact and ROI). They correspond to five levels of objectives and five levels of needs assessment. Together, these make up the results framework. Next is the step-by-step process to plan a study at the impact and ROI levels, collect the data, analyze the data, and report results. This is the process model. Third Next are the operating standards and philosophy, which serve as guiding principles for the process. These twelve standards are conservative, making results CEO and CFO friendly. They also guarantee consistency and efficiency. The fourth component is applications and practice. To date, this is the most used evaluation system in the world, with applications throughout many functions, including EHS. The final component is implementation, which addresses the issues of building capability, developing appropriate processes to manage resources, efficiently evaluating programs and communicating results. This chapter essentially serves as a summary of the rest of the book.

Keywords: Results, framework, levels of evaluation, needs assessment, objectives, process model, guiding principles, converting data to money, roi analysis, benefit-cost ratio

Jack Phillips, Patti Phillips, and Al Pulliam, Measuring ROI in Environment, Health, and Safety, (73–102) 2014 © Scrivener Publishing LLC

4.1 A Brief Overview

The process for showing the value of EHS initiatives, including measuring the ROI, is comprehensive and systematic. It includes five key components: a results framework, a process model, operating standards and philosophy, applications and practice and implementation (Figure 4.1). Together, these five components ensure that a practice of accountability is sustainable. This chapter briefly describes the components of the ROI Methodology that are necessary to achieve the level of accountability demanded for environmental, health and safety. Detailed information on these components is presented throughout the remainder of the book.

4.2 Results Framework

The richness of the ROI Methodology is inherent in the results framework. This framework represents a variety of types of data, categorized by levels, which are measured and monitored during an EHS project's implementation. Each level represents a link in the chain of impact that occurs as projects are launched. People react and acquire the requisite knowledge, skill and information and apply that knowledge and information. As a consequence, positive impact occurs. Figure 4.2 shows the levels of data and

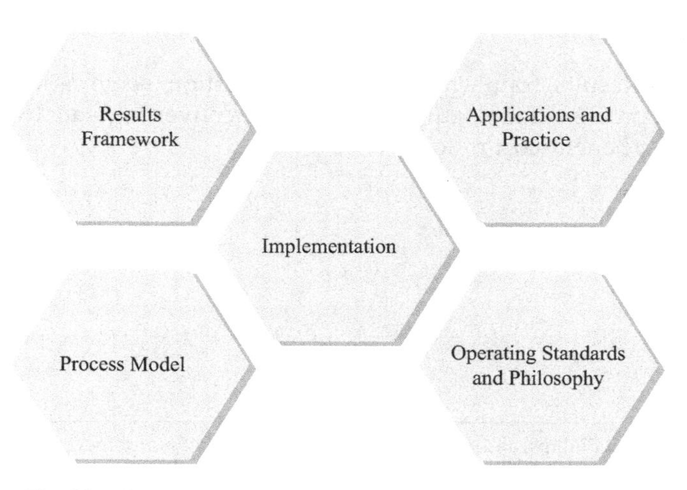

Figure 4.1 The Key Components of the ROI Methodology.

Level	Measurement Focus	Typical Measures
0: Inputs	Inputs into the project, including costs, project scope and duration	Types of projects Number of projects Number of people Hours of involvement Cost of projects
1: Reaction and Perceived Value	Reaction to the project, including the perceived value of the project	Relevance Importance Value Appropriateness Fairness Commitment Motivation
2: Learning and Awareness	Acquisition of knowledge, skill and/or information to prepare individuals to move the project forward	Skills Knowledge Capacity Competencies Confidence Awareness Attitude
3: Application and Implementation	Use of knowledge, skill and/or information and system support to implement the project	Extent of use Actions completed Tasks completed Frequency of use Behavior change Success with use Barriers to application Enablers to application
4: Impact	Immediate and long-term consequences of application and implementation expressed as business measures usually contained in the records	Productivity Accidents / Incidents Quality/Waste Costs / Fines / Penalties Time/Efficiency CO2 emissions Brand / Public image Profits / Growth Customer satisfaction
5: ROI	Comparison of monetary benefits from project to the project costs	Benefit-cost ratio (BCR) ROI (percentage) Payback period

Figure 4.2 Levels and Types of Data.

describes their measurement focus. Subsequent chapters provide more detail on each level, including how to collect and analyze the data and how to report the data so they are meaningful to stakeholders.

4.2.1 Level 0: Input

Level 0 represents the input to a project and includes measures such as the number of people involved, hours of involvement, focus of the project, cost of the project, project duration and project resources. These data represent the activity about a project versus the contribution of the project. Level 0 data also represent the scope of the effort, the extent of commitment and the support for a particular project. For some, this equates to value. However, commitment as defined by involvement expenditures is not evidence that the organization, environment, employees or society are reaping value.

4.2.2 Level 1: Reaction and Perceived Value

Reaction and perceived value (Level 1) marks the beginning of the project's outcome value stream. Reaction data capture the degree to which stakeholders react favorably or unfavorably to the project. The interest in and passion for, EHS initiatives are essential leading indicators of project success. The key is to capture the measures that reflect the content and intent of the project, focusing on issues such as perceived value, relevance, importance and appropriateness.

An adverse reaction to an EHS project usually means that it will not achieve the desired level of success. At this level, project participants identify their intended next actions, make suggestions to advance success and identify potential barriers to success. Data at this level provide the first sign of achievable project success. These data also present project leaders with information they need to make adjustments to project implementation, thereby increasing the chances of positive results.

4.2.3 Level 2: Learning and Awareness

The next level involves measuring learning. For every process, program or project there is a learning component. For some—such as projects for new technology, new systems, new competencies, new processes and new procedures—this component is substantial. Other projects, such as workplace and personal protection, have a learning component. Even the implementation of a new EHS policy includes a learning component to ensure successful execution. Regardless of the initiative, measurement of learning is essential to success. Measures at this level focus on skills, knowledge, capacity, competencies, confidence, attitude and awareness.

4.2.4 Level 3: Application and Implementation

This level measures the extent to which the project is properly applied and implemented. Effective implementation is a must if economic, environmental and societal outcomes are the goals. This is one of the most important data categories because it is here, in the execution of a project, where breakdowns usually occur. Research has consistently shown that in almost half of all projects, participants and users are not doing their parts to make it successful. At this level, measures of success include the extent of the use of technology, task completion, changes in behavior, frequency of use of knowledge, skills and information, success with use, procedures followed and actions completed. Data collection also requires the examination of barriers and enablers to successful implementation of the EHS project. Application and implementation data provide a picture of how well the organizational system supports the successful transfer of knowledge, skills, processes and information to action that leads to the desired outcomes.

4.2.5 Level 4: Impact

Perhaps the most important level of data for understanding the immediate and long-term consequences of the project is collected at Level 4. These data will attract the attention of the sponsor and other executives as well as consumers, suppliers and distributors. This level shows the energy used, waste reduced, time saved, accidents prevented, efficiencies realized, customers satisfied, health costs reduced and employees satisfied, connected to the project. For some, this level reflects the ultimate reason the project exists, to drive environmental health, safety and/or societal impact. Without this level of data, many stakeholders assert, there is no project success.

When this level of measurement is achieved, it is necessary to isolate the effects of the project on the specific measures. Without this extra step, the link between the project and subsequent outcomes is not evident, diminishing the ability to make decisions about project-specific issues.

4.2.6 Level 5: Return on Investment

Impact measures are identified and converted to currency in order to compare the monetary value to the investment, which results in the financial return on investment (ROI). This metric places benefits and costs in equal terms: money. Normalizing benefits and costs enables project owners and other stakeholders to see how resource expenditures compare to benefits. This financial metric is typically stated in terms of a benefit-cost ratio (BCR), ROI (percentage) and/or payback period. This level of measurement requires two important steps: first, the impact data (Level 4) must be converted to monetary values; second, the cost of the project must be captured.

While some stakeholders may suggest that calculating the actual ROI on an EHS initiative diminishes the intangible value of such a project, two important benefits emerge from this process. First, by knowing the ROI, stakeholders concerned with the economic components of EHS see that financial resources are allocated appropriately. Appropriate use of financial resources leads to organization viability and longevity. Second, when an EHS project reaps positive benefits in one plant or region, the results may be used to justify implementation in another plant region. Funds from projects with a positive ROI can be used to help fund the not-so-successful projects or those with longer-term outcomes. It is also helpful to remember that when projects are properly developed and implemented, the chances for a positive ROI are high.

4.2.7 Intangible Benefits

Along with the five levels of outcome results and the initial level of activity (Level 0), there is a sixth type of outcome data—not a sixth level—developed through the processes described in this book. These sixth types of data are the intangible benefits or those impact measures that are purposefully not converted to money. A decision to not convert benefits to money is made when the conversion consumes too many resources or the process is not credible. Yet intangibles are still important measures of success. For initiatives, intangibles may include stress, job engagement, teamwork, brand awareness, reputation, customer satisfaction, employee satisfaction and public image.

4.3 Results Framework and Business Alignment

Our research suggests that the number one reason projects fail is lack of alignment with the business. The results framework supports this alignment by connecting the project needs with its objective and the evaluation of its success. The first opportunity to obtain business alignment is in the initial analysis.

4.3.1 Initial Analysis

Initial analysis of stakeholder needs sets the stage for deciding on the best project(s) to pursue given those needs and the available resources. This initial analysis represents the first phase in aligning projects with the business. It begins with the determination of the payoff needs – the potential opportunity or problem that is worth solving.

4.3.1.1 Payoff Needs

From the start, several steps should be taken to make sure that the project or initiative is necessary. As shown in Figure 4.3, this is the

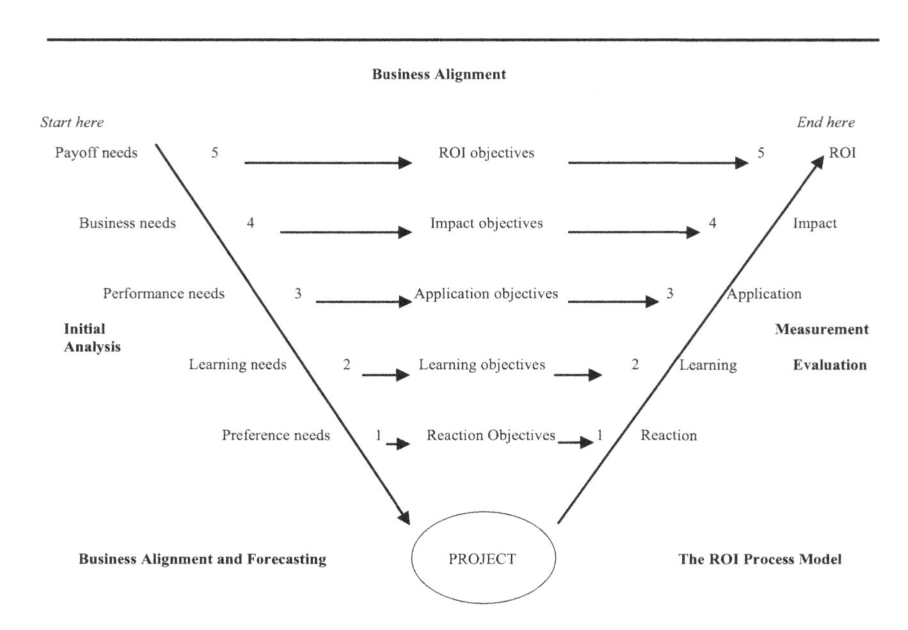

Figure 4.3 The V-Model.

beginning of the complete, sequential model that we often refer to as the V-model. This model is at the heart of the results framework. The first step in this analysis examines the potential payoff of solving a problem or taking advantage of an opportunity. Is this a problem worth solving? Is the project worthy of implementation? For some situations the answer is obviously yes, because of the project's critical nature, its relevance to the issue at hand or its effectiveness in tackling a major problem that affects the organization. An excessive injury rate, a serious CO2 emissions problem or rising health care costs, for example, are worth pursuing. Potential payoff opportunities may be short-term or long-term.

4.3.1.2 Business Needs

The next step is to ensure that the project is connected to one or more key business measures. Key measures that must improve as a reflection of the overall success of the project are defined. Business needs may be long-term, but they often represent more immediate outcomes to the organization, such as cost savings due to accident reduction, health care cost containment and more efficient energy use. These measures are in the system now, in operating reports, key performance indicators, performance scorecards or goals for individuals, departments, functions or organizations.

4.3.1.3 Performance Needs

Next, the performance needs are examined. What must change in terms of behaviors, habits, application or implementation to address the business needs? This step aligns the project with the business and may involve a series of analytical tools to determine the cause of the problem or the changes necessary to take advantage of an opportunity. This step appears to be complex, but it is really a simple approach. A series of questions helps:

- What is keeping the business measure from being where it needs to be?
- If it is a problem, what is its cause?
- If it is an opportunity, what is hindering the measure from moving in the right direction?

This step is important because it provides the link to project or initiative. A simple example may explain. A safety incentive program in a steel company provided cash payments to employees in a work unit for working three months without a medical treatment case. The program focused on the actions and behaviors of employees, including watching out for coworkers. By identifying a simple performance need (the behavior of employees), this initiative addressed a business need (accident costs) with a substantial payoff. The result is the reduction in medical treatment injuries from 116 to 18 in each plant's performance and a savings of almost $400,000 per plant (*Phillips and Phillips, 2007*).

4.3.1.4 Learning Needs

To change performance behaviors and habits, people need to know what they must do, how to do it and when to do it. What specific skills, knowledge or information must be acquired so the performance can change? Sometimes it is just a matter of making people aware of the consequences of their behaviors. Every solution involves a learning component and this step defines what the people involved must know to make the project successful. The required knowledge may be as simple as understanding a policy or as complicated as developing a new set of competencies.

4.3.1.5 Preference Needs

The final step is identifying the structure of the project and the desired reaction to the project. How should the information be presented to ensure that needed knowledge is acquired, performance changes are addressed and the business needs met? This level of analysis involves issues surrounding the perceived value, necessity, importance, scope, timing and budget for project implementation and delivery. It details the desired reaction from stakeholders. Will they perceive it as necessary and important? This step represents preference needs or the preferred approach for the project.

4.3.2 Project Objectives

Collectively, these steps define the issues that lead to project initiation. Still, the actual positioning comes with the development of clear, specific objectives or targets that are communicated to all stakeholders. Objectives represent each level of need and define how stakeholders will know that the need has been met. If the criteria of success are not communicated early and often, project participants will simply go through the motions and there will be little change. Developing detailed objectives with clear measures of success positions each project to achieve its ultimate goal. Objectives provide the connection between organizational needs and project accountability.

4.3.3 Forecasting

Using stakeholder needs and project objectives as the basis, developing a forecast may be useful in making adjustments or choosing alternative solutions. This forecast can be simple, relying on the individuals closest to the situation or it can involve a more detailed analysis of the situation, expected outcomes and potential risks. Recently, forecasting has become a critical tool for project sponsors who need evidence that the project will be successful before they are willing to commit to funding.

4.4 Benefits of Developing the Chain of Impact

Developing data represented in the results framework—including five levels of results along with inputs (Level 0) and intangible measures—provide a variety of benefits, including the following:

- Describing the chain of impact that occurs as people become involved in EHS projects
- Showing project results from multiple perspectives
- Demonstrating how immediate and long-term outcomes are achieved
- Providing information as to why and how outcomes are or are not achieved
- Providing project owners data they can use to make improvements with implementation
- Holding stakeholders accountable for success of all project stages
- Providing stakeholders the data they need to make decisions about the project and the organization

At first, the thought of collecting and analyzing such a comprehensive set of data may seem daunting. However, without this set of information, explaining the basis for an ROI calculation will be a challenge. In addition, decisions about projects and their subsequent success require more than an economic metric. By reviewing data that represent the chain of impact, stakeholders can understand how project implementation evolves, what changes are necessary to improve or sustain success and how the project contributes to the overall good of the organization. Chapter 5 discusses this in greater detail.

To simplify the collection and analysis of data in the results framework, a step-by-step process model is required. This is presented next.

4.5 The ROI Process Model

The second component of the ROI Methodology is the process model. This ten-step process, shown in Figure 4.4, develops the data representing the chain of impact. The process begins with the project objectives and concludes with reporting of data. The model assumes that proper analysis is conducted to define stakeholders' needs prior to project implementation.

4.5.1 Planning the Evaluation

The first phase of the ROI process model is evaluation planning. This phase involves understanding the purpose of the evaluation, determining the feasibility of the planned approach, planning data collection and analysis and outlining the details of the project.

4.5.1.1 Evaluation Purpose

Evaluations of EHS projects are conducted for a variety of reasons:

- To improve the quality of projects and outcomes
- To determine whether a project has accomplished its objectives
- To identify strengths and weaknesses in project implementation
- To enable the cost-benefit analysis
- To assist in the development of future projects or programs
- To determine whether the project was the appropriate solution
- To establish priorities for project funding

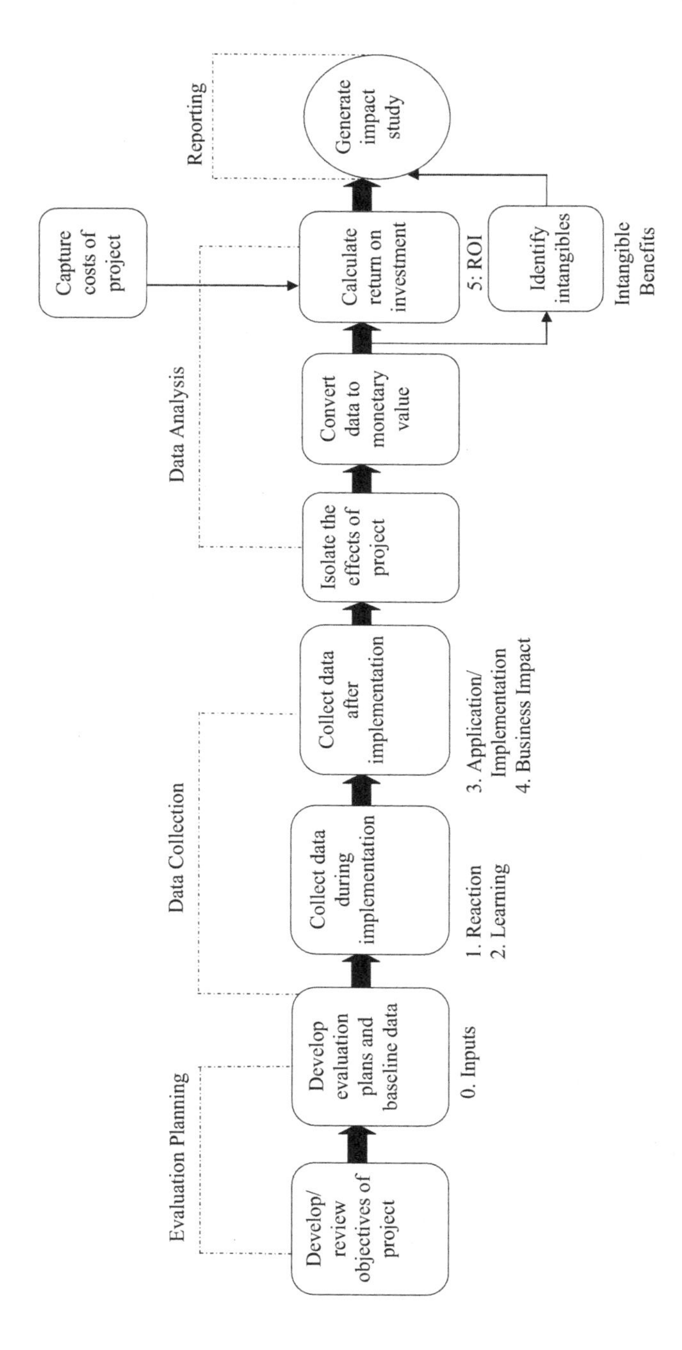

Figure 4.4 The ROI Process Model.

Prior to developing the evaluation plan, the purposes of the evaluation should be considered because they will often determine the scope of the evaluation, the types of instruments used and the kinds of data collected. As with any project, making the purpose of the evaluation clear will give it focus and help to gain support from others.

4.5.1.2 Feasibility

An important consideration in planning an ROI study is the determination of the levels to which the program will be evaluated. Some project evaluations will stop at Level 2, measuring learning and awareness. Other evaluations will stop at Level 3, application, where analysis will determine the extent to which participants are applying what they learned through a project launch. Other projects will be evaluated at Level 4, business impact, where the consequences of application are monitored and measures directly linked to the project are examined. If the ROI calculation is needed, the evaluation will proceed to Level 5, converting project benefits to money and comparing them to project implementation costs. Evaluation at Level 5 is intended for projects that are expensive, high profile and have a direct link to operational and strategic objectives.

The feasibility of an EHS project is determined through the initial analysis and the development of project objectives. The objectives are defined along the same five levels as the needs assessment:

- Reaction objectives (Level 1)
- Learning objectives (Level 2)
- Application and implementation objectives (Level 3)
- Impact objectives (Level 4)
- ROI objectives (Level 5)

Specific objectives take the mystery out of what each project should achieve. They also serve as the basis for comparing results. If application and impact objectives are unavailable, they must be developed using input from a variety of stakeholders.

With a clear purpose and project feasibility, evaluation planning continues with the development of the data collection plan, the ROI analysis plan and the project plan. Appropriate up-front attention to these planning documents will save time later when data are actually collected.

4.5.1.3 Data Collection Plan

Table 4.1 shows a completed safety management program data collection plan for project safety leaders in large construction projects. The two-day program focused on a variety of actions and activities to improve a variety of safety and health measures.

The data collection plan provides a place for the major elements and issues regarding data collection. Broad objectives are appropriate for planning. Specific, detailed objectives are developed later, before the project is designed. Entries in the *measures* column define the specific measure for each objective while entries in the *methods* column describe the technique used to collect the data. In the *sources* column, the source of the data is identified and the *timing* column indicates when the data are collected. Finally, the *responsibilities* column identifies who will collect the data.

4.5.1.4 ROI Analysis Plan

Table 4.2 shows a completed ROI analysis plan for the safety management program for project safety leaders described in the previous section. This planning document captures information on key items that are necessary to develop the actual ROI calculation. In the first column, impact measures are listed. In some cases this column includes application measures. These items will be used in the ROI analysis.

The method employed to isolate the project's effects is listed next to each data item in the second column. Data conversion methods are included in the third column. Cost categories that will be captured for the project are outlined in the fourth column. Normally, the cost categories are consistent from one project to another. Intangible benefits expected from the project are outlined in the fifth column. This list is generated from discussions about the project with sponsors and subject-matter experts. Communication targets are outlined in the sixth column. Finally, other issues or events that might influence project implementation and its outputs are highlighted in the seventh column. Typical items include the capability of participants, the degree of access to data sources, the engagement of stakeholders and unique data analysis issues.

The comments column is for notes and issues important to the team and the evaluation project implementation.

The ROI analysis plan is combined with the data collection plan, illustrating how the evaluation will develop from beginning to end and includes the calculation of the ROI.

4.5.1.5 Project Plan

The final planning document is the project plan for the safety management program, as shown in Table 4.3. A project plan consists of a description of the project and brief details, such as duration, target audience and number of participants. It also shows the timeline of the project, from the planning of the study through the final communication of the results. This plan becomes an operational tool to keep the project on track

Collectively, the three planning documents provide the direction necessary for the ROI study. Most of the decisions regarding the process are made as these planning tools are developed. When the project team spends time up front to plan an evaluation, the project becomes a methodical, systematic process. Planning is a crucial step in the ROI Methodology, in which valuable time allocated to planning will save precious time later.

4.5.2 Collecting Data

Data collection is central to the ROI Methodology. Both hard data (representing output, quality, cost and time) and soft data (including job satisfaction, customer satisfaction and public image) are collected. A variety of data collection methods are employed, including:

- Surveys
- Questionnaires
- Tests
- Observations
- Interviews
- Focus groups
- Action plans
- Performance contracts
- Business performance monitoring

Table 4.1 Data collection plan.

Program: Safety Management Program **Responsibility:**________ **Date:**________

Level	Objective(s)	Measures/Data	Data Collection Method	Data Sources	Timing	Responsibilities

1	**Reaction**
	• Obtain favorable reaction to program and materials on
	  • Need for program
	  • Relevance to project
	  • Importance to project
	• Identify planned actions
		• Average rating of 4.0 out of 5.0 on feedback items
		• 100% submit planned actions
			• Standard questionnaire	• Participant	• End of program	• Facilitator

2	**Learning** After attending this session, participants should be able to: • Establish safety audits • Provide feedback and motivate employees • Measure safety performance • Solve safety problems • Counsel problem employees • Conduct safety meetings	• Achieve an average of 4 on a 5 point scale	• Questionnaire	• Participant	• End of program	• Facilitator

Level	Objective(s)	Measures/Data	Data Collection Method	Data Sources	Timing	Responsibilities
3	**Application/Implementation** • Apply skills in appropriate situations • Complete all steps of action plan	• Ratings on questions (4 of 5) • The number of steps completed on action plan	• Questionnaire • Action plan	• Participant • Participant	• Three months after program • Three months after program	• Safety and Health Team
4	**Business Impact** • Identify three safety and health measures that need improvement	• Varies	• Action plan	• Participant	• Three months after program	• Safety and Health Team
5	**ROI** 20%	Comments: Several techniques will be used to secure commitment to provide data on the questionnaire and action plan.				

Table 4.2

Program: Safety Management Program **Responsibility:**__________ **Date:**______

Data Items (Usually Level 4)	Methods for Isolating the Effects of the Program/ Process	Methods of Converting Data to Monetary Values	Cost Categories	Intangible Benefits	Communication Targets for Finale Report	Other Influences/ Issues During Application	Comments
• Three safety and health measures identified by project safety leader	• All Participant estimation	• Standard values • Expert input • Participant estimation	• Needs assessment • Program development • Program materials • Travel & lodging • Facilitation & coordination • Participant salaries plus benefits while in the program • Extra project expenses related to program • Evaluation	• Job Engagement • Job satisfaction • Stress • Image • Brand	• Project manager • Participants • Safety and Health team • Operating executives • Director, Safety and Health • Senior VP Human resources		
Clie Client Signature: __________ Date: ______							

Table 4.3 Project Plan.

	Month						
	F	M	A	M	J	J	A
Decision to conduct ROI study							
Plan evaluation							
Design and test instruments							
Collect data							
Tabulate data							
Conduct analysis							
Write report							
Print report							
Communicate results							
Initiate improvements							
Complete implementation							

The important challenge in data collection is to select the method or methods appropriate for the setting and the specific project, within the time and budget constraints of the organization. Data collection is covered in more detail in Chapters 6 and 7.

4.5.3 Isolating the Effects of the Project

An often-overlooked issue in evaluation is the process of isolating the effects of the project. In this step, specific strategies are explored that determine the amount of output performance directly related to the project. This step is essential because many factors influence performance data and it is often necessary to identify contribution of certain key factors. The specific strategies of this step pinpoint the amount of improvement directly related to the EHS project, which results in increased accuracy and credibility of ROI calculations. The following techniques have been used by organizations to address this important issue:

- Control groups/comparison groups
- Trendline analysis

- Forecasting and recession models
- Participants' estimates
- Managers' estimates
- Senior manager's estimates
- Experts' input
- Customer input

Collectively, these techniques provide a comprehensive set of tools to handle the important and critical issue of isolating the effects of an EHS project. Chapter 8 addresses this issue in detail.

4.5.4 Converting Data to Monetary Values

To calculate the ROI, Level 4 impact data are converted to monetary values and compared with project costs. This requires that a value be placed on each unit of measure connected with the project. Many techniques are available to convert data to monetary values. The specific technique selected depends on the type of data and the situation. The techniques include:

- Use of standard values for output data
- Use the cost of quality, usually as a standard value
- Time savings converted to participants' wage and employee benefits
- An analysis of historical costs and records
- Use of internal and external experts
- Search of external databases
- Use of participant estimates
- Use of manager estimates
- Soft measures mathematically linked to other measures

This step in the ROI process model is necessary in developing the numerator of the ROI equation. Converting benefits to money normalizes the benefits so a comparison can be made to the costs. The process is challenging, particularly with soft data, but it can be methodically accomplished using one or more of these strategies. Because of its importance, this step in the ROI Methodology is described in detail in Chapter 9.

4.5.5 Identifying Intangible Benefits

In addition to tangible, monetary benefits, intangible benefits (those not converted to money) are identified for most projects. Intangible benefits include items such as:

- Improved teamwork
- Enhanced communications
- Increased brand awareness
- Improved reputation
- Enhanced public image
- Increased employee engagement

During data analysis, every attempt is made to convert all data to monetary values. All hard data such as output, quality, cost and time are converted to monetary values. The conversion of soft data is attempted for each data item. However, if the process used for conversion is too subjective or inaccurate and the resulting values lose credibility in the process, then the data are listed as an intangible benefit with the appropriate explanation. For some projects, intangible, nonmonetary benefits are extremely valuable and these often carry as much influence as the hard data items. Chapter 10 describes in more detail the issue of intangible benefits.

4.5.6 Tabulating Project Costs

An important part of the ROI equation is the calculation of project costs, which make up the denominator of the ROI equation. Tabulating the costs involves monitoring or developing all the related costs of the EHS project targeted for the ROI calculation. Among the cost components to be included are:

- Initial analysis costs
- Cost to design and develop the project
- Cost to acquire equipment and technology, if necessary
- Cost of all project materials
- Cost of the facilities for the project
- Travel, lodging and meal costs for the participants and team members
- Participants' salaries (including employee benefits)

- Administrative and overhead costs, allocated in some convenient way
- Operating costs
- Evaluation costs

The conservative approach is to consider the fully loaded costs of a project. Chapter 10 addresses this step in the ROI Methodology.

4.5.7 Calculating the Return on Investment

Return on investment is reported using a variety of metrics. Standard calculations include the benefit-cost ratio (BCR), ROI percentage and payback period. The BCR is calculated as the project benefits divided by the project costs, shown in formula form below:

$$\text{BCR} = \frac{\text{Project Benefits}}{\text{Project Costs}}$$

The ROI is based on the net project benefits divided by project costs, then multiplied by 100 to develop the percentage. The net benefits are calculated as the project benefits minus the project costs. In formula form, the ROI becomes:

$$\text{ROI (\%)} = \frac{\text{Net Project Benefits}}{\text{Project Costs}} \times 100$$

This is the same basic formula used in evaluating other investments, in which the ROI is traditionally reported as earnings divided by investment. In addition, it may sometimes be necessary to calculate the payback period. Payback period requires that the project costs be compared to annual project benefits. In equation form, the payback period is calculated as:

$$\text{Payback Period} = \frac{\text{Project Costs}}{\text{Annual Project Benefits}}$$

A simple example of the benefit-cost ratio, ROI and payback period illustrate the calculations. An energy savings project for a city's municipal buildings involves replacing current bulbs with energy-saving bulbs. A three-year benefit stream is selected at the beginning of the project based on the expected life of the new bulbs. The project benefits for the three years are $570,000 ($190,000 per year) and the fully loaded cost of replacement is $350,000.

$$\text{BCR} = \$570{,}000/\$350{,}000 = 1.63{:}1$$

$$\text{ROI} = \$220{,}000/\$350{,}000 \times 100 = 63\%$$

$$\text{Payback Period} = \$350{,}000/\$190{,}000 = 1.84 \text{ years or } 22 \text{ months}$$

The ROI calculation of net benefits ($570,000 minus $350,000) divided by total costs brings an ROI of 63 percent. This is what is earned after we get back the $350,000 spent on the project. The ROI calculation accounts for the project costs and shows the resulting net gain.

The BCR calculation uses the total benefits in the numerator. Therefore, the expressed BCR of 1.63:1 does not account for replacing the expended costs. This is why, when using the same values, the BCR will always be 1 greater than the ROI. The BCR of 1.63:1 in this example means that for every dollar spent, $1.63 is gained. One dollar has to pay for the investment, so the net is $0.63 (as expressed in the ROI calculation). The payback period shows that it takes about twenty-two months to pay back the project's investment.

For short-term projects in which an immediate payoff is expected, consider the first-year benefits only. This approach is the most conservative approach to accounting for project costs. With a project investment for which the payoff may not occur for two or three years post-project implementation, consider the time value of the investment and benefits stream. Again, this is a conservative accounting of financial resources. These calculations, along with other issues pertinent to developing the ROI, are described in Chapter 10.

4.5.8 Reporting

The final step in the ROI process model is reporting, a critical step that often lacks the degree of attention and planning required to ensure its success. The reporting step involves developing appropriate information in impact studies and other brief reports. At the heart of this step are the different techniques used to communicate to a wide variety of target audiences. In most ROI studies, several audiences are interested in and need the information. Careful planning to match the communication method with the audience is essential to ensure that the message is understood and that appropriate actions follow. Chapter 11 is devoted to reporting evaluation results developed through the ROI Methodology. Chapter 11 also describes development of a scorecard, a macro-level reporting of success for all sustainability initiatives.

4.6 Operating Standards and Philosophy

An organization's philosophy and standards can have an important influence on how stakeholders perceive the quality of data. This is the third component necessary to create a sustainable evaluation practice. Consistency and replication of studies is the output of evaluation standards. Progress and assumptions inherent in an evaluation process should not vary depending on the individual conducting the evaluation. In addition, instilling a philosophy of conservative assumptions will ensure that results do not overstate the project contribution to outcomes, often, positioning decision makers to make unnecessary and inappropriate overinvestments in a project. Table 4.4 shows the twelve guiding principles that serve as standards of use for the ROI Methodology.

The guiding principles serve not only to consistently address each step of the evaluation process, but also to provide a conservative approach to the analysis. A conservative approach may lower the actual ROI calculation, but it will build credibility and buy-in with the key stakeholders, especially CEOs, managing directors, top administrators and CFOs.

4.7 Case Application and Practice

The fourth component necessary for a sustainable measurement practice is case application and practice. This component puts theory to practice. While the results framework serves as the basis

Table 4.4 Twelve Guiding Principles of the ROI Methodology.

Guiding Principles
1. When conducting a higher-level evaluation, collect data at lower levels.
2. When planning a higher-level evaluation, the previous level of evaluation is not required to be comprehensive.
3. When collecting and analyzing data, use only the most credible sources.
4. When analyzing data, select the most conservative alternative for calculations.
5. Use at least one method to isolate the effects of a project.
6. If no improvement data are available for a population or from a specific source, assume that no improvement has occurred.
7. Adjust estimates of improvement for potential errors of estimation.
8. Avoid use of extreme data items and unsupported claims when calculating ROI.
9. Use only the first year of annual benefits in ROI analysis of short-term projects.
10. Fully load all costs of a project when analyzing ROI.
11. Intangible measures are defined as measures that are purposely not converted to monetary values.
12. Communicate the results of ROI Methodology to all key stakeholders.

for the ROI Methodology and the process model and standards are systematic, it is the practice and use of the process that is important. Application quickly shows the power of this methodology.

The ROI Methodology is the most used evaluation system in the world. Over 3,000 professionals and managers have achieved the designation of Certified ROI Professional (CRP) through the ROI Institute. In addition, 3-5,000 ROI studies are conducted each year from 60 countries, through ROI Institute global partners.

4.8 Implementation

A variety of environmental issues and events must be addressed early to ensure the successful implementation of the ROI process, which is also central to the success of other components described in this chapter. Specific topics or actions important to successful implementation include:

- A policy statement concerning results-based EHS projects
- Procedures and guidelines for different elements and techniques of the evaluation process
- Formal meetings to develop staff skills with the ROI Methodology
- Strategies to improve management commitment to and support for the ROI Methodology
- Mechanisms to provide technical support for data collection, design, data analysis and evaluation strategy
- Specific techniques to place more attention on results

In addition to implementing and sustaining ROI use, the process must undergo periodic review. An annual review is recommended to determine the extent to which the process is adding value. This final element involves checking satisfaction with the process and determining how well it is understood and applied. Essentially, this review follows the process described in this book to determine the ROI on ROI. Chapter 12 is devoted to this important topic.

4.9 Benefits of Applying the ROI Methodology

The approach to evaluating the success of EHS projects presented in this book has been used consistently and routinely by thousands of organizations in the past decade. It has been more prominent in some fields and industries than in others such as performance improvement, quality, human resources, meeting and events and marketing. Much has been learned about the success of this methodology and what it can bring to the organizations using it. Along with the benefits described earlier in the book, specific benefits of applying the ROI Methodology are as follows.

4.9.1 Aligning Projects with the Business

The ROI Methodology ensures alignment with the business, which is enforced in three steps. First, even before the project is initiated, the process ensures that alignment is achieved up front, at the time the EHS project is validated as the appropriate solution. Second, by requiring specific, clearly defined objectives at the impact level, the project focuses on the ultimate outcomes, in essence driving the business measure by its design, delivery and implementation. Third, in the follow-up data, when the outcome measures may have changed or improved, a method is used to isolate the effects of the project on those data, consequently proving the connection to that business measure (i.e., showing the amount of improvement directly connected to the project and ensuring there is business alignment).

4.9.2 Validating the Value Proposition

In reality, most EHS projects are undertaken to deliver value, whether value is defined in business, environmental or societal terms. The definition of value may on occasion be unclear or may not be what a project's various sponsors, organizers and stakeholders desire. Consequently, there are often value shifts. When the values are finally determined, the value proposition is detailed. Using the ROI Methodology, organizations can forecast the value in advance and if the value has been delivered, thereby verifying the value proposition agreed to by the appropriate parties.

4.9.3 Improving Processes

This is a process improvement tool by design and by practice. It collects data to evaluate how projects are, or are not, working. When EHS projects are not progressing as they should, data are available to indicate what must be changed to make the projects more effective. When things are working well, data are available to show what else could be done to make them better. Continuous feedback and process improvement are inherent to the ROI Methodology.

4.9.4 Enhancing Image

Many functions and even entire professions are criticized for being unable, or unwilling, to deliver what is expected. For this, their public image suffers. The ROI Methodology is one way to help build the respect a function organization or profession needs. By showing value defined by all stakeholders and by using evaluation results, EHS project owners communicate to stakeholders their successes and their desire to continuously improve. This methodology shows a connection to the bottom line and the greater good.

4.9.5 Improving Support

Securing support for EHS projects is critical. Many projects enjoy the support of key stakeholders who allocate resources to make the projects viable. Unfortunately, some stakeholders may not support certain projects because they do not see the value the projects deliver in terms they appreciate and understand. Having an accountability approach that shows how a project or program is connected to business goals and objectives can change this support level.

4.9.6 Justifying or Enhancing Budgets

Some organizations have used the ROI Methodology to support existing proposed budgets. Because the process shows the monetary value expected or achieved with specific projects, the data can often be leveraged into budget requests. When a particular function is budgeted, the amount budgeted is often in direct proportion to the value that the function adds. If little or no credible data support the contribution, the budgets are often trimmed, or at least not enhanced. Bringing accountability to the level achieved through use of the ROI process is one of the best ways to secure future funding.

4.9.7 Building Partnerships with Key Executives

Almost every function attempts to partner with operating executives and key managers in the organization. Unfortunately, some managers may not want to be partners. They may not want to waste time and effort on a relationship that does not help them succeed. They want to partner only with groups and individuals who can add value and help them in meaningful ways. Showing the projects' results will enhance the likelihood of building these partnerships, by providing the initial impetus for making the partnerships work.

4.9.8 Earning a Seat at the Table

Many functions are attempting to earn a seat at the table, however defined. Typically, this means participating in the strategy- or decision-making process and in high-level discussions at the top of the organization. Department and project leaders hope to be involved in strategic decision-making, particularly in areas that will affect the projects and programs in which they are involved. Showing the actual contribution and getting others to understand how projects add value can help earn the coveted seat at the table, because most executives want to include those who are genuinely helping the business by providing input that is valuable and constructive. Application of the ROI Methodology may be the most important action toward earning the seat at the table.

4.10 Final Thoughts

The ROI Methodology is an accountability process designed to collect and report multiple types of data that are crucial to the evaluation of environment, health and safety projects:

- Inputs (Level 0)
- Reaction and perceived value (Level 1)
- Learning and awareness (Level 2)
- Application and implementation (Level 3)
- Impact (Level 4)
- ROI (Level 5)
- Intangible benefits

By developing these data and following a step-by-step process grounded in conservative standards, EHS project owners can be confident their results will be perceived as credible. In addition, this process will ensure that projects are aligned with business from the outset. The remainder of the book will describe how to develop data, attribute results to the EHS project and develop the ROI for EHS initiatives.

5

Project Positioning

Beginning with Objectives in Mind

Abstract

This chapter explains how projects are actually initiated by introducing the V Model, the process that links needs assessment and evaluation, with the levels of objectives serving as the transition. First, we introduce the concept of five levels of needs that should be addressed when a major project is initiated: payoff needs, business needs, performance needs, learning needs and preference needs. These needs are ideally addressed in that order to arrive at a particular solution and to develop very specific objectives at each level. Correspondingly, we explore how objectives are written for the five levels: ROI, business impact, application, learning, and reaction. These objectives provide direction and guidance for all stakeholders involved in the project. They correspond with the five levels of evaluation: reaction, learning, application, impact and ROI. This framework is key to understanding clearly how alignment is achieved and the various connections among the parts of a project. The chapter concludes with EHS examples, showing how needs are established, objectives are set and evaluation is accomplished.

Consider the following statements:

1. Most experts agree that organization leaders will fully embrace environment, health and safety projects when there is a perceived business payoff.
2. The vast majority of EHS projects and initiatives result in a significant payoff.
3. EHS projects and initiatives fail primarily because they lack connection to the business need at the onset.

These three conclusions underscore the need for aligning EHS projects to the business.

Jack Phillips, Patti Phillips, and Al Pulliam, Measuring ROI in Environment, Health, and Safety, (103–134) 2014 © Scrivener Publishing LLC

Keywords: Needs assessment, alignment, payoff needs, business needs, performance needs, learning needs, reaction objectives, learning objectives, application objectives, impact applications, ROI objectives

5.1 Creating Business Alignment

As we've noted, aligning EHS projects and initiatives with business needs (as well as environmental and societal needs) serves a variety of purposes. Alignment ensures that an organization not only steps up to the EHS plate, but also serves up a home-run pitch to shareholders, taxpayers, employees and other stakeholders with an interest in the organization's economic vitality. While it may seem obvious that all EHS activities should begin with alignment (e.g., specific environment measures, required health measures or safety performance measures), the reality is that they do not. Many environmental projects begin because of image. Health related programs are initiated because of employee satisfaction and safety programs often focus on the wrong measures.

5.1.1 The Purpose of Alignment

According to approximately two thousand published and unpublished case studies at the ROI Institute, the number one cause of project failure is moving forward without a clearly defined need. The second is misalignment between project and business needs.

Projects must begin with a clear focus on the desired outcome. The end must be specified in terms of business needs and measures so that the outcome—the actual improvement in the measures—and the corresponding ROI are clear. This establishes expectations throughout the project design, development, delivery and implementation stages. It ensures that the right projects are put into place at the right time and involve the right people for the right reason.

Beginning with the end in mind requires pinning down all the details to ensure that the project is properly planned and executed according to schedule. But conducting this up-front analysis is not as simple as one may think, it requires a disciplined approach.

5.1.2 Disciplined Analysis

Proper analysis requires discipline and determination to adhere to a structured, systematic process supported by standards. A standardized approach adds credibility and allows for consistent application so that the analysis can be replicated. A disciplined approach maintains process efficiency through the development and use of various tools and templates. This initial phase of project development calls for focus and thoroughness, with little allowance for major shortcuts.

Not every project should be subjected to the type of comprehensive analysis described in this chapter. Some outcomes and processes are obvious and require little analysis in order to implement the project. For example, in-depth analysis is unnecessary to determine the best approach to changing incandescent light bulbs to fluorescent bulbs in a single building. When over 50 percent of employees are technically obese, a fitness program is in order. A high injury rate in a particular job suggests new safety solutions. Another example is purchasing recycled paper that is less expensive than the current paper, this may not require any analysis other than the cost. Of course, if there is a noticeable difference in the quality of the paper, then users will get involved.

The amount of analysis required often depends on the expected opportunity to be gained if the project is appropriate or the negative consequences anticipated if the project is inappropriate. Usually large-scale and expensive projects need in-depth analysis. When outcomes require involving a large number of people whose perceptions, knowledge and attitude must change, detailed analysis is a must. In essence, if the project is important, strategic, expensive and involves a large number of people, comprehensive analysis at the five levels described in this chapter is appropriate.

Sponsors may react with concern or resistance when analysis is initially proposed. Some indicate concern about the potential for "paralysis by analysis," where requests and directives lead only to additional analyses. These reactions can pose a problem for an organization because analysis is necessary to ensure that a project or an initiative is appropriate for a situation. Unfortunately, the thought of analysis often conjures up images of complex problems, confusing models and a deluge of data along with complicated statistical techniques in an effort to cover all of the bases. In reality, analysis need not be so complicated. Simple techniques can uncover the cause of a problem or the need for a particular project.

Organizations often avoid analysis because:

1. *The specific need appears to point to a particular solution.* Sometimes the information gained from asking individuals what they need appears to point to a legitimate

solution, but in fact the solution is inadequate or inappropriate. For example, when employees are asked what they need to improve the environment, they may identify specific tools, suppliers, materials or equipment. In reality, the solution may be learning to conserve and recycle. Implementing a solution entirely in response to individual input can prove shortsighted and costly.

2. *The solution appears to be obvious.* In the process of examining a problem or identifying a potential opportunity, some seemingly obvious solutions will arise. For example, if employees are having too many injuries, the immediate conclusion may be to implement a program on behavior management of employees. However, although this solution appears obvious, deeper analysis may reveal that another approach, such as the use of engineering or administrative controls, may be appropriate.
3. *Everyone has an opinion about the cause of a problem.* The person requesting a particular project may think that he or she has the best solution. Choosing the solution championed by the highest-ranking or most senior executive is often tempting. When the CEO suggests that a wellness and fitness center should be implemented, the payoff may not be forthcoming. Unfortunately, this person might not be close enough to the situation to offer a solution that will have a lasting effect on the problem.
4. *Analysis takes too much time.* Yes, analysis takes time and consumes resources. However, the consequences of not analyzing can be more expensive. If the implemented solutions do not appropriately address the needs, time and money are wasted and the problem is left unsolved. Ill-advised solutions conceived without proper analysis can have devastating consequences. When designed appropriately and conducted professionally, an analysis can be completed within the budgetary and time constraints of most organizations. The secret is to focus on the right tools for the situation.
5. *Analysis sounds confusing.* Determining a problem's causes may seem complex and puzzling. However, analyses can be simple and straightforward and

achieve excellent results. The challenge is to select the level of analysis that will yield the best solution with minimal effort and the simplest techniques.

In the face of these misconceptions, the difficulty of promoting additional analysis is apparent. But this step is critical and should not be omitted, otherwise the process will be flawed from the outset.

The remainder of the chapter delves into the components of analysis that are necessary for a solid alignment between a project and the business. It may be helpful to first refer to Figure 4.3 in the previous chapter.

5.2 Determining Payoff Needs

The first step in the alignment process is to determine the potential payoff of solving a problem or seizing an opportunity. This step begins with answers to a few crucial questions:

- Is the problem or worth solving?
- Is the opportunity worth pursing?
- Is the investment in a project or solution feasible?
- What is the likelihood of a positive ROI as well as the environmental contribution?

For projects addressing problems or opportunities with high potential rewards, the answers are obvious. The questions may take longer to answer when the expected payoff is less apparent.

Essentially, from an economic perspective, a project will pay off in profit increases and cost savings. Profit increases are generated by projects that drive revenue (e.g., projects that improve sales, drive market share, introduce new products, open new markets, enhance customer service or increase customer loyalty). Other revenue-generating measures such as increasing memberships or donations show an increase in profit after subtracting the cost of doing business.

However, most EHS projects drive cost savings. Cost savings come through cost reduction or cost avoidance. Improved productivity, reduced injury/illness, lowered downtime, reduced energy use, reduced medical expenses and minimized absences are all examples of cost-saving measures.

Cost-avoidance projects are implemented to reduce risks, avoid problems or prevent unwanted events. Some finance and accounting professionals may view cost avoidance as an inappropriate measure used to determine monetary benefits and calculate ROI. However, if the assumptions prove correct, accomplishing an avoided cost (e.g., compliance fines) can be more rewarding than reducing an actual cost. Preventing a problem is more cost-effective than waiting to solve it.

Determining the potential payoff is the first step in the needs analysis process. This step closely relates to the next one, determining the business need, since the potential payoff is often based on a consideration of the business. The payoff depends on two factors: the monetary value derived from the business measure's improvement and the approximate cost of the project. Identifying monetary values in detail usually yields a more credible forecast of what to expect from the chosen project. However, this step may be omitted in situations where the problem (i.e., business need) must be resolved regardless of the cost, or if it becomes obvious that it is a high-payoff activity. For example, if the problem involves a safety concern related to toxic materials, a regulatory environmental compliance issue or a serious injury incidence rate, a detailed analysis may not be needed.

The target level of detail may also hinge on the need to secure project funding. If the potential funding source does not recognize the value of the project compared with the potential costs, more detail may be necessary to provide a convincing case for funding.

5.2.1 Obvious Versus Not-So-Obvious Payoff

The potential payoff is obvious for some projects and not so obvious for others. Examples of opportunities with obvious payoffs include:

- Injury costs are 47 percent higher than industry average
- Use of sick leave has doubled each year for three years
- Environmental friendly rating of 3.89 on a 10-point scale
- Employee productivity (revenue per employee) is the lowest in the industry
- A cost to the city of $12,000 annually per person for landfill operation

- Patient injury rate is twice the hospital average
- Noncompliance OHSA fines totaling $1.2 million, up 82 percent from last year
- Employee medical costs are 35 percent above benchmark figure
- Worker compensation costs are increasing 25 percent per year
- Carbon equivalent are 58 percent above average industry
- Retention rates are 66 percent among the lowest in the industry

Each item appears to reflect a serious problem that needs to be addressed by executives, administrators or politicians.

For other projects, issues are sometimes unclear and may arise from political motives or bias. These potential opportunities are associated with payoffs that may not be so obvious. Examples of such opportunities may include:

- Become a recognized leader
- Provide safety management training to all project safety leaders
- Become a green company
- Implement behavior based safety for all employees
- Improve safety leadership competencies for all managers
- Improve the health for all employees
- Create a safe workplace
- Create a great place to work
- Establish a project management office
- Build a wellness and fitness center
- Implement green recruiting
- Create a culture of zero harm
- Establish a recycling program

With each of these opportunities, there is a need for more specific detail regarding the measure. For example, if the opportunity is to become a zero harm company, one might ask: What is a zero harm company? What are the advantages of becoming a zero harm company? How is it measured? Projects with not-so-obvious payoffs require greater analysis than those with clearly defined outcomes.

5.2.2 The Cost of a Problem

The potential payoff establishes the fundamental reason for pursuing new or enhanced projects. But the payoff—whether obvious or not—is not the only reason for moving forward with a project. The cost of a problem is another factor. If the cost is excessive, it should be addressed. If not, then a decision must be made as to whether the problem is worth solving.

Problems are expensive and their solution can result in high returns, especially when the solution is inexpensive. Problems may encompass time, quality, productivity and team or customer issues. All of these factors must be converted to monetary values if the cost of the problem is to be determined. Injuries (a quality measure) are directly associated with the cost of medical care as well as with the cost of investigation, reporting and prevention. Time can easily be translated into money by calculating the fully loaded cost of an individual's time spent on unproductive tasks. Calculating the time for completing a project, task or cycle involves measures that can be converted to money. Productivity problems and inefficiencies, equipment damage and equipment underuse are other items for which conversion to monetary value is straightforward.

In examining costs, considering *all* the costs and their implications is crucial. For example, the full cost of disposing of hazardous materials includes not only the disposal fee, but also transportation, record keeping, time required for investigations, damage to image and time spent by all involved employees who are addressing the issue. The cost of an OSHA complaint includes not only the cost of the time spent resolving the complaint but also the cost of the possible penalty or adjustment because of the complaint. The costliest consequences are the potential loss of business and goodwill from a negative image and from potential customers who learn about the complaint. Placing a monetary value on a problem helps in determining if the problem's resolution is economically feasible.

5.2.3 The Value of an Opportunity

Just as the cost of a problem can be easily tabulated in most situations, the value of an opportunity can also be calculated. Examples of opportunities include implementing green cost-saving projects, exploring new technology for accident investigation, increasing research and development efforts to prevent injury and upgrading the workforce to create a safe working environment. In these situations, a problem may not exist but an opportunity to get ahead of the competition or to prevent a problem's occurrence by taking immediate action does. Assigning a proper value to this opportunity requires considering what may happen if the project is not pursued or acknowledging the potential windfall if the opportunity is seized. The value of an opportunity is determined by following the different possible scenarios to convert business impact measures to money. The difficulty in this process is conducting a credible analysis. Forecasting the value of an opportunity entails many assumptions compared to the value of a known outcome.

5.2.4 To Forecast or Not to Forecast?

The need to seek and assign value to opportunities leads to an important decision, whether or not to forecast ROI. If the stakes are high and support for the project is not in place, a detailed forecast may be the only way to gain the needed project backing and funding or to inform decision makers about the choice between multiple potential projects. When developing the forecast, the rigor of the analysis is an issue. In some cases, an informal forecast is sufficient, given certain assumptions about alternative outcome scenarios. In other cases, a detailed forecast is needed that uses data collected from a variety of experts, previous studies from another project or a more sophisticated analysis. Other references provide techniques for developing forecasts (Phillips & Phillips, 2010).

5.3 Determining Business Needs

When the potential payoff, including financial value, has been determined, the next step is to clarify business needs. This requires identifying specific measures so that the business situation can be clearly assessed.

The concept of business needs refers to the need for gains in productivity, quality, efficiency, time and cost. This is true for the private sector as well as in government, nonprofit, non-governmental and academic organizations.

5.3.1 The Opportunity

A business need is represented by a business measure. For example, let's say you have been receiving an extraordinary number of OSHA complaints regarding safety conditions. The need is to reduce complaints. The specific measurement is the number of complaints regarding safety conditions. Any process, item or perception can be measured and such measurement is critical to this level of analysis. If the project focuses on solving a problem, preventing a problem or seizing an opportunity, the measures are usually identifiable. The important point is that the measures are present in the system, ready to be captured for this level of analysis. The challenge is to define the measures and to find them economically and swiftly.

5.3.2 Hard Data Measures

To focus on the desired measures, distinguishing between hard data and soft data may be helpful. Hard data are primary measures of improvement presented in the form of rational, undisputed facts that are usually gathered within functional areas throughout an organization. These are the most desirable type of data because they are easy to quantify and are easily converted to monetary values. The fundamental criteria for gauging the effectiveness of an organization are hard data items such as revenue, productivity and profitability, as well as measures that quantify such processes such as cost control and quality assurance.

Hard data are objective and credible measures of an organization's performance. Hard data can usually be grouped into four categories, as shown in Table 5.1. These categories—output, quality, costs and time—are typical performance measures in any organization.

Hard data from a particular project involve improvements in the output of the work unit, section, department, division or the entire organization. Every organization, regardless of the type, must have basic measures of output, such as number of pages printed, tons produced or packages shipped. Since these values are monitored, changes can easily be measured by comparing "before" and "after" outputs.

Quality is another important hard data category. If quality is a major priority for the organization, processes are likely in place to measure and monitor quality. The rising prominence of quality

Table 5.1 Examples of Hard Data.

Output	**Quality**	**Costs**	**Time**
Energy use	Accidents	Workers compensation costs	Investigation time
Units produced	Disabling injuries	Accident costs	Cycle time
Carbon emissions	Sick leave used	Investigation cost	Equipment downtime
Recycle volume	Turnover	Medical costs	Overtime
Items assembled	Failure rates	Landfill costs	On-time shipments
Money collected	Scrap/waste	Energy costs	Time to project completion
Items sold	Rejects	Supplies costs	Processing time
Materials consumed	Error rates	Fuel costs	Supervisory time
New accounts generated	Rework	Budget variances	Time to proficiency
Forms processed	Shortages	Unit costs	Repair time
Inventory turnover	Deviation from standard	Cost by account	Efficiency
Applications processed	Product	Variable costs	Work stoppages
Tasks completed	Inventory adjustments	Fixed costs	Order response
Output per hour	Incidents	Overhead costs	Late reporting
Productivity	Compliance discrepancies	Project cost savings	Lost-time days
Work backlog	Agency fines	Material costs	Time for safety check
Shipments	Absenteeism	Healthcare costs	Time for permit

improvement processes (such as total quality management, continuous process improvement and Six Sigma) has contributed to the tremendous recent successes in pinpointing the proper quality measures—and assigning monetary values to them. Most EHS measures are in this category.

Another important hard data category is cost. Many projects are designed to lower, control or eliminate the cost of a specific process or activity. Achieving cost targets has an immediate effect on the bottom line. Some organizations focus narrowly on cost reduction. Consider Wal-Mart, whose tagline is "Always low prices. Always." All levels of the organization are dedicated to lowering costs for processes and products and to passing the savings along to customers. Sometimes the EHS measures are converted to costs, such as landfill costs, employee healthcare costs and accident investigation costs.

Time is another critical measure in any organization. Hundreds of publications, workshops and seminars focus on saving time. Some organizations gauge their performance almost exclusively in relation to time. When asked what business FedEx is in, company executives say, "We engineer time." These are important for EHS, such as the time to obtain an environmental permit, the time to investigate an accident, lost days of work and the time to perform a safety check.

5.3.3 Soft Data Measures

Soft data are probably the most familiar measures of an organization's effectiveness, yet their collection can present a challenge. Values representing attitude, motivation and satisfaction are examples of soft data. Soft data are more difficult to collect and analyze and therefore, they are used when hard data are unavailable or to supplement hard data. Soft data represent qualitative measures, which make them more difficult to convert to monetary values than hard data. They are less objective as performance measurements and are usually behavior related, yet, organizations place great emphasis on them. Improvements in these measures represent important business needs, but many organizations omit them from the ROI equation because of their subjectivity. However, soft data can contribute to economic value to the same extent as hard data measures. Table 5.2 shows typical examples of soft data by category. The key is to avoid focusing too much on the hard versus

Table 5.2 Examples of Soft Data.

Work Habits
Excessive socialization
Wasteful activities
Visits to the dispensary
Violations of rules
Communication breakdowns
Work Climate/Satisfaction
Grievances
Discrimination charges Employee complaints
Job satisfaction
Organization commitment
Employee engagement
Employee loyalty
Intent to leave
Stress
Customer Service
Customer complaints
Customer satisfaction
Customer dissatisfaction
Customer impressions
Customer loyalty
Customer retention
Lost customers
Employee Development/Advancement
Promotions

(Continued)

Table 5.2 (*cont.*)

Capability
Intellectual capital
Requests for transfer
Performance appraisal ratings
Readiness
Networking
Initiative/ Innovation
Creativity
Innovation
New ideas
Suggestions
New products and services
Trademarks
Copyrights and patents
Process improvements
Partnerships/alliances
Image
Brand awareness
Reputation
Leadership
Social responsibility
Environmental friendliness
Social consciousness
External awards

soft data distinction. A better approach is to consider data as tangible or intangible.

5.3.4 Tangible Versus Intangible Benefits

A challenge with regard to soft versus hard data is converting soft measures to monetary values. The key to addressing this challenge is to remember that, ultimately, all roads lead to hard data. Although stress may be categorized as a form of soft data, a less stressful workplace can lead to less sick time, which lead to, for example, reduction in healthcare costs—clearly a hard data measure. Although it is possible to convert the measures listed in Table 5.2 to monetary amounts, it is often more realistic and practical to leave them in nonmonetary form. This decision is based on considerations of credibility and the cost of the conversion. According to the standards of the ROI Methodology, an intangible measure is defined as a measure that is intentionally not converted to money. If a soft data measure can be converted to a monetary value credibly using minimal resources, it is considered tangible, reported as a monetary value and incorporated in the ROI calculation. If a data item cannot be converted to money credibly with minimal resources, it is listed as an intangible measure. Therefore, when defining business needs, the key difference between measures is not whether they represent hard or soft data, but whether they are tangible or intangible. In either case, they are important contributions toward the desired payoff and important business impact data.

5.3.5 Impact Data Sources

Sources of EHS impact data, whether tangible or intangible, are diverse. Data come from routine reporting systems in the organization, city, community or industry. In many situations, these measures have led to the need for the project; therefore, the source is evident. A vast array of documents, systems, databases and reports can be used to select the specific measure or measures to be monitored throughout the project. Impact data sources include—but are not limited to—quality reports, service records, suggestion systems and employee engagement data.

Some EHS project planners and project team members assume that corporate data sources are scarce because the data are not readily available to them. However, data can usually be located by investing a small amount of time. Rarely do new data collection systems or processes need to be developed in order to identify measures representing the business needs of an organization.

When searching for the proper measures to connect to a project and to identify business needs, it is helpful to consider all the possible measures that could be influenced. Sometimes, collateral measures move in harmony with a project. For example, efforts to reduce injuries may also improve quality and increase job satisfaction. Weighing adverse impacts on certain measures may also help. For example, when using recycled materials, quality may suffer, or when delivery schedules are altered to save fuel, customer satisfaction may deteriorate. Finally, project team members must anticipate unintended consequences and capture them as other data items that might be connected to or influenced by the project.

In the process of settling on the precise business measures for the project, it is useful to examine various "what if" scenarios. For example, what if the organization does nothing? The potential consequences of inaction should be made clear. The following questions may help in understanding the consequences of inaction:

- Will the situation deteriorate?
- Will operational problems surface?
- Will our workplace be safe?
- Will budgets be affected?
- Will we lose influence or support?
- Will our image suffer?

Answers to these questions can help the organization identify a precise set of measures and can provide a hint of the extent to which the measures may change as a result of the project.

5.4 Determining Performance Needs

The next step in the needs analysis is to understand what led to the business need. If the proposed project addresses a problem, this step focuses on the cause of the problem. If the project makes use of an opportunity, this step focuses on what is inhibiting the organization from taking advantage of that opportunity. Answers to the following questions help determine changes in performance necessary to address business needs:

- What is happening or not happening within the organization that is causing the business measure to be at its current level?
- What behaviors or habits need to change in order to improve the business measure?
- What systems or processes need to change in order to support the change in behavior?
- What workplace alterations are necessary?
- What barriers prevent people from employing behaviors or habits necessary to improve the business measure?

Changing habits and behavior are often the critical success factors in implementing EHS projects. People often get stuck in their comfort zones and prefer not to change. Analyzing the behavioral aspects of performance ensures that the right behaviors are targeted for business needs or opportunities. For example, if your business measure is to reduce landfill costs, analysis may reveal that employees do not recycle. The behavior that needs to change is placing recyclable materials in the appropriate bins.

5.4.1 Analysis Techniques

Uncovering the causes of the problem or the inhibitors to success with key business measures that can be influenced by EHS projects requires a variety of analytical techniques. These techniques—such as interviews, focus groups, problem analysis, nominal group technique, force field analysis and brainstorming—clarify job performance needs. The technique employed depends on the organization or community setting, the apparent depth of the problem and the funding available for such analysis. Multiple techniques can be used since job performance may be lacking for a number of reasons. The details of applying these techniques can be found in many sources (Whiteside & McKenna, 1999).

5.4.2 A Sensible Approach

Analysis takes time and adds to an EHS project's cost. Examining records, researching databases and observing individuals can provide important data, but a more cost-effective approach may include employing internal and/or external experts to help analyze the problem. Performance needs can vary considerably and may include ineffective behavior, dysfunctional habits, inadequate processes, a disconnected process flow, improper procedures, a non-supportive culture, outdated rules and methods and a non-accommodating environment, to name a few. When needs vary, and with many techniques from which to choose, the possibility of over analysis and excessive costs exists. Consequently, a sensible approach is needed.

5.5 Determining Learning Needs

Changing behaviors, habits and supporting processes often requires acquiring new knowledge, skills and/or information. For example, participants and team members may need to learn about the EHS solutions, how to perform a task differently or how to use a process, system or technology. In some cases, learning is the principal solution, as in competency or capability development for EHS projects in major production and system installations. For many EHS projects, however, learning may be a minor aspect of an overall solution and may involve simply understanding the process, procedure or policy. For example, in the implementation of a new safety audit process for an organization, the learning component requires understanding why the process is necessary, how it works and the employees' role in the process. In short, a specific learning solution is not always needed, but all EHS solutions have a learning component.

A variety of approaches are available for measuring specific learning needs. Often, multiple tasks and jobs are involved in a project and should be addressed separately. One of the most useful ways to determine learning needs is to ask individuals who understand the process. Subject matter experts can often best determine what skills and knowledge are necessary to address performance issues. This may be the appropriate time to find out the extent to which knowledge and skills already exist.

Job and task analyses are effective when a new job is created or when an existing job description changes significantly. As jobs are redesigned and new tasks must be identified, this type of analysis offers a systematic way of detailing the job and task. Essentially, a job analysis is the collection and evaluation of work-related information. A task analysis identifies the specific knowledge, skills, tools and conditions necessary for the safe performance of a particular job.

Perhaps the most effective way to assess learning needs is to conduct informal discussions, surveys and self-assessments with the planned participants of green projects. Understanding what they do or do not know about issues and their roles and responsibilities can provide insights into the best approach to addressing the performance issue.

Sometimes, the demonstration of knowledge surrounding a certain task, process or procedure provides evidence of what capabilities exist and what is lacking. Such demonstration can be as simple as skill practice or role-playing a safety procedure, or as complex as an extensive mechanical or electronic simulation. The point is to use demonstrations as a way of determining if employees know how to perform a particular process.

Testing as an assessment process for learning needs is not used as frequently as other methods, but it can be helpful. Employees are tested to reveal what they know about the safety, safety procedures, accident prevention, wellness, fitness, nutrition, environment, climate change and sustainability. This information helps guide learning issues.

Input from managers or team leaders may provide a good assessment of knowledge, skills and information gaps. Input can be solicited through surveys, interviews or focus groups. It can be a rich source of information about what the users of the project, if implemented, will need to know to make the plan successful.

Where new knowledge, skill and/or information are minor components, learning needs are simple. Determining learning needs can be time-consuming for major projects in which new procedures, technologies and processes must be developed. As in developing performance needs, it is important not to spend excessive time analyzing learning needs but rather to collect as much data as possible with minimal resources.

5.6 Determining Preference Needs

The final level of needs analysis determines the preferences that drive the project requirements. Essentially, individuals prefer certain processes, schedules or activities for the structure of the project. These preferences define how the particular project will be perceived and launched. If the project is a solution to a problem, this step defines how the solution will evolve. If the project takes advantage of an opportunity, this step outlines how the opportunity will be addressed, considering the preferences of those involved in the project.

Perhaps the most important aspect of preference is the desired reaction for the key stakeholders in the project. Stakeholders include participants, their supervisors, managers and the client funding the project, among others. The EHS project must be perceived as important, relevant, useful, needed, appropriate and valuable. Perhaps the most powerful reaction occurs when employees report their intent to do something about the issue. This indicates commitment and has a strong correlation with actual action.

Preference needs also define the parameters of the project in terms of scope, timing, budget, staffing, location, technology, deliverables and the degree of allowable disruption. Preference needs are developed from the input of several stakeholders rather than from one individual. For example, participants in the project (i.e., those who must make it work) may have a particular preference, but that preference could exhaust resources, time and budgets. The immediate manager's input may help minimize the amount of disruption and maximize resources.

The urgency of project implementation may introduce a constraint to the preferences. Those who support or own the project often impose preferences on the project in terms of timing, budget and the use of technology. Because preferences represent a Level 1 need, the project structure and solution will relate directly to the reaction objectives and to the initial reaction to the project.

When determining the preference needs, there can never be too much detail. Projects often go astray and fail to reach their full potential because of misunderstandings and differences in expectations surrounding the project. Preference needs should be addressed before the project begins. Pertinent issues are often outlined in the project proposal or planning documentation.

5.7 Developing Objectives for EHS Projects and Programs

EHS projects are driven by objectives. Objectives position the project or program for success if they represent the needs of the business and include clearly defined measures of achievement. Developing project objectives is the second phase of alignment. A project may be aimed at implementing a solution that addresses a particular need, problem or opportunity. In other situations, the initial project is designed to develop a range of feasible solutions, with one specific solution selected prior to implementation. Regardless of the project, multiple objective levels are necessary. These levels define precisely what will occur as a project is implemented. Project objectives correspond to the levels of evaluation and the levels of need presented.

5.7.1 Reaction Objectives

For a project to be successful, the stakeholders immediately involved in the process must react favorably—or at least not negatively—to the project. Reaction objectives come from the preference needs. Ideally, those directly involved should be satisfied with the project and see its value. This feedback must be obtained routinely during the project in order to make adjustments, keep the project on track and to redesign certain aspects as necessary. Unfortunately, for many projects, specific objectives and data collection mechanisms are not developed and put in place to allow channels for feedback.

Developing reaction objectives should be straightforward and relatively easy. The objectives reflect the degree of immediate as well as long-term satisfaction and explore issues important to the success of the project. They also form the basis for evaluating the chain of impact, emphasizing planned action, when necessary. Typical issues addressed in the development of reaction objectives are relevance, usefulness, importance, necessity, appropriateness and motivation. The most important reaction objective is for participants in an EHS project to indicate intent to apply, use, or implement the concepts in the success of green projects, often using a 5-point scale.

5.7.2 Learning Objectives

Every EHS project involves at least one learning objective, and usually more. With projects entailing major change, the learning component is very important. In situations narrower in scope, such as the implementation of a new policy, the learning component is minor but still necessary. To ensure that the various stakeholders have learned what they need to know to make the project successful, learning objectives are developed. Here are some examples of learning objectives supporting successful implementation of an EHS project.

After completing the learning component, participants should be able to:

- Identify the six features of the new green policy
- Identify the top five hazardous materials
- Demonstrate the use of each safety procedure within the standard time
- Score 10 or better on the healthy living test
- Explain the value of recycling to a group
- Complete the safety simulation with a score of 80 percent or greater within the five-minute time limit
- Correctly complete a confined space entry permit

Objectives are critical to the measurement of learning because they communicate the expected outcomes from the learning component and define the competency or level of performance necessary to make project implementation successful. They provide a focus to allow participants to clearly identify what it is they must learn and do—sometimes with precision.

5.7.3 Application and Implementation Objectives

Application and implementation objectives clearly define what is expected of the project and often the target level of performance. Application objectives are similar to learning objectives but relate to actual performance. They provide specific milestones indicating when one part or all of the process has been implemented. Here are some typical application objectives for when a project or program is implemented.

After completing the project and returning to work, participants should:

- Follow the correct safety sequences after three weeks on the job
- Develop or revise job safety analyses at a rate of 5 per month

- Participate in health and wellness programs
- Report at least 95 percent of near misses
- Wear personal protective equipment every day
- Routinely recycle waste products
- Conduct a five-minute meeting with the team at the beginning of each shift
- Conduct two safety audits each month
- Monitor and report safety hazards
- Follow up with at least one green exhibitor from the conference

Application objectives are critical because they describe the expected outcomes in the intermediate area—between the learning of new information tasks, procedures and skills and the delivery of the impact of this learning. Application and implementation objectives describe how things should be or the desired state of the workplace when the project solution has been implemented. They provide a basis for evaluating on-the-job changes and performance.

5.7.4 Impact Objectives

Impact objectives indicate key business measures that should improve as the application and implementation objectives are achieved. The following are typical impact objectives.
After the project is completed:

- Total accident costs at the foundry should decrease by 20 percent within the next calendar year.
- The number of safety complaints will be reduced by twelve next year.
- Energy costs will be reduced by 30 percent at the tire plant in nine months.
- The company-wide safety index should increase by 10 percent during the next calendar year.
- Employee sick leave use should be reduced by 20 percent in six months.
- First aid treatments should be reduced 40 percent in three months.
- Employee healthcare costs should increase no more than five percent next year.

- Water usage at the headquarters building should decrease by 25 percent within three years.

Impact objectives are critical to measuring business performance because they define the ultimate expected outcome from the project, describing the business unit performance that should result. Above all, impact objectives emphasize achievement of the bottom-line results that key sponsor groups expect and demand.

5.7.5 ROI Objectives

The fifth level of objectives for EHS projects represents the acceptable ROI—the economic impact. Objectives at this level define the expected payoff from investing in a green project. An ROI objective is typically expressed as an acceptable ROI percentage, which is expressed as annual monetary benefits minus cost, divided by the actual cost and multiplied by one hundred. A 0 percent ROI indicates a breakeven project. A 50 percent ROI indicates recapture of the EHS project cost and an additional 50 percent "earnings" (fifty cents for every dollar invested).

For capital projects, such as the purchase of a new company, a new building or major equipment, the ROI objective may be larger relative to the ROI of other expenditures. However, the calculation is the same for both. For many organizations, the ROI objective for an EHS project is set slightly higher than the ROI expected from other "routine investments" because of the relative newness of applying the ROI concept to EHS projects. For example, if the expected ROI from the purchase of new equipment is 20 percent, the ROI from a new EHS project might be around 25 percent. The important point is that the ROI objective should be established up front, ideally in discussions with the project sponsor. Excluding the ROI objective leaves stakeholders questioning the economic success of a project. If a project reaps a 25 percent ROI, is that successful? Not if the objective was a 50 percent ROI.

5.8 Case Study Examples

The following two case studies may prove helpful in describing the alignment process. One case study involves a medium-sized sales and service organization, focusing on a variety of green initiatives, principally involving employees. The second case study includes a food processor that is addressing a different type issue, the safety and health of employees.

5.8.1 Progressive Specialty Company

Progressive Specialty Company (PSC) sells and distributes a variety of specialty advertising products such as pens, calendars, mugs, T-shirts, umbrellas and calculators. In total, 150 different products are sold. Most of the items are produced by suppliers and imprinted at PSC. However, most of the paper-based products are printed at PSC. Top executives are interested in protecting the environment. However, the headquarters office has done little to focus on the green projects and sustainability efforts. To improve their image in the community and to ensure that customers and other stakeholders are aware of their commitment to the environment, PSC decided to implement a variety of green projects at company headquarters.

Approximately 500 people work in two buildings at the headquarters and are, for the most part, professional administrative and operations employees. Executives could see possible opportunities to save costs with energy savings and procurement as a result of the initiatives; these savings would be in addition to the benefit of enhancing the corporate office image. After their needs were clearly defined, the project was positioned around a specific set of objectives. Table 5.3 shows the framework for the project assessment, objectives and evaluation.

Level 5 analysis determined the rationale for pursuing the project: Is it needed? Is it necessary? Is it a problem worth solving? Is it an opportunity worth pursuing? Two issues clearly indicated that the opportunities were worth pursuing. First, the potential environmental impact was a valuable opportunity to pursue. PSC wants to be seen as an organization supportive of environmental sustainability. In addition, cost-saving opportunities existed in a variety of operational and supply categories. The cost savings should be more than the cost of the project. This is important from an organizational perspective. Three specific business measures include:

1. Energy costs
2. Operations costs
3. Materials costs

Table 5.3 Create a Green Organization: Progressive Specialty Company.

Level	Needs	Objectives	Evaluation	Level
5	• Help protect the environment. • Net cost savings.	• Reach ROI target of 10 percent.	• Compare project benefits to costs.	5
4	• Improve image as a green company. • Reduce energy costs. • Reduce costs of operations. • Address increasing costs of materials/supplies.	• Improve image with employees, customers and by 1 point on a 5 point scale. • Reduce energy costs by: • Reduce operating costs ten percent. • Reduce materials/supplies by twenty percent.	• Conduct external survey. • Examine organization records.	4
3	• Increase recycling of materials. • Change consumption habits. • Use less materials and supplies. • Start making environmentally friendly choices.	• After the project is implemented, employees will: • Recycle in eight categories. • Alter consumption patterns. • Reduce usage. • Increase conservation. • Use environmentally friendly supplies.	• Conduct self-assessment questionnaire. • Check recycle records. • Check records of purchasing eco-friendly products.	3

Level	Needs	Objectives	Evaluation	Level
2	• Lack of knowledge or how actions effect the environment. • Specific green actions. • Environmental issues.	All employees will be able to: • Environmental issues. • Take specific green actions. • Make eco-friendly choices.	• Conduct self-assessment questionnaire. • Conduct environmental quiz.	2
1	Employees must see project as: • Necessary • Important • Relevant • Feasible.	Project receives favorable rating of 4 out of 5 on: • Necessary to PSC. • Relevant to our work. • Important to adhering. concepts in support of public good. • Feasible to implement.	• Administer reaction questionnaire to all project participants.	1

Improvement in these measures lead to a reduction in another measure: carbon emissions.

A brief Level 3 analysis revealed that employees needed to do more to help the environment and save costs. Specifically the actions observed involved not recycling, failing to change consumption habits and not making enough environmentally friendly choices.

At the learning level (Level 2), several specific needs were identified through brief conversations with employees and with a quiz that was promoted in the company newsletter as a contest to offer an incentive for people to complete it. Executives noticed a lack of understanding about green projects, their effects on the environment and why they were necessary.

With the learning needs defined, the project was defined around preference needs, which covered when projects would be implemented, how they would be implemented and the specific target audiences. Details about how participants should view these projects were also addressed. Because there is often apathy about green initiatives, the preference needs analysis considered the overall participant perspective of perceived necessity, importance and relevance to the organization's goals, as well as feasibility of pursuing the project.

With these needs in mind, the objectives were set at multiple levels and the evaluation system was employed based on these objectives. Aligning green initiatives using this approach is especially valuable when multiple projects are undertaken at the same time.

5.8.2 Western Food Services

Western Food Services (WFS) is a regional food processor, serving most of the western states. WFS prepare fresh, frozen and canned foods for supermarkets, specialty shops and restaurants. WFS purchases food from farmers and processes the food at a network of eight plants, employing about 2,000 people. Food purchases follow quality standards to ensure that the food is healthy and will remain fresh until consuming, freezing or canning.

The food is grown in fertile soil in appropriate high-yield climates. The plants are near the growing areas to minimalize transportation costs and ensure freshness. Usually, food supplies are sufficient to meet needs at a quality level appropriate to WFS standards.

Although automation is used in the plants, much of food processing is labor intensive. With pressure for high production and low costs, the challenge is to maintain a safe and healthy workplace.

In the last two years, the safety and health record has not been acceptable. Worker compensation costs are excessive and rising. First aid visits, medical treatment cases and disabling injuries are too high. OSHA citations are excessive, prompting more frequent visits by inspectors. To improve these measures, a program is needed that will encourage employees in the plants to work safely in a healthy environment.

There can be many reasons for excessive accidents and injuries, such as inadequate protection, faulty equipment, high risk processes, unsafe tools exist, carelessness and knowledge about how to work safely. Through an analysis of accidents, audits, work places, behavior, knowledge and attitudes, a program, labeled WORKSAFE, was implemented at all plants.

Table 5.4 shows the alignment between the needs and objectives for the WORKSAFE program. The payoff need is obvious and broad. The plants should be operating safely to improve costs and job satisfaction. This should be accomplished in a cost effective way so that cost savings are not less than the cost of the program. Expenses are excessive, as are first aid visits, medical treatment cases, disabling injuries, OSHA citations and violations in the use of hazardous materials.

An analysis of what is and isn't occurring in the plants identified six actions that need to be in place.

To address these performance and process needs, employees must learn how to operate and maintain a safe work process, behavior and attitude and the six learning topics we identified. This requires 16 hours of training.

Finally, for this program to be successful, the employees must see that it is necessary to have a meaningful safety program and that it is imperative for the survival of the plants. They must perceive all elements of the program as relevant. Employees must view what they learn as immediately useful and that they will commit to using it.

With these needs clearly defined, the specific objectives are developed as outlined in Table 5.2, which represents a simplified version of a complex program. The important point is that for any project, the multiple levels of analysis are an excellent way to understand the needs so that objectives can be developed and used to drive success in the process. With objectives at Level 3, 4 and 5 and appropriate data collection at Level 3 and 4, the project is not only positioned for ultimate success, but it can easily be evaluated at the impact and ROI level.

Table 5.4 WORKSAFE Program at WFS.

Level	Needs	Objectives	Evaluation	Level
5	• Improve safety performance and the impact on employees in a cost effective way.	• Break even (BCR = 1:1). ROI of 0 percent	• Compare program money benefits to program costs	5
4	• Worker compensation costs must be reduced. • First aid visits must be lowered. • Medical treatment cases must reduce. • Disabling injuries must be lowered. • OHSA citations must be reduced. • Hazmat violations must be reduced.	• Reduce costs by 25 percent in one year. • First aid visits reduced by 25 percent. • Medical treatments reduced by 15 percent. • Disabling injuries reduced by 20 percent. • OSHA citations reduced by 50 percent. • Hazmat violations reduced by 300 percent.	• Check safety records. • Check industry records.	4
3	At the workplace: • OSHA standards must be followed. • Protective equipment must be utilized. • Unsafe citations must be reported. • Violations and unsafe equipment must be reported. • Co-workers must be protected. • Checklists must be followed.	Employees will: • Follow standards. • Utilize protective equipment. • Correct unsafe situations. • Report violations. • Report unsafe equipment. • Look after co-workers. • Follow checklists.	• Use interviews. • Check lists. • Distribute questionnaires. • Examine WFC records.	3

Level	Needs	Objectives	Evaluation	Level
2	Employees must know how to work safely: • Safe practices • Safe equipment • Safe behavior • Safe attitudes • Safe conditions	Employees will demonstrate their: • Safe practices • Safe equipment • Safe behavior • Safe attitudes • Safe conditions	• Prepare a simple quiz. • Offer checklists. • Provide demonstrations. • Promote self-assessment.	2
1	Employees must see program as: • Necessary • Important to their survival • Relevance to their work • Something they will use	Program receives favorable rating of 4 out of 5 on • Necessary. • Importance. • Relevance. • Usefulness. • Commitment to follow processes.	• Administer reaction questionnaire to employees.	1

5.9 Final Thoughts

The alignment of any EHS initiative with relevant needs and objectives positions the project for success. This level of detail ensures that the green project remains results-focused throughout its implementation. Without upfront analysis, the project runs the risk of failing to deliver the value that it should, or of not being aligned with one or more business needs. The outputs of the analysis are objectives, which provide a focus for project designers, developers and implementers, as well as participants and users who must make the project successful. The third and final phase of aligning projects to organization needs and strategies occurs through the evaluation process. By evaluating projects against objectives that are representative of stakeholder needs, project owners can report, with confidence, the contribution their project makes to the organization and the environment.

6
Measuring Reaction and Learning

Abstract
Here we begin the journey of data collection, a key part of the ROI Methodology. Chapter 6 focuses on the measurement of reaction and learning, Levels 1 and 2 of the methodology. These two levels of evaluation provide the first wave of data. Participant feedback supplies powerful information to use in making adjustments to a project as it rolls out. Project participants' reactions, value perceptions and acquired learning, with regard to the EHS project, provide indications of its potential for success. The chapter outlines the most common approaches to collecting reaction and learning data and explores ways to use the information for maximum value.

Keywords: Reaction, learning, surveys, questionnaires, interviews, focus groups, data source, simulations, case studies, informal assessment

6.1 Why Measure Reaction?

It is difficult to imagine a project being conducted without the collection of feedback from those involved. The collection of reaction and perceived value data represents the first level of outcomes from the launch of an EHS project. Participant feedback is critical to a project's success so that adverse reactions can be identified and corrected before they derail an otherwise successful project. Here are a few specifics to consider as an organization works to measure reaction.

Jack Phillips, Patti Phillips, and Al Pulliam, Measuring ROI in Environment, Health, and Safety, (135–150) 2014 © Scrivener Publishing LLC

6.1.1 Customer Satisfaction

Reaction data are essentially measures of customer satisfaction with the project. Without sustained, favorable reactions from participants—the employees, volunteers, members, citizens, suppliers or customers involved in an EHS project—the project success may be in jeopardy. Individuals who have a direct role in the project are immediately affected; they often must change processes or procedures or make other adjustments in response to the project's initiation. Participant feedback regarding preferences is necessary to make positive adjustments and changes in the project as it unfolds. The opinions of project supporters are also important because this group will be in a position to influence the project's continuation and development. Sponsors—who approve budgets, allocate resources and ultimately live with the project's success or failure—must be satisfied with the project and their overall approval must be verified early and often.

6.1.2 Immediate Adjustments

Projects can go astray quickly and at times, a project can end up being the wrong solution for the specified problem. A project can also be mismatched to the solution from the beginning. Securing feedback early in the process allows for immediate adjustment. This can help prevent misunderstandings, miscommunications and, more importantly, misappropriations. Collecting and using reaction data promptly can enable an improperly designed project to be altered before more serious problems arise.

6.1.3 Predictive Capability

A relatively recent application of reaction data involves predicting the success of a project using analytical techniques. Project participants are asked about their reaction to the project in terms of its potential utility. The measures of utility are compared to the project's measures of success with application. The reaction data thus become a predictor of application. Figure 6.1 demonstrates the correlation between reaction feedback and application data.

In this analysis, reaction measures are taken as the EHS project is introduced and the success of the implementation is later judged using the same scale (i.e., a 1 to 5 rating). When significant positive

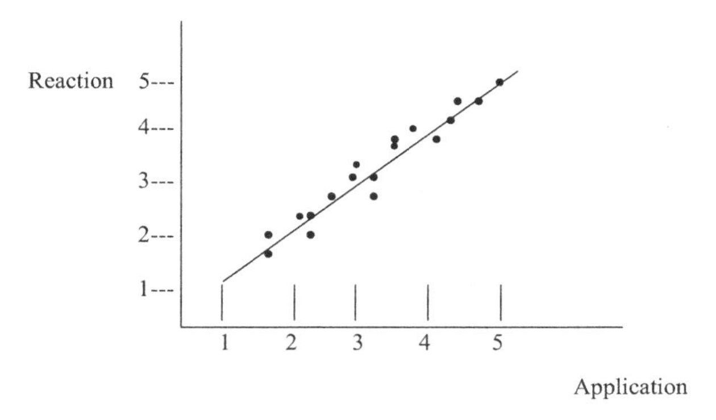

Figure 6.1 Correlations Between Reaction and Application.

correlations are present, reaction measures can have predictive capability. Some reaction measures shown to have predictive capability include statements such as these:

- The project is relevant to my work.
- The project is necessary for a safe workplace.
- The project is important to my job success.
- The project represents a valuable investment for the organization.
- I intend to make the project successful.
- I recommend this project to others in similar jobs.

These measures consistently lead to strong positive correlations and consequently represent more powerful feedback than typical measures of overall satisfaction with the project. Some organizations collect these or similar reaction measures for every project.

6.1.4 Important but Not Exclusive

Feedback data are critical to a project's success and should be collected for every project. Unfortunately, in some organizations, feedback alone has been used to measure project success. For example, at a particular trucking firm, the traditional method of measuring the effectiveness of a project is to rely entirely on feedback data from employees by asking them if the safety program project is appropriate, important and necessary. Positive feedback is critical to the project's acceptance, but it is no guarantee the project will be successfully executed. Executives become interested in the extent to which employees change their approach or behavior and implement the project in their work (application) and the effectiveness of the project in reducing the impact on the environment, health and safety (impact). Only when these additional measures are taken can the full scope of success be identified.

6.2 Sources of Data for Measuring Reaction

Possible sources of reaction and perceived value data concerning the success of a project can be grouped into six distinct categories. We address each category here.

6.2.1 Participants

The most widely used data source for EHS project evaluation is the participants who are directly involved in the project. The participants must take the knowledge and skills they acquire during the project and apply them in their work or personal lives and they may be asked to explain the potential impact of that application. Participants are a rich source of data for almost every aspect of a project. They are the most credible source of reaction and perceived value data.

6.2.2 Managers and Supervisors

For organizations, another key source of data are the individuals who directly supervise or lead participants. Managers and supervisors have a vested interest in the project and are often in a position to observe the participants as they attempt to apply the knowledge and skills acquired in the project. Consequently, they can report on the successes associated with the project as well as the difficulties and problems.

6.2.3 Other Team Members

When entire teams are involved in the implementation of an EHS project, all team members can provide useful information about the perceived changes prompted by the project. It's important to note that input from this group is pertinent only to issues directly related to their work; otherwise, the potential exists for introducing inaccuracies to the feedback process. Data collection should be restricted to those team members who are capable of providing meaningful insight into the value of the EHS project.

6.2.4 Sponsors and Senior Managers

One of the most useful data sources is the sponsor group, usually a senior management team. The perception of the sponsor, whether an individual or a group, is critical to project success. Sponsors can provide input on all types of issues and are usually available and willing to offer feedback. Sponsors and senior managers are a preferred source for reaction data, since these data usually indicate what is necessary to make adjustments and to measure success.

6.3 Areas of Feedback

When capturing reaction data, it is important to focus on the content of the EHS project. Too often, feedback data reflect aesthetic issues that may not be relevant to the project's substance. Table 6.1 distinguishes content and non-content issues explored in a reaction questionnaire from the launch of a safety leadership project. A traditional way to evaluate activities is to focus on non-content issues

Table 6.1 Content Versus Non-Content Issues.

Non-Content Issues	Content Issues
• Parking	• Necessity of project
• Location	• Practicality of topics
• Learning environment	• Relevance of project
• Registration	• Importance of topics
• Logistics	• Timing of project
• Hotel service	• Use of my time
• Food/Refreshments	• Amount of new information
• Facilitator	• Intent to use the content
• Coordination	• Value of content
• Communications with participants	• Appropriate level of content

(experience) or inputs. The column on the left in the table represents areas important to the activity surrounding the project, but contains nothing about the content. The column on the right reflects a focus on content, when it is necessary to drive business value. This is not to suggest that the service, the atmosphere of the event and the quality of the facilitator are not important but it is assumed that these issues will be addressed appropriately if there are problems. A more important set of data, focused on results, incorporates detailed information about the perceived value of the project, the importance of the content and the planned use of material—indicators that successful results were achieved.

Many topics are critical targets for feedback, which is needed in connection with almost every major issue, step or process to make sure things are advancing properly. Stakeholders will provide reaction input as to the appropriateness of the project planning schedule and objectives and the progress made with the planning tools. If the project is perceived as irrelevant or unnecessary to the participants, it is more than likely that it will not succeed in the workplace. Support for the project—including resources—represents an important area of feedback.

Participants must be assured that the project has the necessary commitment. Issues important to project management and the organization sponsoring the project include project leadership, staffing, coordination and communication. Also, it is important to collect feedback on how well the project team is working to address such issues as motivation, cooperation and capability.

Finally, the perceived value of the project is often a critical parameter. Major funding decisions are made based on perceived value when stronger evidence of value is unavailable.

6.4 Data Collection Timing for Measuring Reaction

As discussed previously, feedback during the early stages of implementation is critical. Ideally, this feedback validates the decision to go forward with the project and confirms the alignment with business needs. Notation of problems in initial feedback means that adjustments can be made early in its implementation. In practice, however, many organizations omit this early feedback, waiting until significant parts of the project have been implemented, at which point feedback may be more meaningful.

For longer projects, concerns related to the timing of feedback may require data collection at multiple moments. Measures can be taken at the beginning of the project and then at routine intervals once the project is underway.

6.5 Data Collection Methods for Measuring Reaction

A variety of methods can be used to collect reaction data. Instruments range from simple surveys to comprehensive interviews. The appropriate measure depends on the type of data needed (quantitative vs. qualitative), the convenience of the method to potential respondents, the culture of the organization and the cost of a particular instrument.

6.5.1 Questionnaires and Surveys

The questionnaire or survey is the most common method of collecting and measuring reaction data. Questionnaires and surveys come in all sizes, ranging from short forms to detailed, multiple-page instruments. They can be used to obtain subjective data about participants' reactions as well as to document responses for future use in a projected ROI analysis. Proper design of questionnaires and surveys is important to ensure versatility.

There are several basic types of questions such as yes/no and numerical scale (e.g., 1 to 5). Essentially with numerical scores, the individual is indicating the extent of agreement with a particular statement or is giving an opinion of varying conviction on an issue.

6.5.2 Interviews

Interviews, although not used as frequently as questionnaires to capture reaction data, may be conducted by the project team or a third party to secure data that are difficult to obtain through written responses. Interviews can uncover success stories that may help to communicate early achievements of the project. The interview is versatile and is appropriate for soliciting reaction data as well as application data. A major disadvantage of the interview is that it consumes time, which increases the cost of data collection. It also requires interviewer preparation to ensure that the process is consistent.

6.5.3 Focus Groups

Focus groups are particularly useful when in-depth feedback is needed. The focus group format involves a small-group discussion conducted by an experienced facilitator. It is designed to solicit qualitative judgments on a selected topic or issue. All group members are required to provide input, with individual input building on group input.

Compared with questionnaires, surveys and interviews, the focus group approach has several advantages. The basic premise behind the use of focus groups is that when quality judgments are subjective, several individual judgments are better than one. The group process, in which participants often motivate one another, is an effective method for generating and clarifying ideas and hypotheses. Focus groups are inexpensive and can be quickly planned and conducted. The flexibility of this process allows exploration of a project's unexpected outcomes or applications.

6.6 Use of Reaction Data

Unfortunately, reaction and perceived value data are sometimes collected and then disregarded. Too many project leaders use the information to feed their egos and then allow it to quietly disappear into their files, forgetting the original purpose behind its collection. In an effective evaluation, the information collected must be used to make adjustments or verify successes; otherwise, the exercise is a waste of time. Because this input is the principal measure supplied by key stakeholders, it provides an indication of their reaction to and satisfaction with the project. More important, these data provide evidence relating to the potential success of the project. Data collected at this level (Level 1) should be used to:

- Identify the strengths and weaknesses of the project and make adjustments
- Evaluate project team members
- Evaluate the quality and content of planned improvements

- Develop norms and standards for benchmarking and comparison
- Link with follow-up data, if feasible
- Market future projects based on the positive reaction

6.7 Why Measure Learning?

Several key principles illustrate the importance of measuring learning during the course of a project. Collectively, these principles provide an indication of the full range of benefits that result from measuring the changes in knowledge and skills information provided during the project.

6.7.1 Compliance Issues

Organizations face an increasing number of environmental and safety regulations with which they must routinely comply. These regulations involve all aspects of business and are considered essential by governing bodies to protect customers, investors and the environment. Employees must have a certain amount of knowledge about the regulations to maintain compliance. Consequently, an organization must measure the extent of employee learning and understanding with regard to regulations to ensure that compliance is not a problem.

Some projects are compliance driven. For example, a steel manufacturer may have to implement a major project to ensure that its employees were all familiar with the requirements of hazardous waste disposal regulation. This project was precipitated by the fact that the firm's process generates a USEPA "listed" hazardous waste and the firm's internal audit procedures revealed what appeared to be a lack of knowledge of the rules. When projects such as this are initiated, learning must be measured.

6.7.2 Use and Development of Competencies

The use of competencies has dramatically increased in recent years. In the struggle for a competitive advantage, many organizations have focused on people as the key to success. Competency models are used to ensure that employees do the right things, clarifying and articulating what is required for effective performance.

These models help organizations align behavior and skills with the strategic direction of the company. A competency model describes a particular combination of knowledge, skills and characteristics necessary to perform a role in an organization. With the increased focus on competencies, measuring learning is a necessity, particularly when environmental and sustainability issues are a part of the competencies.

6.7.3 Role of Learning in EHS Projects

Although some EHS projects involve new equipment, processes and technology, the human factor remains critical to project success. Employees must understand green issues and learn how to work in a new way and this requires the development of new knowledge and skills. Simple tasks and procedures do not always come with new processes. Sometimes, complex process procedures and tools must be used in an intelligent way to reap the desired benefits. Employees must learn in different ways—not just in a formal meeting, but also through technology-based learning and informal processes. Team leaders and managers may serve as coaches or resource experts in some projects. In a few cases, learning coaches or subject matter experts are used in conjunction with a project to ensure that learning is transferred to the job and is implemented as planned.

Participants do not always fully understand what they must do. Although the chain of impact can be broken at any level, a common place for such a break is at Level 2 (learning), when participants do not know what to do or how to do it properly. When the application and implementation does not go smoothly, project leaders can determine if a learning deficiency is the problem and if so, they may be able to correct it. In other words, learning measurement is necessary to contribute to the leaders' understanding of why employees are or are not, performing the way they should.

6.8 Challenges and Benefits of Measuring Learning

Measuring learning involves major challenges that may inhibit a comprehensive approach to the process. The good news is that a comprehensive approach is not necessary for most EHS projects and sustainability concerns. While measuring learning is an essential part of the ROI Methodology, this measurement provides many benefits that help ensure project success.

6.8.1 Challenges

The greatest challenge in measuring learning is to maintain objectivity without crossing ethical or legal lines while keeping an eye on time and costs. A common method of measuring learning is testing, but this approach generates additional challenges.

The first challenge is the "fear" factor. Few people enjoy being tested and some are offended by it. They may feel that their professional prowess is being questioned. Others are intimidated by tests, which bring back memories of their third-grade math teacher, red pen in hand.

Another testing challenge involves the legal and ethical repercussions of basing decisions involving employee status on test scores. Therefore, organizations use other techniques to measure learning, such as surveys, questionnaires and simulations. The difficulty with these methods, however, is the potential for inaccurate measures and the financial burden they impose. Consequently, there is a constant tradeoff between additional resources and the accuracy of the learning measurement process.

6.8.2 Benefits

Learning measurement checks the progress of the project against the learning objectives. Learning objectives are critical to a project in terms of participant readiness to execute the project. Fundamentally, the measurement of learning reveals the extent to which knowledge, skill and/or information is acquired during project roll out. This knowledge is necessary to fully understand the project and make it successful. Learning measurements provide data to project leaders so that adjustments can be made. They can identify strengths and weaknesses in the project presentation and may point out flaws in the design or delivery.

Learning measures enhance participant performance. Verification and feedback concerning the knowledge and skills acquired can encourage participants to improve in certain areas. When employees excel, feedback motivates them to enhance their performance even further.

Measuring learning also helps to maintain accountability. Because projects are aimed at making the environment better and workplaces safer, learning is an important part of any project and its measurement is vital in confirming that improvement has in fact occurred.

6.9 Learning Measurement Issues

Several items affect the nature and scope of measurement at the learning level. These include project objectives, the measures themselves and timing.

6.9.1 Project Objectives

The starting point for any level of measurement is the project objective. The measurement of learning builds on the learning objectives. For EHS projects, the first step is to ensure that objectives are in place. Typically, the objectives are broad and indicate only major information or general knowledge areas that should be acquired as the project is implemented. These are sometimes called *key project learning objectives.* They can be divided into subcomponents that provide more detail, which is necessary when a tremendous number of tasks, procedures or new skills must be learned to implement an EHS project. For other projects, this level of detail may not be needed. Identifying the major objectives and indicating what must be accomplished to meet each one is often sufficient.

6.9.2 Typical Measures

Measuring learning focuses on knowledge, skills and attitudes as well as the individual's confidence in applying or implementing the project or process as desired. Typical measures collected at this level concern the following areas:

- Knowledge
- Awareness
- Understanding
- Information
- Skills
- Capability
- Capacity
- Readiness
- Contacts
- Confidence

Obviously, the more detailed the knowledge area, the greater the number of objectives. The concept of knowledge is quite general and often includes the assimilation of facts, figures and ideas. Instead of knowledge, terms such as *awareness, understanding and information* may be used to denote specific categories. Sometimes skills are improved for comprehensive EHS projects. In some cases, the issue involves developing a reservoir of knowledge and related skills toward improving capability, capacity or readiness. Networking is often part of a project and developing internal or external contacts that may be valuable later is important. For example, a safety management initiative may include managers from different areas to contact each other at particular times in the future. For projects that involve different organizations, such as recycling programs offered to the public, new contacts that result from the program can be important and ultimately pay off with shared approaches.

6.9.3 Timing

The measurement of learning can occur at various times. If formal learning sessions connected with the project are offered, the measure is taken at the end of those sessions to ensure that participants are ready to apply their newly acquired knowledge. If a project has no formal learning sessions, measurement may occur at different intervals. For long-term projects, as skills and knowledge grow, routine assessment may be necessary to measure both the acquisition of additional skills and knowledge and the retention of the previously acquired skills. The timing of measurement is balanced with the need to know the new information; this is offset by the cost of obtaining, analyzing and responding to the data. In an ideal situation, the timing of measurement is part of the data collection plan.

6.10 Data Collection Methods for Measuring Learning

One of the most important considerations with regard to measuring learning is the specific way in which data are collected. Learning data can be collected using many different methods. Below are a few of these data collection methods.

6.10.1 Questionnaires and Surveys

Questionnaires and surveys, introduced earlier in the chapter with the focus on measuring reaction, are also used to collect learning data. These questionnaires may include similar types of questions used in collecting reaction data including yes/no questions, agree/disagreement questions and rating scales. Multiple choice is probably the most common question type, where participants are asked to choose one or more items from a series of alternative answers. Matching exercises are also useful, where participants match particular items. Short-answer questions can be easy to develop, but they are difficult to score. Developing questions in an attempt to measure learning can be fairly simple. The key is to ensure that the questions asked are relevant to the knowledge or information presented.

6.10.2 Simulations

Another technique for measuring learning is simulation. This method involves the construction and application of a procedure or task that mimics or models the work involved in the project. The simulation is designed to represent, as closely as possible, the actual job situation. Participants attempt the simulated activity and their performance is evaluated based on how well they accomplish the task or understand the process. For example, a simple simulation shows the energy use of light bulbs or small appliances as they are connected to a wattmeter. This vividly shows people who are involved in projects the most energy-saving appliances and bulbs. This simple simulation is helpful to understand the effects of making changes or selecting the proper ones.

Although the initial development can be expensive, simulations can be cost effective in the long run, particularly for large projects or situations where a project may be repeated. For example, Luminant, the Texas-based energy production and distribution company, has developed large simulation of their power plants. These simulations, which are located at the Luminant Academy, adjacent to a community college in Texas, provide power plant operators an opportunity to simulate different sequences, schedules and possibilities. It not only teaches them the correct way to run the power plant, but it also helps them understand the most efficient and safe way to operate the plants. Although this is expensive, it has provided savings over the long run.

6.10.3 Case Studies

A popular technique of measuring learning is the case study. A case study presents a detailed description of a problem and usually contains a list of several questions posed to the participant. The participant is asked to analyze the case and determine the best course of action. The problem should reflect conditions in the real world and in the content of the project. This approach is helpful for green projects.

The difficulty in using a case study lies in objectively evaluating participant performance. Many possible courses of action are available, making an objective, measurable performance rating of successful knowledge and understanding difficult.

6.10.4 Informal Assessments

Many projects include activities, exercises or problems that must be explored, developed or solved. Some of these are constructed in the form of interactive exercises, while others require individual problem-solving skills. When these tools are integrated into the learning activity, they can be effective in gathering learning data.

A commonly used informal method is participant self-assessment. Participants are provided with an opportunity to assess their acquisition of skills and knowledge. In some situations, a project leader or a facilitator provides an assessment of the learning that has taken place. Although this approach is subjective, it may be appropriate when project leaders or facilitators work closely with participants.

6.11 Use of Learning Data

Data must be used to add value and improve processes. Appropriate uses of learning data include the following:

- Provide individual feedback to build confidence
- Validate that learning has been acquired
- Provide additional support to ensure successful implementation
- Evaluate project leaders and facilitators
- Build a database for project comparisons
- Improve the project, program or process

6.12 Final Thoughts

This chapter discusses data collection during the first two levels of evaluation: reaction and learning. Measuring reaction is a component of every study and is a critical factor in a project's success. Projects fail because of a negative reaction. Learning must be assessed to determine the extent to which the participants in a project learn new information, knowledge, processes and procedures. By measuring learning, project leaders can ascertain the degree to which participants are capable of successfully executing the EHS project plan. Measuring learning provides an opportunity to make adjustments quickly so that improvements can be made.

Reaction and learning data are collected using a variety of techniques, although surveys and questionnaires are most often used because of their cost-effectiveness and convenience. The data are important in allowing immediate adjustments to be made to the project. While reaction and learning data are important, the value of data to executives increases as the evaluation moves up the chain of impact. Data collection at the next two levels, application and impact, are discussed in the next chapter.

7

Measuring Application, Implementation and Impact

Abstract

This chapter completes the topic of data collection by exploring application and impact, Levels 3 and 4 in the ROI Methodology. Application focuses on what individuals are doing to apply or implement the EHS project. Unfortunately, projects often break down at this level. Individuals just don't follow through with what is needed to be successful, and there are many barriers and enablers to application. Impact is the number one measure desired by executives. This connects the EHS project to impact measures such as workers' compensation costs, accidents, environmental fines and penalties, health care costs, and many other EHS measures. In addition, impact connects the EHS program to productivity, quality, cost and time. The specific methods of data collection are detailed with examples, along with explanation of how the data can be used to make future adjustments. The chapter ends with guidelines for selecting the appropriate data collection method, ensuring that process improvement is always the goal of data collection, analysis and evaluation.

Many EHS projects fail because of breakdowns in implementation. In these cases, project participants simply do not do what they should on a timely basis. Measuring application and implementation is critical to understanding the success of project implementation. Without successful implementation, positive impact will not occur—and no positive return will be achieved.

Most sponsors regard business impact data as the most important data type because of its connection to business success. For many projects, the opportunity to improve business measures (i.e., the business need) is what has initiated the project. Impact evaluation data close the loop by showing a project's success in meeting business needs.

This chapter explores the most common ways to measure two of the most important levels of data: application and implementation data and business impact data. The possibilities vary from using questionnaires to

---

Jack Phillips, Patti Phillips, and Al Pulliam, Measuring ROI in Environment, Health, and Safety, (151–174) 2014 © Scrivener Publishing LLC

monitoring business performance records. In addition to describing the techniques to evaluate these two levels, this chapter addresses the challenges and benefits of each approach.

Keywords: Application measurement, input measurement, focus groups, interviews, action plans, scorecards, response rate

7.1 Why Measure Application and Implementation?

Measuring application and implementation is absolutely necessary if a project is intended to improve EHS. For some projects, it is the most critical data set because it provides an understanding of the extent to which successful project implementation occurs. It also provides evidence of the barriers and enablers that influence success.

7.1.1 Focus on the Project

Because many EHS projects focus directly on implementation and application, a project sponsor often speaks in these terms and has concerns about these measures of success. For example, the sponsor of a recycling project will want to know if people are engaged in the process: Is the target audience naturally recycling? A sponsor of a safety program will want to know if employees are operating safely. A sponsor of a wellness and fitness program will want to know if employees are involved in the various program activities. By measuring and monitoring the extent to which people are actively participating, a focus remains on the project and its ultimate intent.

7.1.2 Identify Problems and Opportunities

If the chain of impact breaks at this level, little or no corresponding impact data will be available. Without improvement in impact measures, there is no ROI. This breakdown most often occurs because participants in the project encounter barriers, inhibitors and obstacles that deter implementation. A dilemma arises when reactions to the project are favorable and participants learn what is intended, but then they fail to apply necessary tools, knowledge or skills, thereby, failing to accomplish the objective.

When an EHS project goes astray, the first question usually asked is, "What happened?" More important, when a project appears to add no value, the first question should be, "What can we do to change its direction?" In either scenario, it is important to identify the barriers to success, the problems in implementation and the obstacles to application. At this level of evaluation, these issues are addressed, identified and examined. In many cases, the stakeholders directly involved in the process can provide important recommendations for making changes or using a different approach in the future.

When a project is successful, the obvious question is, "How can we repeat this or improve it in the future?" The answer to this question is also found at Level 3. Identifying the factors that contribute directly to the success of the project is critical. Those same items can be used to replicate the process and produce enhanced results in the future. When key stakeholders identify those issues, they make the project successful and provide an important case history of what is necessary for success.

7.1.3 Reward Effectiveness

Measuring application and implementation allows the sponsor and project team to reward those who do the best job of applying the processes and implementing the project. Measures taken at this level provide clear evidence of success and achievement and they provide a basis for performance reviews. Rewards often have a reinforcing value, helping keep participants on track and communicating a strong message for future improvement.

7.2 Application Measurement Issues

Collecting application and implementation data brings issues into focus that must be addressed for success at this level. These challenges often inhibit an otherwise successful evaluation.

7.2.1 Sufficient Amount of Data

Whether collecting data by questionnaires, interviews or focus groups, insufficient response rates are a problem in most organizations. Having individuals participate in the data collection process is a challenge. To ensure that adequate amounts of high-quality data are available, a serious effort is needed to improve response rates.

When a follow-up evaluation is planned, a wide range of issues will be covered in a detailed questionnaire. Asking for too much detail in either the reaction questionnaire or the follow-up questionnaire can reduce the response rate. The following actions can help maximize response rates:

- Provide advance communication regarding the need for data
- Identify who will see the data
- Describe the data integration process
- Design the instrument for simplicity and ease of response
- Use local management support, if feasible
- If applicable, let the participants know they are part of the sample
- Consider the use of incentives
- Have an executive sign the introductory letter or memo
- Issue at least two follow-up reminders
- Send a copy of the results to the participants
- Make sure the survey or questionnaire looks professional
- Introduce the questionnaire or survey in the early stages of the project
- Collect the data anonymously or confidentially

Because many projects are planned on the basis of the ROI Methodology, it is expected that sponsors will collect impact data, monetary values and the project's actual ROI. This need to "show the money" sometimes results in less emphasis being placed on measuring application and implementation. In many cases, it may be omitted or slighted in the analysis. But it is through focused effort on process and behavior change that business impact will occur. Therefore, emphasis must be placed on collecting application and implementation data.

7.2.2 Application Needs

A needs assessment (detailed in Chapter 5) asks, "What is being done, or not being done, that is inhibiting the business measure?" When this question is answered adequately, a connection is made between the solution and the business measure. When this issue is addressed, the activities or behaviors that need to change are identified, serving as the basis of the data collection. The bottom line is that too many evaluations focus on either impact measures, which define the business measure to collect, or on learning, which uncovers what people need to know. More focus is needed at Level 3, which involves the tasks, processes, procedures and behaviors that need to be in place for successful implementation on the job.

7.2.3 Objectives

As with the other levels, the starting points for data collection are the objectives set for project application and implementation. Without clear objectives, collecting data would be difficult. Objectives define what activity is expected. Chapter 5 discusses the basic principles for developing these objectives.

7.2.4 Coverage Areas

To a certain extent, the area of coverage for this level focuses on activity or action, not on the ability to act (Level 2) and not on the consequences of acting (Level 4). The sheer number of activities to measure can be mind-boggling. Table 7.1 shows examples of coverage areas for application, which will vary from project to project.

7.2.5 Data Sources

Essentially, all key stakeholders are potential sources of data. Perhaps the most important sources of data are the users of the solutions—those directly involved in the application and implementation of the project. Good sources may also be the project team or team leaders charged with the implementation. In some cases, the source may be the organizational records or system.

7.2.6 Timing

The timing of data collection can vary significantly. Because data collection occurs as a follow-up after the project launch, the key issue is determining the best time for a post-implementation evaluation. The challenge is to analyze the nature and scope of the application and implementation and to determine the earliest time

Table 7.1 Examples of Coverage Areas for Application.

Action	Explanation	Example
Increase	Increasing a particular activity or action	Increase the frequency of use of a particular skill
Decrease	Decreasing a particular activity or action	Decrease the number of times a particular safety process must be checked
Eliminate	Stopping a particular task or activity	Eliminate the formal monthly meeting and replace it with a brief daily briefing
Maintain	Keeping the same level of activity for a particular process	Continue to monitor the process with the same schedule previously used
Create	Designing or implementing a new procedure, process or activity	Create a procedure for the safe handling of a hazardous material
Use	Using a particular process, procedure, skill or activity	Use the new work safe skill in situations for which it was designed
Perform	Carrying out a particular task, process or procedure	Conduct a safety audit at the end of each week
Participate	Becoming involved in various activities, projects or programs	Submit a suggestion for reducing accident costs
Enroll	Signing up for a particular process, program or project	Enroll in a fitness program
Respond	Reacting to groups, individuals or systems	Respond to customer inquiries within fifteen minutes
Network	Facilitating relationships with others who are involved in or have been affected by the project	Continue networking with project safety leaders on a quarterly basis (at minimum)

that a trend and pattern will evolve. This occurs when the application becomes routine and the implementation is making significant progress. Identifying this timing is a judgment call. Going in as early as possible is important so that potential adjustments can still be made. At the same time, leaders must wait long enough so that behavior changes are allowed to occur and so implementation can be observed and measured. In EHS projects that span a considerable length of time, several measures may be taken at three- to six-month intervals. Using effective measures at well-timed intervals will provide successive input on implementation progress and clearly show the extent of improvement.

Convenience and constraints also influence the timing of data collection. If the participants are conveniently meeting to observe a milestone or special event, this would be an excellent opportunity to collect data. Sometimes, constraints are placed on data collection. Consider, for example, the time constraint that sponsors may impose. If they are anxious to have the data to make project decisions, they may request that the data collection is moved to an earlier time than would otherwise be ideal.

7.3 Data Collection Methods for Measuring Application

Some of the techniques previously mentioned that are available to collect application and implementation data are easy to administer and provide quality data. Other techniques are more robust, providing greater detail about success but raising more challenges in administration.

7.3.1 Questionnaires

Questionnaires have become a mainstream data collection tool for measuring application and implementation because of their flexibility, low cost and ease of administration. One of the most difficult tasks is determining the specific issues to address in a follow-up questionnaire. Figure 7.1 presents content items necessary for capturing application, implementation and impact information (Level 3 and Level 4 data).

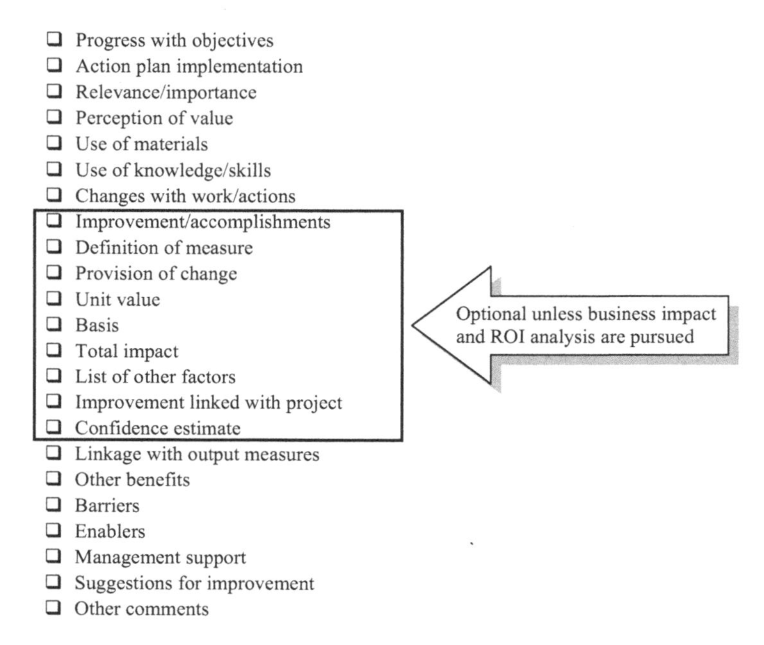

Figure 7.1 Questionnaire Content Checklist.

7.3.2 Interviews, Focus Groups and Observation

Interviews and focus groups can be used during implementation or on a follow-up basis to collect data on implementation and application. Observing participants and recording any changes in behavior and specific actions taken is also an approach to collect application data. When observation is used, the observer must be invisible or unnoticeable, such as when using a supervisor to observe safe work practices. Technology lends itself as a tool to assist with observations. Recorders, video cameras and computers play an important role in capturing application data. Call centers are a classic forum for gathering observation data using technology. Supervisors or other observers routinely listen in while call center representatives respond to calls.

7.3.3 Action Plans

In some cases, action plans can be used to develop implementation and application data. A typical action plan requires the participant to meet a goal or complete a task or steps by a set date. A summary of results of the completed action plan provides further evidence of the project's success. Figure 7.2 shows an example of an action plan for Level 3 (Application) with a hint of Level 4 (Impact).

Action Plan

Name ____________ Facilitator's Signature ____________ Follow-Up Date ____________

Objective ____________ Evaluation Period ______ to ______

ACTION STEPS: *I will do this*	END RESULT: So that
1.	
2.	
3.	
4.	
5.	
6.	
7.	

Figure 7.2 Action plan example with a focus on application.

7.4 Barriers to Application

One of the important reasons for collecting application and implementation data is to uncover barriers and enablers. Although both groups are important, barriers, a serious problem in every project, can lead to failure. The barriers must be identified in order to become important reference points for change and improvement. Subsequently, actions must be taken to minimize, remove or circumvent them so that the project can be implemented. Typical barriers that will stifle the success of an EHS project include:

- We have no opportunity to use this knowledge or information.
- The appropriate materials are not available for the project.
- Resources are not available to implement the project.
- The project is not appropriate for our work unit.
- Another project or issue has gotten in the way.
- The culture in our work group does not support the project.
- We have no time to implement the project.
- We don't see a need to implement the project.

As noted, the important point is to identify any barriers and to use the data in meaningful ways to make the obstacles less of a problem.

7.5 Use of Application Data

Data are meaningless if they are not used properly. As we move up the chain of impact, the data become more valuable in the minds of sponsors, key executives and others who have a strong interest in the project. Data also become more useful in ensuring successful project implementation. The principal uses of application data are the following:

- Report and review results with various stakeholders
- Adjust project design and implementation
- Identify and remove barriers
- Identify and enhance enablers
- Recognize individuals who have contributed to project success
- Reinforce in current and future project participants the value of desired actions
- Improve management support for green projects
- Market future projects

The key difference in using reaction and learning data and using application data is that the use of application data reaches beyond the boundaries of the EHS project implementation team. Application data provide evidence that the system (i.e., organization) is supporting implementation. When that support isn't there, these data can be leveraged to help other departments or functions that may be impeding progress with the project. Following the assumption that higher-level data create more value for key stakeholders, business impact measures often offer the most valuable data for stakeholders.

7.6 Why Measure Impact?

For most projects, business impact data represent the initial drivers for the project. The problem of deteriorating performance (e.g., workers' compensation costs) or the opportunity for improvement of a business measure (e.g., employee healthcare costs) usually leads to a project. If the business needs defined by business measures are the drivers for a project, then the key measure for evaluating the project is the business measure. The extent to which measures have changed is the principal determinant of project success.

7.6.1 "Show the Money" to Sponsors

From the sponsor's perspective, business impact data reflect key payoff measures that the sponsor desires or wants to see changed or improved. They often represent hard, indisputable facts that reflect performance that is critical to the business and operating unit level of the organization. Business impact leads to "the money"—the actual ROI in the project. Without credible business impact data linked directly to the project, it would be difficult, if not impossible, to establish a credible monetary value for the project. This makes this level of data collection one of the most critical.

7.6.2 Easy to Measure

One unique feature of business impact data is that they are often easy to measure. Hard and soft data measures at this level often reflect key assessments that are plentiful throughout an organization. It is not unusual for an organization to have hundreds or even thousands of calculations reflecting specific business impact measures including employee healthcare, fines due to safety and environmental violations, accidents, product, failure rates, lost time, equipment downtime and safety violations. The challenge is to connect the objectives of the project to the appropriate business measures. This is more easily accomplished at the beginning of a project.

7.7 Impact Measurement Issues

Chapter 5 defined impact measures as hard data, soft data, tangible or intangible. However they are categorized, these measures are plentiful. The key is to identify relevant impact measures specifically linked to the project.

7.7.1 Metric Fundamentals

When determining the type of measures to use, reviewing metric fundamentals can be helpful. The first important issue is identifying what makes an effective measure. Table 7.2 shows some of those criteria. These are issues that should be explored when examining any type of measure.

These criteria serve as a screening checklist as measures are considered, developed and ultimately added to the list of possibilities. In addition to meeting criteria, the factual basis of the measure should be stressed. In essence, the measure should be subjected to a fact-based analysis, a level of analysis never before applied to decisions about many projects, even when these decisions have involved huge sums of money. Distinguishing between the various "types" of facts is beneficial. As the following list shows, the basis for facts ranges from common sense to what employees "say", to actual data:

- *Omits facts.* Common sense tells us that employees will become involved in fitness programs if they are convenient for them.
- *Includes unreliable facts.* Employees say they are more likely to stay with a company if it has a good record for workplace safety.
- *Offers irrelevant facts.* We have benchmarked three world-class companies for green project measures: a bank, a hotel chain and a defense contractor. All reported good results.
- *Is fact-based.* Excessive accidents are increasing operational costs.

7.7.2 Scorecards

In recent years, interest has increased in developing documents that reflect appropriate measures in an organization. Scorecards, like those used in sporting events, provide a variety of measures for top executives. In their landmark 1996 book, *The Balanced Scorecard,* Robert Kaplan and David Norton explore the concept of the scorecard for use by organizations. They suggest that data can be organized in four categories: finances, customers, learning and growth

Table 7.2 Criteria for Effective Measures (Kerr, 1995; Mayo, 2003).

Criteria: Effective Measures Are	Definition: The Extent to Which a Measure . . .
Important	Connects to strategically important business objectives rather than to what is easy to measure
Complete	Adequately tracks the entire phenomenon rather than only part of the phenomenon
Timely	Tracks at the right time rather than being held to an arbitrary date
Visible	Is visible, public, openly known and tracked by those affected by it, rather than being collected privately for management's eyes only
Controllable	Tracks outcomes created by those affected by it who have a clear line of sight from the measure to results
Cost-effective	Is efficient to track using existing data or data that are easy to monitor without requiring new procedures
Interpretable	Creates data that are easy to make sense of and that translate into employee action
Simplicity	Is simple to understand from each stakeholder's perspective
Specific	Is clearly defined so that people quickly understand and relate to the measure
Collectible	Can be collected with no more effort than is proportional to the usefulness that results
Team-based	Will have value in the judgment of a team of individuals, not in the judgment of just one individual
Credible	Provides information that is valid and credible in the eyes of management

and process. Measures driven by green projects exist in all four categories.

The scorecard approach is appealing because it provides a quick comparison of key business impact measures and examines the status of the organization. As a management tool, scorecards can be important in shaping and improving or maintaining the performance of the organization through the implementation of projects. Scorecard measures often link to EHS projects. In many situations, it is a scorecard deficiency measure that initially prompts a project.

7.7.3 Specific Measures Linked to Projects

An important issue that often surfaces when considering ROI applications is the understanding of specific measures that are frequently driven by specific EHS projects. Although no standard answers are available, most projects are driving business measures. The monetary values are based on what is being changed in the various business units, divisions, regions and individual workplaces. These are the measures that matter to senior executives. The difficulty often comes in ensuring that the connection to the program exists. This is accomplished through a variety of techniques to isolate the effects of the project on the particular business measures, as discussed in Chapter 8.

7.7.4 Business Performance Data Monitoring

Data are available in every organization to measure business performance. Monitoring performance data enables measure performance in terms of environmental and safety compliance, accidents, employee health and other measures. When determining the source of data in the evaluation, the first consideration should be existing databases, reports and scorecards. In most organizations, EHS performance data that is suitable for measuring project-based improvement is available. If a particular data item is not available, additional record-keeping systems will have to be developed for measurement and analysis. It is at this moment that economics questions may begin to surface. Is it economical to develop the record-keeping systems necessary to evaluate an EHS project? If costs will be greater than expected benefits, then developing those systems is pointless.

7.7.5 Appropriate Measures

Existing performance measures should be thoroughly researched to identify those related to the proposed objectives of the project. Often, several performance measures are related to the same item. For example, the safety performance of a work team can be measured in several ways:

- First aid treatments
- Medical treatment cases
- Disabling injury rate
- Workers' compensation costs
- OSHA incidents
- Near misses
- Property damages

Each of these measurements, in its own way, evaluates the safety performance of the team. All related measures should be reviewed to determine those most relevant to the project.

7.8 Data Collection Methods for Measuring Impact

For many projects, business data are readily available to be monitored. However, at times, data will not be easily accessible to the project team or to the evaluator. Sometimes data are maintained at the individual, work unit or department level and may not be known to anyone outside that area. Tracking down all those data sets may be too expensive and time consuming. When this is the case, other data collection methods may be used to capture data sets and make them available for the evaluator. Three other options described in this book are the use of action plans, performance contracts and questionnaires.

7.8.1 Action Plans

Action plans can capture application and implementation data, as discussed earlier. They can also be a useful tool for recording business impact data. For business impact data, the action plan is more focused and credible than using a questionnaire. The basic design principles and the issues involved in developing and administering action plans are the same for business impact data as they are for application and implementation data (Phillips & Phillips, 2007).

For example, a hospital implemented a safety program to reduce the number of slips and falls for each unit as shown in Figure 7.3. Slips and falls involved patients and employees. The target audience is nurses. The number of slips and falls were excessive and increasing. The program involved one day of training to prepare nurses to prevent the incidents. As part of the program, an action plan was created, which was used to drive the project and collect data. Time was allotted to complete the action plan and a facilitator managed the process. It detailed specific actions nurses could take to prevent slips and falls. The specific measure the project was intended to influence, monthly slips and falls, was on the action plan. Not only did this step-by-step process become an application tool for the participants, but it also became an important input for the program evaluation. Essentially the monetary value was a collection of the impacts for the various participants. This allowed the hospital to evaluate the projects at the impact and ROI levels, minimizing the normal resources that would be needed to evaluate them.

7.8.2 Performance Contracts

Another technique for collecting business impact data is the performance contract, which is a slight variation of the action plan. Based on the principle of mutual goal setting, a performance contract is a written agreement between a participant and the participant's manager. It states the goal for the participant to accomplish during the project or after the project's completion and details what is to be accomplished, at what time and with what results.

Although the steps can vary according to the organization and the specific kind of contract, a common sequence of events usually follows:

- The employee (participant) becomes involved in project implementation.
- The participant and his or her immediate manager agree on a measure or measures for improvement related to the project.
- Specific, measurable goals for improvement are set.
- In the early stages of the project, the contract is discussed and plans are developed to accomplish the goals.

Safe Workplace Action Plan

Name: *Ellie Hightower* Facilitator Signature: Follow-Up Date *2 June*

Objective: *Improve workplace safety* Evaluation Period: *December* to *May*

Improvement Measure: *Monthly slips and falls* Current Performance *11/six months* Target Performance *2/six months*

Action Steps	
1. *Meet with team to discuss reasons for slips and falls*	*2 Dec*
2. *Review slip and fall records for each incident with safety – look for trends and patterns.*	*18 Dec*
3. *Make adjustments based on reasons for slips and falls*	*22 Dec*
4. *Counsel with housekeeping and explore opportunities for improvement..*	*5 Jan*
5. *Have safety conduct a brief meeting with team members*	*11 Jan*
6. *Provide recognition to team members who have made extra efforts for reducing slips and falls .*	*As needed*
7. *Follow-up with each incident and discuss improvement or lack of improvement and plan other action.*	*As needed*
8. *Monitor improvement and provide adjustment when appropriate.*	*As needed*

Intangible Benefits: *Image, risk reduction*

Analysis

A. What is the unit of measure? *1 slip and fall*

B. What is the value (cost) of one unit? $1750

C. How did you arrive at this value? *Safety and Health —Frank M.*

D. How much did the measure change during the evaluation period? (monthly value) 8

E. What other factors could have caused this improvement? *A new campaign from safety and health.*

F. What percent of this change was actually caused by this program? *70%*

G. What level of confidence do you place on the above information? (100% = Certainty and 0% - No Confidence) 80%

Figure 7.3 Completed Action Plan Example for Impact.

- During project implementation, the participant works to meet the deadline set for contract compliance.
- The participant reports the results of the effort to his or her manager.
- The manager and participant document the results and forward a copy, with appropriate comments, to the project team.

The process of selecting the area for improvement is similar to the process used in an action plan. The topic can cover one or more of the following areas:

- Routine performance related to the project, including specific improvement in measures such as safety and health of employees.
- Problem solving focused on such problems as an unexpected increase in compliance violations, an increase in workplace fatalities or a public image issue.
- Innovative or creative applications arising from the project, which could include the initiation of environmental practices, methods, procedures, techniques and processes

The topic of the performance contract should be stated in terms of one or more of the following objectives:

- Written
- Understandable by all involved
- Challenging (requiring an unusual effort to achieve)
- Achievable (something that can be accomplished)
- Largely under the control of the participant
- Measurable and dated

The performance contract objectives are accomplished.

7.8.3 Questionnaires

As described in the previous chapters, the questionnaire is one of the most versatile data collection tools and it can be appropriate for collecting data at Levels 1 through 4. Essentially, the design principles and content issues for a business impact evaluation are the same as at other levels, except that questionnaires will include additional questions to capture those data specific to business impact.

Using questionnaires for impact data collection has advantages and disadvantages. The good news is that questionnaires are easy to implement and low in cost. Data analysis is efficient and the time required to provide the data is often minimal, making questionnaires among the least disruptive of data collection methods. The bad news is that the data can be distorted and inaccurate and are sometimes missing. The challenge is to take all the steps necessary to ensure that questionnaires are complete, accurate and clear and that they are returned.

As noted, questionnaires are popular, convenient and low-cost; for these reasons, they have become a way of life. Unfortunately, questionnaires are among the weakest methods of data collection. Paradoxically, they are the most commonly used because of their advantages. Of the first 300 case studies published on the ROI Methodology, roughly 50 percent used questionnaires as a method of data collection. The philosophy in the ROI Methodology is to take processes that represent the weakest method and make them as credible as possible. Here the challenge is to make questionnaires credible and useful by ensuring that they collect all the data needed, that participants provide accurate and complete data and that return rates are in at least the 70 to 80 percent range.

The reason that return rates must be high is explained in the sixth guiding principle of the ROI Methodology that was outlined in Table 4.4. If an individual provides no improvement data, it is assumed that the person had no improvement to report. This is a conservative principle, but it is necessary to bring the required credibility. Consequently, using questionnaires will require effort, discipline and personal attention to ensure proper response rates as described earlier. There are other references that present suggestions for ensuring high response rates for data collection. For example, "Return to Sender: Improving Response Rates for Questionnaires and Surveys" (Phillips & Phillips, 2004) can be found at www.roiinstitute.net/publications/articles/?page=7.

7.9 Considerations for Selecting Data Collection Methods

The data collection methods presented in this and earlier chapters offer a wide range of opportunities for collecting data in a variety of situations. Seven aspects of data collection should be considered when deciding on the most appropriate method of collecting any type of data.

7.9.1 Type of Data

One of the most important issues to consider when selecting the data collection method is the type of data to be collected. Some methods are more appropriate for business impact. Follow-up questionnaires, observations, interviews, focus groups and action planning are best—sometimes exclusively—suited for application data. Performance monitoring, action planning, performance contracting and questionnaires can easily capture business impact data.

7.9.2 Investment of Participant Time

Another important factor when selecting the data collection method is the amount of time participants must spend with data collection and evaluation systems. Time requirements should always be minimized and the method should be positioned so that it is a value-added activity. Participants must understand that data collection is a valuable undertaking and not an activity to be resisted. Sampling can be helpful in keeping total participant time to a minimum. Methods such as performance monitoring and observation require no participant time, whereas others, such as interviews and focus groups, require a significant investment in time.

7.9.3 Cost of Method

Cost is always a consideration when selecting the method. Some data collection methods are more expensive than others. For example, interviews and observations are expensive, because they are one-on-one activities. Surveys, questionnaires and performance monitoring are usually inexpensive. The balance between accuracy and cost is always an issue.

7.9.4 Disruption of Normal Activities

For organizations, the issue that generates perhaps the greatest concern among managers is the degree of work disruption that data collection will create. Routine processes should be disrupted as little as possible. Data collection techniques such as performance monitoring require little time and cause little distraction from normal activities. Questionnaires generally do not disrupt the work environment and can often be completed in just a few minutes, perhaps even during normal work hours. At the other extreme, techniques such as the focus group and interviews may disrupt the work unit.

7.9.5 Accuracy of Method

The accuracy of the technique is another factor to consider when selecting the method. For example, performance monitoring is usually accurate, whereas questionnaires are subject to distortion and may be unreliable. If on-the-job behavior must be captured, observation is clearly one of the most accurate methods. There is often a tradeoff in the accuracy and costs of a method.

7.9.6 Utility of an Additional Method (Source or Timeframe)

Because many different methods to collect data exist, using too many methods is tempting. Multiple data collection methods add time and cost to the evaluation and may result in little added value. Utility refers to the value added by each additional data collection method. When more than one method is used, the utility should always be addressed. Does the value obtained from the additional data warrant the extra time and expense of the method? If the answer is no, the additional method should not be implemented. The same issue must be addressed when considering multiple sources and time frames.

7.9.7 Cultural Bias of Data Collection Method

The culture or philosophy of the organization can dictate which data collection methods are best to use. For example, questionnaires will work well in an organization or audience that is accustomed to using questionnaires. If, however, an organization tends to overuse questionnaires, this may not be the best choice for collecting project data. Some organizations routinely use focus groups. However, others view the technique as invasive.

7.10 Measuring the Hard to Measure

Impact measures are typically easy to collect and easy to measure. They represent the classic definitions of hard data and soft data and tangible and intangible data. For green projects, much attention today is focused on the hard to measure—that is, on some of the classic soft items such as image, reputation, social responsibility and environmental friendliness. Although this subject is discussed in more detail in Chapter 9, a few brief comments are appropriate here.

7.10.1 Everything can be Measured

Contrary to the views of some professionals, any item, issue or phenomenon that is important to an organization can be measured. Even image, perception and ideas can be measured. The thorny issue usually lies in identifying the best way and the available resources to measure. Although the community's image of an organization or the way customers become aware of a brand can be measured accurately, doing so takes time and money.

Organizations collect these types of data in various time frames from different sources. For example, Starbucks has been implementing sustainability projects and green initiatives for years. They make it a point for the customers, the public and the employees to fully understand their commitment to helping and protecting the environment. Starbucks is interested in feedback related to their corporate social responsibility position, sustainability success and environmental stewardship. They routinely collect data on their website and in customer surveys. They also collect similar data from employees so executives will fully understand the success of these projects, which are aimed at image building.

7.10.2 Perceptions are Important

Many measures are based on perceptions. For example, corporate social responsibility is an important component of a company's image. Concepts such as brand awareness are based strictly on perception (i.e., on what a person knows or perceives about an item, product or service). At one time, perceptions were considered irrelevant and not valuable, but today many decisions are based on perceptions. For example, customers' perceptions about service quality often drive tremendous organizational changes. Similarly, employees' perceptions of their employer often drive huge investments in projects to improve job satisfaction organizational commitment and engagement. The image of BP as being environment and safety conscious eroded after the Gulf oil spill. Because perceptions are important, they must be part of the measurement plan for the hard to measure.

7.10.3 All Measures can be Converted to Money, but Not All Measures should be

Just as everything can be measured, so can every measure be converted to monetary value. The concern involves credibility and resources. As the eleventh guiding principle of Table 4.4 indicates, some measures (intangibles) cannot credibly be converted to money with minimum resources.

Important emphasis must be placed on measuring the hard to quantify and valuing the hard to rate – the intangibles. Intangible measures are often the principal drivers for green projects. Knowing when to pursue conversion to money and when to avoid it is important. More details and examples of intangibles are presented in Chapter 9. The chapter includes techniques to measure the hard to measure and addresses the issue of converting to money.

7.11 Final Thoughts

Measuring application and implementation is critical in determining the success of a project or program. This essential measure not only determines the success achieved but also identifies areas where improvement is needed and where success can be replicated in the future.

Business impact data are critical to address an organization's business needs. These data lead the evaluation to the money. Although perceived as difficult to find, business impact data are readily available and credible. This chapter presented a variety of techniques to collect application and impact data, ranging from questionnaires to monitoring records and systems. The method chosen must match the scope of the project. Understanding success is important to the ability to provide evidence that business needs are being met.

8

Isolating the Impact of EHS Projects

Abstract

This chapter focuses on perhaps the most credible issue: isolating the impact of EHS projects from other influences. Whenever a measure is improved, such as lost-time accidents, health care costs, workers' compensation claims or penalties for environmental violations, multiple influences are usually at work. It is unlikely that a single EHS project will make all the difference; consequently, there must always be a step to isolate the effects of the EHS project from other influences. This chapter shows how this is accomplished and underscores the fact that it must always be accomplished. The range of techniques includes the classic experimental vs. control group, various analytical processes such as trend analysis and forecasting, estimates from a variety of individuals with the ability to sort out the influences. The chapter begins with showing the importance of this step and outlining what is necessary to accomplish it for every project.

Reporting improvement in business impact measures such as energy consumption, accident costs and employee health costs is an important step in an EHS project evaluation. Invariably, however, one essential question arises: How much of this improvement was the result of the EHS project or which EHS project? Unfortunately, the answer is rarely given with an acceptable degree of accuracy and confidence. Although the change in performance may in fact be linked to the EHS project, other, non–project–related factors may have contributed to the improvement as well. If this issue is not addressed, the results reported will lack credibility. This chapter explores useful techniques for isolating the effects of an EHS project.

Keywords: Isolation, isolating project impact, control groups, trendline analysis, forecasting, estimates, credibility

Jack Phillips, Patti Phillips, and Al Pulliam, Measuring ROI in Environment, Health, and Safety, (175–200) 2014 © Scrivener Publishing LLC

8.1 Why the Concern About Isolating Project Impact?

Multiple factors influence the business measures targeted by almost any project. Isolating the effect of an EHS project is necessary for accuracy and credibility. Without this isolation, the project's success cannot be confirmed; moreover, the effects of the project may be overstated if the change in the business impact measure is attributed entirely to the project. If this issue is ignored, the impact study may be considered invalid and inconclusive. This places appropriate pressure on EHS project leaders to take credit only for what their projects have accomplished.

8.1.1 Reality

Isolating the effects of projects on business measures has led to some important conclusions. First, in almost every situation, multiple factors generate business results. The rest of the world does not stand still while an EHS project is being implemented. Other processes and programs are also operating to improve the same metrics targeted by a particular EHS.

Next, if the project effects are not isolated, no business link can be established. Without steps taken to document the project's contribution, there is no proof that the project actually influenced the measures. Instead, the evidence will show that the project *might* have made a difference. Results have improved, but other factors may have influenced the data. Proof is much better than evidence.

Also, the outside factors and influences have their own protective owners, who will insist that it was their processes that made the difference. Some of them will probably be certain that the results are due entirely to their efforts. They may present a compelling case to management, stressing their achievements. For example, a green project at a large hospital chain focused on eliminating or minimizing paper, saving millions of dollars and thousands of trees. The green project focused on employee attitudes, habits and practices. At the same time, a lean six-sigma program targeted the same area, using process improvement techniques and technology enhancement. When the savings were tabulated, the lean six-sigma team claimed all of the results in paper reduction, leaving the green team disappointed.

Finally, isolating the effects of the project on impact data is a challenging task. For complex projects in particular, the process is not easy, especially when strong-willed owners of other processes are involved. Fortunately, a variety of approaches are available to facilitate the procedure. These approaches are presented in this chapter.

8.1.2 Myths

The myths surrounding the isolation of project effects create confusion and frustration with the process. Some researchers, professionals and consultants go so far as to suggest that such isolation is not necessary. The most common myths are listed here:

1. *Our project is complementary to other processes; therefore, we should not attempt to isolate the effects of the project.* A project often complements other factors at work (or other EHS projects), all of which together drive results, sometimes even the same measure. For example, when there is a serious safety issue, several projects may be implemented to correct it quickly. If the sponsor of a particular project needs to understand its relative contribution, the isolation process is the only way to do it. If accomplished properly, the isolation process will reveal how the complementary factors interact to drive improvements.
2. *EHS project leaders do not address this issue.* Some project leaders do not grapple with the isolation problem because they wish to make a convincing case that all of the improvement is directly related to their own processes. For example, when implementing a safety project to correct a problem with excessive accidents, project leaders may assume that any improvement is caused by the project alone. In practice, other factors can affect accidents, such as work schedules, overtime, volume of work, equipment maintenance, down time, workflow changes and supervision. A step must be taken to know which project, process or system influenced the safety performance.
3. *If we cannot use a research-based control group, we should not attempt this procedure.* Although an experimental

research design using randomly assigned control and experimental groups is the most reliable approach to identifying causes and effects, it is not applicable in the majority of situations.

If the experimental vs. control group is not appropriate, other methods must be used to isolate the effects of a project. The challenge is to find a method that is effective—with reproducible results—even if it is not as credible as the group comparison method.

4. *The stakeholders will understand the link to business impact measures; therefore, we do not need to attempt to isolate the effects of the project.* Unfortunately, stakeholders try to understand only what is presented to them. The absence of information about the connection to the project makes it difficult for them to understand the business linkage, particularly when others are claiming full credit for the improvement.
5. *Estimates of improvement provide no value.* It may be necessary to tackle the isolation process using estimates from those who understand the process best. Although this should be pursued only as a last alternative, the technique can provide value and credibility, particularly when the estimates have been adjusted for error to reduce subjectivity. A tremendous amount of research from the past hundred years supports the validity of estimates (Surowieki, 2004).
6. *If we ignore the issue, maybe the others will not think about it.* Audiences are becoming more sophisticated about this topic and they are aware of the presence of multiple influences. If no attempt is made to isolate the effects of the project, the audience will assume that the other factors have had a major effect—perhaps the only effect. A project's credibility can deteriorate quickly.

These myths underscore the importance of addressing the isolation step. The emphasis on isolation is not meant to suggest that a project is implemented independently and exclusively of other processes. Obviously, all groups should be working as a team to produce the desired results. Multiple EHS projects are often implemented at the same time. However, when funding is parceled among different functions, departments or project leaders, there is always a struggle to show, and often to understand, the connection between their activities and the results. If you do not undertake this process, others will—leaving your project with reduced budgets, resources, influence and respect.

8.2 Preliminary Issues

The cause-and-effect relationship between an EHS project and EHS performance measures can be confusing and difficult to prove, but it can be demonstrated with an acceptable degree of accuracy. The challenge is to develop one or more specific techniques to isolate the effects of the project early in the process, usually as part of an evaluation plan conducted before the project begins. Up-front attention ensures that appropriate techniques will be used with minimal cost and time commitments. Two important issues in isolating the effects of a project are covered next, followed by specific methods.

8.2.1 Chain of Impact

Before presentation of the methods, reflecting on the chain of impact described in earlier chapters is important. Measurable impact data from an EHS project (Level 4 data) should be derived from the project's application (Level 3 data). Successful application of the project should stem from the project participants' learning (Level 2 data) of new skills or techniques. Successes with learning will usually occur when project participants react favorably to the EHS project's intent, content and objectives (Level 1 data). Without this preliminary evidence, isolating the effects of a project is difficult. From a practical standpoint, this requires data collection at four levels for an ROI calculation (the first guiding principle in Table 4.4) as a prerequisite to isolating the project's effects.

8.2.2 Identify Other Factors: A First Step

As a first step to isolate a project's impact on performance, all key factors that may have contributed to the performance improvement should be identified. This step communicates to interested parties that other factors may have influenced the results, underscoring that the project is not the sole source of improvement. Consequently, the credit for improvement is shared among several possible variables and sources—an approach that is likely to garner the sponsor's respect. Several potential sources are available for identifying major influencing variables:

- Project sponsor
- Participants in the project (employees, customers, suppliers, citizens, etc.)
- The immediate managers of participants
- The EHS team
- Subject matter experts
- Other process owners
- Experts on the situation, issue or project
- Middle and top management

The importance of identifying all of the factors is underscored by an example. A low-cost hotel chain decided to pursue a variety of green projects. While the principal driver was image with customers, much of it was an attempt to save operating costs. Several different functions and departments were involved, including marketing, maintenance, engineering, staff training, procurement and quality. Although there were several impact measures representing the payoff for the green projects, one important measure was energy consumption, a number that is routinely monitored for each property. A year into the program, the electrical energy savings had been impressive, representing a significant amount of cost savings. To determine what caused the improvement, several factors were identified:

- The marketing department communicated directly with the customers, requesting permission to not change the linens and wash the towels every day, asking them to conserve by keeping the air conditioning thermostat higher and the heating thermostat lower and reminding them to turn the lights off whenever possible. The maintenance function installed more efficient lighting systems, a system to automatically shut off the lights when a person leaves and automatic adjustments for the heating and air during the daytime.
- Engineering reviewed the electrical bill and made consolidations, changed some purchasing arrangements,

improved some efficiency on connections to the property and upgraded some of the generators and pumps for more energy-efficient ones.
- Procurement purchased materials that were friendly to the environment and some of these also caused less energy consumption. For example, the housekeeping department purchased laundry detergent that used cold water only, thus saving energy when washing the linens and towels.
- The quality function worked on several projects, including ways to cut down on waste. The waste reduction meant less use of the trash compactor, which saved some electrical energy.
- Staff training had a brief one-day session on the entire green issue and taught all employees to take specific steps to conserve wherever possible.

The challenge was to determine how much electrical energy savings is connected to each factor.

A method is available to sort out the cause-and-effect relationship. Project team leaders should go beyond assuming the contribution of their efforts. They should use one or more of the following techniques to isolate the impact of their project.

8.3 Methods to Isolate the Impact of Projects

Just as there are many data collection methods available for collecting data at different levels, a variety of methods are also available to isolate the effects of an EHS project. Some are more credible than others. The most credible approaches are presented first.

8.3.1 Control Groups

The most accurate approach for isolating the project's impact is an experimental design with control groups. This approach involves the use of an experimental group that experiences the implementation of the EHS project and a control group that does not. The two groups should be as similar in composition as possible and, if feasible, participants for each group should be randomly assigned. When this is achievable and the groups are subjected to the same

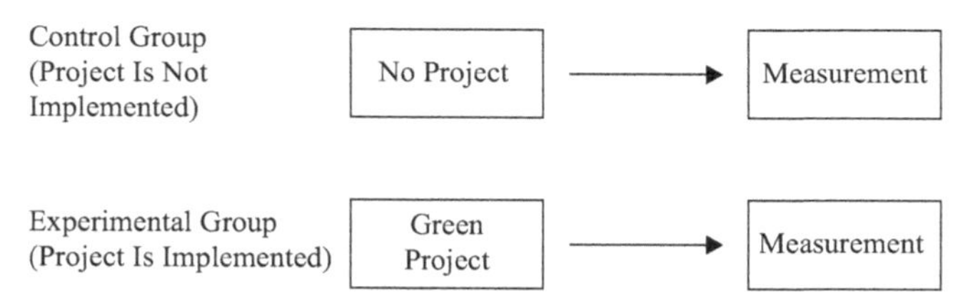

Figure 8.1 Use of Control Groups.

environmental influences, any difference in performance between the two groups can be attributed to the project.

As illustrated in Figure 8.1, the control group and experimental group do not necessarily require pre-project measurements. Measurements can be taken during the project and after the project has been implemented, with the difference in performance between the two groups indicating the amount of improvement that is directly related to the project.

For EHS initiatives and sustainability projects, the use of a comparison group analysis should be feasible. It is a matter of comparing one situation to another, such as the disabling frequency rate (DFR) in one plant compared to another similar plant with no project. The plants would have to be similar in terms of the factors that affect the disabling frequency rate such as plant size, type of work, age of plant, type of project and same DFR natural sets occur when considering the vast opportunities for green projects. Some examples include stores, buildings, branches, houses, subdivisions, employees, customers, plants, neighborhoods and cities. There are often many natural comparisons that are possible because the green projects and initiatives should be, can be, and are being implemented everywhere.

One caution should be observed: The use of control groups may create the impression that the project leaders are reproducing a laboratory setting, which can cause a problem for some executives and administrators. To avoid this perception, some organizations conduct a pilot project as the experimental group. A similarly constituted comparison group is selected but is not involved in the project. The terms *pilot project* and *comparison group* are less threatening to executives than *experimental group* and *control group*.

It's also important to recognize that the control group approach has some inherent challenges that can make it difficult to apply in practice. The first involves the selection of the groups. From a theoretical perspective, having identical control and experimental groups is next to impossible. Dozens of individual and contextual factors can affect performance, so on a practical basis, it is best to select the four to six variables that will have the greatest influence on performance. Essentially, this involves the 80/20 rule (the Pareto Principle). This rule is aimed at selecting the 20 percent of variables that might account for 80 percent of the difference. It requires working from the most important variable and moving to the next to cover four to six issues that have the greatest influence on the improvement in the impact measure in question.

Another issue is that the control group process is not suited to many situations. For some types of EHS projects, withholding the initiative from one particular group while implementing it with another may not be appropriate. This is particularly true where critical solutions are needed immediately; management is usually not willing to withhold a safety solution from one area to see how it works in another. This limitation keeps control group analyses from being implemented in many instances. However, in practice, many opportunities arise for a natural control group to develop in situations where a solution is implemented throughout an organization. If it takes several months for the solution to be implemented in the organization, enough time may be available for a parallel comparison between the first group and the last group for implementation. In these cases, ensuring that the groups are matched as closely as possible is critical. These naturally occurring control groups can often be identified in the case of major enterprise-wide project implementations. For example, a restaurant chain is implementing a new safety program for all of the restaurants, one at a time with a safety team. The project will take four months to complete, which offers enough time to have a comparison of the first stores to begin the implementation with the last. The challenge is to address this possibility early enough to influence the implementation schedule to ensure that similar groups are used in the comparison.

Another issue is contamination. This occurs when the control group realizes a program is being implemented that improves important EHS measures and their performance is being compared to those involved in the project. Ensuring that the control and project groups are in different locations or different buildings can minimize this problem. When this is not possible, it should be explained to participants in both groups that one group will be involved in the project now and the other will be involved at a later date. Appealing to participants' sense of responsibility and asking them not to share information with others may help prevent contamination.

A closely related issue involves the passage of time. The longer a control versus experimental group comparison operates, the greater the likelihood that other influences will affect the results and more variables will enter into the situation, contaminating the results. On the other end of the scale, enough time must pass to allow a clear pattern of success to emerge, distinguishing the two groups. Thus, the timing of control group comparisons must strike a delicate balance between waiting long enough for performance differences to show, but not so long that the results become contaminated.

Still another issue occurs when the different groups are subject to different environmental influences during the experiment. This is usually the case when groups are at different locations. Sometimes selection of the groups can prevent this problem from occurring. Another tactic is to use more groups than necessary and discard those groups that show some environmental differences.

A final issue is that the use of control and experimental groups may appear too research-oriented for most business organizations. In addition to the selective withholding problem discussed earlier, management may not want to take the time to experiment before proceeding with a project. Because of these concerns, some project managers will not entertain the idea of using control groups.

Because the use of control groups is an effective approach for isolating impact, it should be considered when an ROI study is planned. In these situations, isolating the project impact with a high level of accuracy is essential and the primary advantage of the control group process is accuracy.

For example, a hospital chain was experiencing excessive use of sick days. A program was implemented with one hospital and compared to another similar hospital. Figure 8.2 shows an experimental and control group comparison. Both groups are experiencing about 40 days per week of sick days, too much for these two nursing units. A sick leave reduction project involving nurse managers was implemented with the experimental group. The control group was not involved. The criteria used to select the two groups were current performance on sick leave, staffing levels, type of care and overtime use. The control group experienced a reduction from 40 days to 28 days. The experimental group moved from 40 days to 18 days. The improvement, connected to the reduction project, is 10 (28-18) days per week.

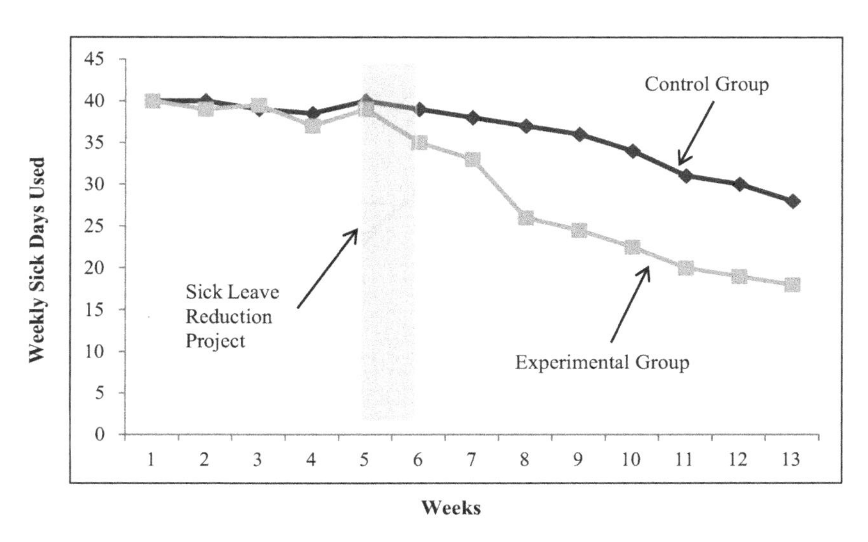

Figure 8.2 Experimental-versus-Control Group Comparison for Sick Leave Reduction.

8.3.2 Trendline Analysis

Another useful technique for approximating the project's impact is trendline analysis. In this approach, a trend line is projected in the future, using previous performance as a base. When the project is fully implemented, actual performance is compared with the trendline projection. Any improvement in performance beyond what the trend line predicted can be reasonably attributed to project implementation when certain conditions are met. While this is not a precise process, it can provide a reasonable estimate of the project's impact.

Figure 8.3 shows a trendline analysis from a medium sized manufacturing plant. The vertical axis reflects the level of disabling injury rate. The horizontal axis represents time in months. Data reflect conditions before and after the safe operations practice (SOP) project was implemented in May. As shown in the figure, an upward trend for the data existed prior to project implementation. However, the project apparently had an effect on the DFR, as the trend line is much greater than the actual. Project leaders may have been tempted to measure the improvement by comparing the one-year average for the DFR prior to the project to the one-year average after the project. However, this approach understates the improvement because the measure in question is moving in the wrong direction and the SOP turns the DFR in the right direction.

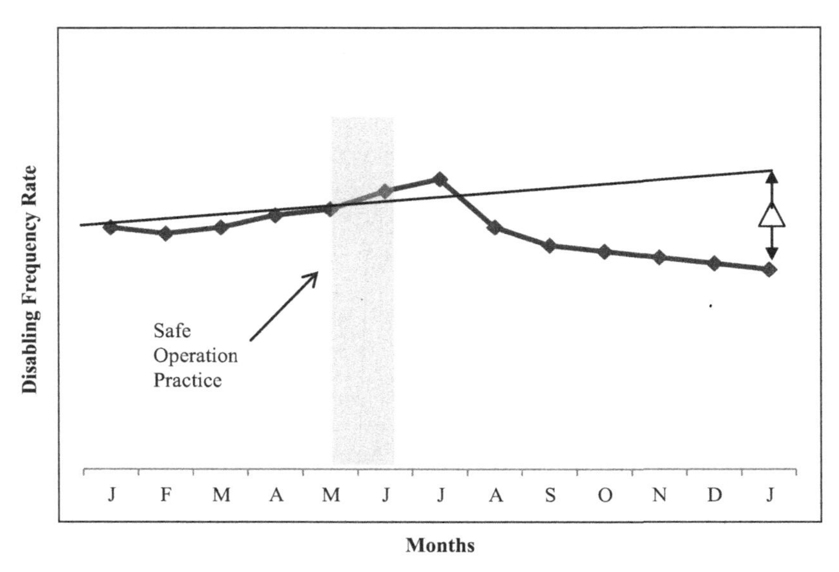

Figure 8.3 Trend Line Example.

A more accurate comparison is the actual value after the impact has occurred (the last month, January) versus the trend line value for the same period. Using this measure increases the accuracy and credibility of the process in terms of isolating the project's impact.

To use this technique, two conditions must be met:

1. It can be assumed that the trend that developed prior to the project would have continued if the project had not been implemented to alter it (i.e., had the project not been implemented, this trend would have continued on the same path). The experts who understand the situation best should be able to provide input to confirm this assumption. If the assumption does not hold, trendline analysis cannot be used. If the assumption is valid, the second condition is considered.
2. No other new variables or influences entered the process during project implementation. The key word here is *new;* the understanding is that the trend has been established from the influences already in place and no additional influences have entered the process beyond the green project. If this is not the case, another method will have to be used. Otherwise, the trendline

analysis presents a reasonable estimate of the impact of this project.

In the example, the downward trend prior to the project implementation was caused by the emphasis on safety by the plant manager and was still there in the post-project period. Also, this group stated that no other new influence entered the process during the post-project period. Nothing else had caused this improvement. Thus, the improvement difference was attributed to the safe operation project.

Pre-project data must be available in order for this technique to be used and the data should show a reasonable degree of stability. If the variance of the data is high, the stability of the trend line will be an issue. If the stability cannot be assessed from a direct plot of data, more detailed statistical analyses can be used to determine if the data are stable enough to allow a projection. The trend line can be projected directly from historical data using a simple formula that is available in many calculators and software packages, such as Microsoft Excel.

A primary disadvantage of the trendline approach is that it will not work much of the time. As noted, it only works if the trends established prior to the project will continue in the same relative direction and if no new influences have entered the situation during the course of the project. This may not be the case.

The primary advantage of this approach is that it is simple and inexpensive. Top executives appreciate this type of data, which can be represented with charts. If historical data are available, a trend line can quickly be drawn and the differences estimated. While not exact, it does provide a quick general assessment of project impact.

8.3.3 Forecasting Methods

A more analytical approach to isolation is the use of forecasting methods that predict a change in performance variables. This approach represents a mathematical interpretation of the trendline analysis when other variables enter the situation at the time of implementation. With this approach, the output measure targeted by the project is forecast based on the influence of variables that have changed during the implementation or evaluation period for the project. The actual value of the measure is compared with the forecast value and the difference reflects the contribution of the project.

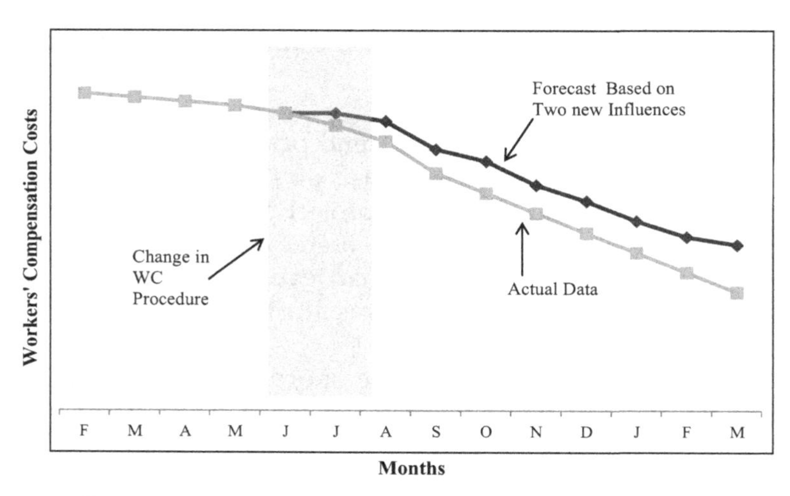

Figure 8.4 Forecasting Example.

An example will help illustrate the effect of the forecasting. A trucking firm was focusing on decreasing workers' compensation costs. In July, a new project began, changing several procedures that made diagnosis, treatment and rehabilitation faster, with various ways to quickly recognize improvement and make decisions and adjustments accordingly. All this was aimed at reducing the average workers' compensation costs. Figure 8.4 shows that the workers' compensation costs prior to the change in procedures and the actual data shows a significant downward improvement in the ten months since the program was implemented. However, two important changes occurred about the same time as the new project was implemented. A regulatory authority issued new procedures for worker compensation costs. This influence has a tendency to cause organizations to focus more intensely on getting employees back to work as quickly as possible. At the same time, the equipment for lifting the truck's content was installed. Those types of improvements should affect worker compensation costs. The analysts in the business process improvement department developed a forecast showing the effects of the new regulation and the change in the lifting equipment. They were able to generate a multiple variable analysis to forecast the workers' compensation costs, as shown in the figure. The data from March shows the difference in the forecasted value and the actual value. That difference represents the impact of the new worker WC procedures since they were not included in the forecasted value.

A major disadvantage to this approach emerges when several variables enter the process. The complexity multiplies and the use of sophisticated statistical packages designed for multiple variable analyses is necessary. Even with this assistance, however, a good fit of the data to the model may not be possible. This technique may be appropriate for large, expensive projects when it is essential to know more precisely the impact of the EHS project. Still, some organizations have not developed mathematical relationships for output variables as a function of one or more inputs and without them the forecasting method is difficult to use.

8.3.4 Estimates

The most common method of isolating the effects of a project is to use estimates from a group of individuals. Although this is a weak method, it is feasible in all situations and its credibility can be enhanced if adequate precautions are taken. The first step in using this method is ensuring that the estimates are provided by the most credible source, which are often the project participants or other experts (the third guiding principle in Table 4.4). The individuals who provide this information must understand the different factors and, particularly, the influence of the project on those factors. Essentially, there are four common groups for input:

1. The participants directly involved in the project are the first source considered.
2. Managers are another possible source.
3. Customers provide credible estimates in particular situations
4. Technical experts may provide insight into causes for improvement.

These sources are described in more detail next.

8.3.4.1 Participants' Estimate of Impact

An easily implemented method of isolating the project's impact is to obtain information directly from participants involved in project implementation. The participants are the employees, suppliers, citizens or other individuals who should be able to make the project successful. The usefulness of this approach rests on the assumption that participants are capable of determining or estimating how much of the performance improvement is related to the project implementation. Because their actions have led to the improvement, participants may provide accurate data. Although an estimate, the value they provide is likely to be acceptable to management because they know that the participants are at the center of the change or improvement. The estimate is obtained by defining the improvement and then asking participants the series of questions shown in Table 8.1.

Participants who do not provide answers to the questions in Table 8.1 are excluded from the analysis. Erroneous, incomplete and extreme information should also be discarded before the analysis. To obtain a conservative estimate, the confidence percentage, which is a reflection of the error of the estimate, can be factored into each of the values. Thus, an 80-percent confidence level equates to a potential error range of plus or minus 20 percent. In this approach, the estimate is multiplied by the level of confidence in order to adjust for (i.e., remove) the error.

An example will help to describe the situation. In an effort to increase recycling in the community, three actions were taken. Recycling had been available but because of the apathy of the community, the inconvenience with the location and a lack of incentive to do it, the results were not acceptable. The community implemented three new approaches. One approach was to conduct awareness sessions in the schools, neighborhoods, community groups and churches to make people aware of the recycling program and what it means to them and the environment. In addition,

Table 8.1 Questions for Participant Estimation.

What is the link between these factors and the improvement?
What other factors have contributed to this improvement in performance?
What percentage of this improvement can be attributed to the implementation of this project?
How much confidence do you have in this estimate, expressed as a percentage? (0 percent equals no confidence; 100 percent equals complete confidence.)
What other individuals or groups could provide a reliable estimate of this percentage to determine the amount of improvement contributed by this project?

recycling was made more convenient so it was easier for residents to conserve. Essentially, they could place three different containers on the street and have them picked up. In addition, when citizens participated in recycling, a discount was provided to their regular waste management bill. With these three services implemented, it was important to understand the effects of each of the processes. On a questionnaire, a sample of participants were asked to allocate the percentage that each of these services led to their increased participation.

Moreover, the participants were told the actual amount of recycling volume increase and were asked to indicate if other factors, besides the new recycling processes, could have caused that increase. Residents mentioned only a few other processes. Table 8.2 shows one participant's response. In the example, the participant allocates 60 percent of the improvement to the awareness program and has a level of confidence in the estimate of 80 percent.

The confidence percentage is multiplied by the estimate to produce a usable project value of 48 percent. This adjusted percentage is then multiplied by the quantified amount of the improvement in recycling volume (post-project minus pre-project value) to isolate the portion attributed to the project. For example, if volume increased by 50 percent, 24 percent would be attributed to the awareness program. The adjusted improvement is then ready to convert to a monetary value and, ultimately, to use in the ROI calculation. Although the reported contribution is an estimate, this approach

Table 8.2 Example of a Participant's Estimation.

Fact: Recycling volume has increased by 50% percent			
Factor That Influenced Improvement	**Percentage of Improvement Caused By Project**	**Confidence Expressed as a %**	**Adjusted % of Improvement Caused by**
Green awareness	60%	80%	48%
Convenience for participation	15%	70%	10.5%
Discounts for participating	20%	80%	16%
Other	5%	60%	3%
Total	**100%**		

offers accuracy and credibility. Five adjustments are effectively applied to the participant estimate to produce a conservative value:

1. Participants who do not provide usable data are assumed to have observed no improvements.
2. Extreme data values and incomplete, unrealistic or unsupported claims are omitted from the analysis, although they may be included in the "other benefits" category.
3. For short-term projects, it is assumed that no benefits are realized from the project after the first year of full implementation. For long-term projects, several years may pass after project implementation before benefits are realized.
4. The portion directly related to the project, expressed as a percentage, adjusts the amount of improvement.
5. The improvement value is multiplied by the confidence level, expressed as a percentage, to reduce the amount of the improvement in order to reflect the potential error.

As an enhancement of this method, the level of management above the participants may be asked to review and concur with each participant's estimate if it is an employment, as opposed to community, situation.

In using participants' estimates to measure impact, several assumptions are made:

1. The project encompasses a variety of different activities, practices and tasks all focused on improving the performance of one or more business measures.
2. One or more business measures were identified prior to the project and have been monitored since the implementation process. Data monitoring has revealed an improvement in the business measure.
3. There is a need to associate the project with a specific amount of performance improvement and determine the monetary impact of the improvement. This information forms the basis for calculating the actual ROI.

Given these assumptions, the participants can specify the results linked to the project and provide data necessary to develop the

ROI. This can be accomplished using a focus group, an interview or a questionnaire.

8.3.4.2 Manager's Estimate of Impact

In lieu of, or in addition to, participant estimates, the participant manager may be asked to provide input concerning the project's role in improving performance if the project participant is an employee of an organization. In some organizational settings in, managers may be more familiar with other factors influencing performance and may therefore be better equipped to provide estimates of impact. The questions to ask managers, after identifying the improvement ascribed to the project, are similar to those asked of the participants.

Managers' estimates should be analyzed in the same manner as the participant estimates, and they may also be adjusted by the confidence percentage. When participants' and managers' estimates have both been collected, the decision of which estimate to use becomes an issue. Each estimate source provides a unique perspective. If there is a compelling reason to believe that one estimate is more credible than the other, then that estimate should be used. If both are equally credible, the most conservative approach is to use the lowest value and include an appropriate explanation. This is in the fourth guiding principle shown in Table 4.4.

In some cases, higher levels of management may provide an estimate of the percentage of improvement attributable to a project. After considering other factors that could contribute to the improvement—such as technology, procedures and process changes—they apply a subjective factor to represent the portion of the results that should be attributed to the project. However, a word of caution may be in order. Since higher levels of management are farther from the action, their input can be subjective, yet those who provide or approve funding for the project may still accept it. Sometimes, their comfort level with the process becomes the most important consideration.

8.3.4.3 Customers' Input on Project Impact

An approach that is useful in some situations is to solicit input regarding the project's impact directly from customers. This may be helpful in situations involving product safety. In this scenario, customers are asked how an EHS project has influenced them to operate a product safely. This technique focuses directly on the safe use of the product. For example, an electric utility company produced a video for customers to learn to install an electric water heater. Market research data showed that the customers attributed a 5.1 percent of the reduction of product complaints to the video, with an average confidence of 83 percent. Consequently, a 4.23-percent savings (5.1 x .83) is attributed to the safety project.

Routine customer feedback provides an excellent opportunity to collect input directly from customers concerning their reactions to new or improved products and services and their safe operation. Pre- and post-project data can pinpoint the improvements spurred by a new project.

Customer input should be secured using existing data collection methods; the creation of new surveys or feedback instruments should be minimized. Ideally, this measurement process should not add to the data collection systems in use. Customer input may constitute the most powerful and convincing data if it is convenient, complete, accurate and valid.

8.3.4.4 Technical Experts' Input

External or internal experts are usually available to estimate the portion of results that can be attributed to an EHS project. With this technique, experts must be carefully selected based on their knowledge of the process, project and situation. For example, an expert in green lighting may be able to provide estimates of how much electricity is saved with a change in bulbs. Other experts may know how much can be attributed to other factors. Also, this industry has many experts who have documented this type of information (MacKay, 2009).

This approach has its drawbacks, however. It can yield inaccurate data unless the project and the setting where the estimate is made are familiar to the expert. Also, this approach may lack credibility if the estimates come from sources that do not have credibility.

This process has the advantage that its reliability is often a reflection of the reputation of the expert or independent consultant. It is a quick and easy form of input from a reputable expert or consultant and as noted, many experts are often available. Sometimes top management has more confidence in external experts than in members of its own staff.

8.3.5 Estimate Credibility: The Wisdom of Crowds

The following story is a sample of the variety of research showing the power of input from average individuals. It is taken from James Surowieki's bestselling book *The Wisdom of Crowds* (2004).

In the fall of 1906, British scientist Francis Galton left his home in the town of Plymouth and headed for a country fair. Galton was 85 years old. While he was beginning to feel his age, he was still brimming with the curiosity that had won him renown and notoriety for his work on statistics and the science of heredity.

Galton's destination was the annual West of England Fat Stock and Poultry Exhibition, a regional fair where the local farmers and townspeople gathered to appraise the quality of each other's cattle, sheep, chickens, horses and pigs. Wandering through rows of stalls examining workhorses and prize hogs may seem like a strange way for a scientist to spend an afternoon, but there was certain logic to it. Galton was obsessed with the measurement of physical and mental qualities and breeding. The livestock show was no more than a large showcase for the effects of good and bad breeding?

Galton was interested in breeding because he believed that only a few people had the characteristics necessary to keep societies healthy. He had devoted much of his career to measuring those characteristics, in fact, in an effort to prove that the vast majority of people did not possess them. His experiments left him with little confidence in the intelligence of the average person, "the stupidity and wrong-headedness of many men and women being so great as to be scarcely credible." Galton believed, "Only if power and control stayed in the hands of the select, well-bred few, could a society remain healthy and strong."

As he walked through the exhibition that day, Galton came across a weight-judging competition. A fat ox had been selected and put on display and many people were lining up to place wagers on what the weight of the ox would be after it was slaughtered and dressed. For sixpence, an individual could buy a stamped and numbered ticket and fill in his or her name, occupation, address and estimate. The best guesses would earn prizes.

Eight hundred people tried their luck. They were a diverse group. Many of them were butchers and farmers, who were presumably experts at judging the weight of livestock, but there were also quite a few people who had no insider knowledge of cattle.

"Many non-experts competed," Galton wrote later in the scientific journal *Nature*. "The average competitor was probably as well fitted for making a just estimate of the dressed weight of the ox, as an average voter is of judging the merits of most political issues on which he votes."

Galton was interested in figuring out what the "average voter" was capable of because he wanted to prove that the average voter was capable of very little. So he turned the competition into an impromptu experiment. When the contest was over and the prizes had been awarded, Galton borrowed the tickets from the organizers and ran a series of statistical tests on them. Galton arranged the 787 legible guesses in order from highest to lowest and plotted them to see if they would form a bell curve. Then, among other things, he added up all of the contestants' estimates and calculated the mean. That number represented, you could say, the collective wisdom of the Plymouth crowd. If the crowd was viewed as a single person, that would be the person's guess as to the ox's weight.

Galton had no doubt that the average guess of the group would be way off the mark. After all, mix a few smart people with some mediocre people and a lot of dumb people and it seems likely that you would end up with a dumb answer. But Galton was wrong. The crowd had guessed that the slaughtered and dressed ox would weigh 1,197 pounds. In fact, after it was slaughtered and dressed, the ox weighed 1,198 pounds. In other words, the crowd's judgment was essentially perfect. The "experts" were not even close. Perhaps breeding did not mean so much after all. Galton wrote later, "The result seems more creditable to the trustworthiness of a democratic judgment than it might have been expected." Considering the results of the experiment, Galton's statement is an understatement.

On that fateful day in Plymouth, Francis Galton discovered a simple but powerful truth. Under the right circumstances, groups, of even the most average of people, are remarkably intelligent and are often smarter than the smartest group members.

Groups do not need to be dominated by exceptionally intelligent people in order to be smart. Even if most of the people within a group are not especially informed, collectively they can reach a wise decision. Although estimates are based on input from average individuals, the combined effort of those people often leads to an accurate assumption of a condition.

8.3.6 Calculate the Impact of Other Factors

It is often possible to calculate the impact of factors, besides the EHS project in question, that account for part of the improvement and then credit the project with the remaining part. The project assumes credit for improvement that cannot be attributed to other factors.

Another example taken from the low-cost hotel chain's situation will help explain this approach. As noted, the hotel had implemented several different types of green projects and initiatives, which were owned by different parts of the organizations (see Table 8.3).

Green initiatives such as these can drive several different measures in an organization. In this particular example, the improvement measure was electricity use. When considering the impact of the various factors on the reduction of electricity, some can be calculated based on a variety of studies and previous analyses. For example, in the maintenance project, many—if not all—of the items could be estimated or could be developed. For engineering, changes in purchase, distribution and equipment could be easily translated into electricity cost reductions. Procurement could estimate how the new materials are affecting electricity based on previous analysis examples. The same is true for quality, but this may not be as accurate. What is not possible is to show the impact of

Table 8.3 The Contributing Factors.

Green Initiatives Contributing to the Improvement	Is the Impact a Known Factor?
Customer Green Project (e.g., requesting guests' permission not to wash towels)	No
Maintenance Green Project (e.g., changing systems to turn off lights)	Yes, part of it
Engineering Green Project (e.g., replacing low-efficiency pumps)	Yes
Procurement Green Project (e.g., purchasing environmentally friendly detergents)	Yes
Quality Green Project (e.g., focusing on waste reduction)	Yes, maybe
Green Training Project (e.g., teaching employees conservation techniques)	No

the customers' efforts and the green training—these will have to be estimated.

The important point is that whenever a factor's contribution can be calculated, it is developed and taken out of the total pie. The total electricity savings is the complete pie; only part of it goes to each of these different green initiatives.

This method is appropriate when the other factors can be easily identified and the appropriate mechanisms or experts are in place to calculate their impact on the improvement. In some cases, estimating the impact of outside factors is just as difficult as estimating the impact of the project, limiting this approach's applicability. However, the results can be reliable if the procedure used to isolate the impact of other factors is sound.

8.4 Considerations When Selecting Isolation Methods

With all of these techniques available to isolate the project's impact, selecting the most appropriate ones for a specific project can be difficult. Some techniques are simple and inexpensive; others are time-consuming and costly. In choosing among them, the following factors should be considered:

- Feasibility of the technique
- Accuracy associated with the technique, compared to the necessary need
- Credibility of the technique with the sponsor audience
- Specific cost to implement or use the technique
- Amount of disruption in normal work activities resulting from the technique's implementation
- Participant, staff and management time required to use the technique

The use of multiple techniques or multiple sources of data input should be considered since two sources are usually better than one.

When multiple sources are used, a conservative method should be used to combine the inputs. That means using the technique that provides the lowest ROI (as noted in the fourth guiding principle of Table 4.4), because a conservative approach builds acceptance.

The sponsor should always be provided with an explanation of the process and the subjective factors involved.

Multiple sources allow an organization to experiment with different strategies and build confidence in the use of a particular technique. For example, if management is concerned about the accuracy of participants' estimates, the combination of a control group arrangement and participant estimates could be useful for checking the accuracy of the estimation process.

It is not unusual for the ROI of a project to be extremely large. Even when a portion of the improvement is allocated to other factors, the magnitude can still be impressive in many situations. The audience should understand that even though every effort has been made to isolate the project's impact, it remains an imprecise figure that is subject to error, although every attempt has been made to adjust for the error. The result is the most accurate amount of the contribution given the constraints, conditions and resources available. Chances are, it is more accurate than other types of analysis regularly used in other functions within the organization. More information on how to isolate the results is found in other sources (Phillips & Aaron, 2008).

8.5 Final Thoughts

Isolating the effects of an EHS project is an important step in answering the question of how much of the improvement in a business measure was caused by the project. The techniques presented in this chapter are the most effective approaches available to answer this question and are used by some of the most progressive organizations. Too often, results are reported and linked to a project with no attempt to isolate the specific portion of the outcome associated with the project. This leads to a suspicious report about project success. If managers and professionals in the field wish to improve the success of green projects and are committed to obtaining results, the need for isolation must be addressed early in the process for all major projects. When this important step is completed, the impact data must be converted to monetary values to prepare for the ROI calculation. The process for converting data to monetary values is detailed in the next chapter.

9

Converting Impact Data to Money

Abstract
Here we discuss a necessary step for ROI study—converting data to money. With impact data collected, the challenge remains to convert them to credible, reliable values. The good news is that most of the measures that matter to organizations, particularly those linked to EHS projects, are already converted to money. This chapter underscores the importance of monetary value, as it is the response to a client's request to "Show Me the Money." This value is a common denominator, even when several individuals are driving different measures. The chapter shows where to find the values in an organization; how to calculate them, if necessary; and how to use external experts to find the value if it's possible. If converting to monetary value cannot be accomplished, then the data set is considered intangible. This chapter explores a variety of intangible measures, defined as those not converted to money because the conversion is not credible or requires excessive resources.

To show the economic contribution of an EHS project, the improvement in business measures that is attributable to the project (after the effects of the project have been isolated from other influences) must be converted to monetary values, which are then compared with project costs. This represents the ultimate level of project success in the five-level evaluation framework presented in Chapter 2. This chapter explains how business leaders and EHS project owners develop the monetary values of impact measures used to calculate ROI.

In addition to showing stakeholders the money, it is important to also account for the intangible benefits of the project. Sometimes these benefits are just as powerful as the monetary benefits derived from sustainability initiatives. As mentioned earlier, intangible measures are those measures not converted to money. In addition to describing how to convert measures to money, this chapter explores the role of intangibles, how to measure them, when to measure them and how to report them.

Keywords: Show me the money, standard values

Jack Phillips, Patti Phillips, and Al Pulliam, Measuring ROI in Environment, Health, and Safety, (201–226) 2014 © Scrivener Publishing LLC

9.1 Why the Concern About Converting Data to Monetary Values?

Placing monetary values on measures is not new. Monetary values have been placed on some of the most difficult measures, such as human life, for centuries. But why do it? Because money is the ultimate normalizer. It places different types of measures in a similar unit, thereby allowing for comparison among multiple measures, including project costs.

A project can be shown to be a success just by providing business impact data, showing the amount of change directly attributable to the project. For example, a change in first aid treatments, medical treatment cases, disabling injuries, environmental complaints and unplanned absenteeism can represent a significant improvement linked directly to an EHS project. For some EHS projects, it may be sufficient to report improvements. However, many sponsors want to relate that improvement to tangible, monetary values. There are five fundamental reasons why it is important to convert improvement in impact measures to money:

1. Normalizes the definition of value
2. Makes impact more meaningful
3. Aligns the evaluation process to the budgeting process
4. Aligns project implementation to operations
5. Clarifies the magnitude of an issue

Each of these reasons is discussed in more detail in the following sections.

9.1.1 Normalize the Definition of Value

As described earlier, there are many different types of value. However, monetary value is becoming one of the primary criteria of success for all types of projects. Executives, sponsors, administrators and other leaders are concerned in particular with the allocation of funds. Unless the value of processes is represented by a common measure, resource allocation is much more difficult. Value is ultimately an indicator of financial gains. Rather than leaving that financial gain to guesswork, converting a measure to money defines value in meaningful terms.

9.1.2 Make Impact More Meaningful

For some projects, the impact is more understandable when it is stated in terms of monetary value. Consider for example, the impact of a major project to implement a wellness and fitness center for employees and thereby enhance the organization's image as a great employer. This project involves all employees and has an impact on all parts of the organization. The only way to understand the value of such a project would be to convert the individual efforts to become healthier and their consequences to monetary values. Totaling the monetary values of all the outcomes would provide a sense of the project's value. Then the project's impact could be compared to the project's costs to show the actual return on investment.

9.1.3 Align Budgeting Process

Professionals and administrators are typically occupied with budgets and they are expected to develop budgets for projects with an acceptable degree of accuracy. They are also comfortable with handling costs. When it comes to benefits, however, many are not comfortable, even though some of the same techniques used in developing budgets are used to determine benefits. Some of the project's benefits will take the form of cost savings or cost reductions and this can make identification of the costs or value easier. The monetary benefit resulting from a project is a natural extension of the budget.

9.1.4 Align Operations

With global competitiveness, the drive to improve the efficiency of operations and the drive to sustain Mother Earth, awareness of the costs related to particular processes and activities is essential. In the 1990s this emphasis gave rise to activity-based costing (ABC) and activity-based management. (ABM), processes used to identify and cost business activities. ABC is used to assign overhead costs to activities so these costs are more precisely allocated to products and services. ABM focuses on managing activities to reduce costs and increase customer value. ABC is not a replacement for traditional, general ledger accounting. Rather, it is a translator or medium between cost accumulations or the specific expenditure account balances in the general ledger and the end users who must apply cost data in decision making. In typical cost statements, the actual cost of a process or problem is not readily discernible. ABC converts inert cost data to relevant, actionable information. ABC has become increasingly useful for identifying improvement opportunities and measuring the benefits realized from performance initiatives on an after-the-fact basis (Cokins, 1996). Most EHS projects contribute to cost savings for the organization. The measures listed in Chapter 5 (Tables 5.1 and 5.2) can all be converted to money. Consequently, understanding the cost of a problem and the payoff of the corresponding solution is essential to proper management of the business.

9.1.5 Clarify the Magnitude of an Issue

In any business, costs are essential to understanding the magnitude of a problem. Consider, for example, the cost of unhealthy employees. Traditional records and even those available through activity-based costing will not indicate its full value or cost. If employees lose weight, lower blood pressure, stop smoking, exercise and eat nutritional foods, productivity will improve, accidents will decrease, absenteeism will decrease and sick leave usage will decrease. A variety of estimates and expert inputs may be necessary to supplement immediately available data to arrive at a definite value of improvements. The good news is that organizations have developed a number of standard procedures for identifying the costs of these measures.

9.2 Five Steps to Convert Data to Money

Converting measures to monetary value involves five steps for each data item:

1. Focus on the unit of measure
2. Determine the value of each unit
3. Calculate the change in performance

4. Determine the annual change in performance
5. Calculate the annual value of improvement

9.2.1 Focus on the Unit of Measure

First, a unit of measure must be defined. For output data, the unit of measure is the item produced (e.g., one item assembled), service provided (e.g., one package shipped) or sale completed. Time measures could include the time to complete a project or time to investigate a claim and the unit here is usually expressed in terms of minutes, hours or days. Quality is another common measure, with a unit defined as one accident, disabling injury or one OHSA citation, one ton of carbon emissions, or one kilowatt-hour. Soft data measures vary, with a unit of improvement expressed in terms of absences, turnover or a change in the employee satisfaction index. Specific examples of units of measure are:

- One OHSA citation
- One toxic material exposure
- One visit to the first aid center
- One medical treatment case
- One disabling injury
- One property damage
- One fatality
- One hour of downtime
- One hour of employee time
- One near miss
- One gallon of water
- One pound of solid waste
- One employee complaint
- One point increase in job satisfaction
- One kilowatt-hour of electricity

9.2.2 Determine the Value of Each Unit

The second step is to place a monetary value *(V)* on the unit identified in the first step. For measures of productivity, quality, cost and time, the process is relatively easy. Most organizations maintain records or reports that can pinpoint the cost of one accident or one OHSA citation. Soft data are more difficult to convert to money. For example, the monetary value of one employee complaint or a one-point change in job satisfaction may be difficult to determine. The techniques described in this chapter provide an array of approaches for making this conversion. When more than one value is available, the most credible or conservative is generally used in the calculation.

9.2.3 Calculate the Change in Performance Data

The change in impact data is calculated after the effects of the project have been isolated from other influences. This change (Δ) is the performance improvement that is directly attributable to the project, represented as the Level 4 impact measure. The value may represent the performance improvement for an individual, a team, a group of participants or several groups of participants.

9.2.4 Determine the Annual Amount of Change

The Δ value is annualized to develop a value for the total change in the performance data for one year (ΔP). Using annual figures is a standard approach for organizations seeking to capture the benefits of a particular project, even though the benefits may not remain constant throughout the year. For a short-term project, first-year benefits are used even when the project produces benefits beyond one year. This approach is considered conservative; therefore, it presents a more credible (and likely) outcome.

9.2.5 Calculate the Annual Value of the Improvement

The total value of improvement is calculated by multiplying the annual performance change (ΔP) by the unit value (V) for the complete group in question. For example, if one group of participants is involved in the project that is being evaluated, the total value will include the total improvement for all participants who are providing data in the group. This value for annual project benefits is then compared with the costs of the project to calculate the BCR, ROI or payback period.

9.3 The Five Steps to Convert Data in Practice

A simple example will demonstrate these five steps described in the previous section. Suppose a large retail store chain pilots an effort to replace all traditional lighting with energy-efficient LED bulbs. Prior to project implementation, the 90,000-square-foot store in which the pilot program occurs uses approximately 14 kilowatt-hours (kWh) per square foot annually for a total of 1,260,000 kWh per year. On average this is 105,000 kWh per month. After implementing the project, the store monitors electricity usage for six months, showing a new average monthly usage of 73,500 kWh. This is a decrease of 31,500 kWh per month. By comparing this store's usage to that of another store with comparable characteristics, the 31,500 reduction is attributed to the change in bulbs. Given a monthly change in performance, the annual change is 378,000 kWh. The value of a kWh is approximately 10 cents per kWh. The total annual savings to the store is $37,800. Table 9.1 shows the example using the five steps.

9.4 Methods to Convert Impact Measures to Money

The steps to convert a measure to money are straightforward. The challenge, however, is determining the value (Step 2) of a particular measure. A variety of techniques are available that run the gamut from standard values to the use of conservative estimates.

Table 9.1 Converting Kilowatt-Hours to Monetary Values.

Setting: Retail store piloting replacement of traditional lighting to LED lighting.
Step 1: Define the unit of measure. The unit of measure is defined as one kWh.
Step 2: Determine the value (V) of each unit. According to historical data (i.e., the cost per kWh paid per month), the cost is 10 cents per kWh (V = 10 cents).
Step 3: Calculate the change (Δ) in performance data. Six months after the project was completed, electricity usage decreased an average of 31,500 kWh per month. The isolation technique used was a control group (see Chapter 8).
Step 4: Determine an annual amount of the change (ΔP). Using the six-month average of 31,500 kWh per month yields an annual improvement of 378,000 ($\Delta P = 31{,}500 \times 12 = 378{,}000$).
Step 5: Calculate the annual monetary value of the improvement. $(\Delta P \times V) = 378{,}000 \times .10 =$ $37,800 cost savings

9.4.1 Standard Values

A standard value is a monetary value assigned to a unit of measurement that is accepted by key stakeholders. Most hard data items (output, quality, cost and time) have standard values. Standard values have been developed because these are often the measures that matter to the organization. They reflect problems and their conversion to monetary values shows their impact on the organization's operational and financial well being. Standard values are used to convert measures of output, quality and time to money. Virtually every major function within an organization has standard monetary values, particularly for measures they monitor on a routine basis. Standard values can typically be found in the following:

- Finance and Accounting
- Production
- Operations
- Engineering
- IT
- Administration
- Sales and Marketing
- Customer Service
- Procurement
- Logistics
- Compliance
- Research and Development
- HR
- Environment, Health and Safety
- Risk Management

The following discussion describes how measures of output, quality and time can be converted to standard values.

9.4.1.1 Converting Output Data to Money

When a project results in a change in output, the value of the increased output can usually be determined from the organization's accounting or operating records. For projects intended to drive profit, this value is typically the marginal profit contribution of an additional unit of production or service provided. For projects that are performance driven rather than profit driven, this value is usually reflected in the savings realized when an additional unit of output is realized for the same input. For example, United Parcel Service (UPS) implemented a new mapping system that enabled a "no left turn" rule, which enabled drivers to deliver the same number of packages in less time and with less fuel costs by avoiding left-hand-turns (Myrow, 2007). One of the more important measures of output is productivity, particularly in a competitive organization. Today, most organizations competing in the global economy do an excellent job of monitoring productivity and placing a value on it.

9.4.1.2 Calculating the Cost of Quality

Quality and the cost of quality are important issues in manufacturing and service organizations. Because many EHS projects are designed to improve quality (e.g., prevent accidents), the project team may have to place a value on the improvement of certain quality measures. For some quality measures, the task is easy. For example, if quality is measured in terms of the defect rate, the value of the improvement is the cost to repair or replace the product. The most obvious cost of poor quality is the amount of scrap or waste generated by mistakes. Defective products, spoiled raw materials and discarded paperwork, rework and repair are all the result of poor quality and represent monetary costs to the organization.

Quality costs can be grouped into six major categories: (Campanella, 1999)

1. *Internal failure* represents costs such as those related to reworking and retesting.
2. *Penalty costs* are fines or penalties incurred as a result of unacceptable quality.
3. *External failure* refers to problems detected after product shipment or service delivery, such as technical support, complaint investigation, remedial upgrades and fixes.
4. *Appraisal costs,* such as product quality audits, assess the condition of a particular product or service.
5. *Prevention costs* include service quality administration, inspections, process studies and improvement costs.
6. *Customer dissatisfaction,* which is perhaps the costliest element of inadequate quality, may be difficult to quantify and arriving at a monetary value may be impossible using direct methods. More and more quality experts are measuring customer and

client dissatisfaction with the use of market surveys (Rust, Zahorik and Keiningham, 1994).

The good news is that many quality measures have been converted to standard values. Some of these measures are:

- Incidents
- Accidents
- Disabling Injury
- Sick Day
- Violations
- System downtime
- Solid waste

9.4.1.3 *Converting Employee Time Using Compensation*

A third type of standard value is the value of time. Reducing the workforce or saving employee time is a common objective for projects. In a team environment and on an individual basis, a project may enable an organization to reduce time in completing tasks. The value of the time saved is an important measure and determining a monetary value for it is relatively easy.

The most obvious timesaving stems from reduced labor costs for performing a given amount of work. The monetary savings are found by multiplying the hours saved by the labor cost per hour. For example, Tecmotiv's Lean and Energy Environment Initiatives (LE2), which include value stream mapping, help the team to identify cost savings opportunities in their processes. One such cost savings opportunity was over-processing. By addressing this unnecessary effort in their process, inhibitor labor hours were reduced (Chapman & Green, 2010).

The average wage, with a percentage added for employee benefits, will suffice for most calculations of the cost of time. However, employee time may be worth more. For example, additional costs for maintaining an employee (office space, furniture, telephones, utilities, computers, administrative support and other overhead expenses) could be included in calculating the average labor cost. However, for most projects, the conservative approach of using salary plus employee benefits is recommended.

A word of caution concerning time savings. Savings are realized only when the amount of time saved translates into a cost reduction or a profit contribution. Even if a project produces savings in time, monetary value is not realized unless the time saved is put to productive use. Having individuals estimate the percentage of time saved that is devoted to productive work may be helpful, if it is followed up with a request for examples of how the extra time was used. An important preliminary step in figuring out time savings is determining whether the expected savings will be genuine.

9.4.2 Historical Costs

Sometimes standard values are unavailable. When this occurs, the next best measure of monetary value is the through the use of historical costs. This classic approach to placing monetary value on measures considers how much performance a measure has cost in the past.

Take the earlier example of converting kilowatt-hours to money. Because this is not standard, the next best technique is to consider how much has been paid for kWh usage in the past.

The challenges with using historical cost include the time to sort through databases, cost statements, financial records and activity reports to calculate the cost of performance in a measure; the accuracy of the cost, accounting for all related costs; accessing the data, particularly those that are somewhat sensitive; and accuracy in the analysis of the data that result in historical cost of a measure. Because of these issues it is important that two conditions exist:

1. The sponsor approves the use of additional time, effort and money to develop cost.
2. The measure is simple and can be formed by searching only a few records.

9.4.3 Expert Input

When it is necessary to convert data items for which historical cost data are not available, input from experts might be a consideration. Internal experts can provide the cost, or value, of one unit of improvement in a measure. Individuals with knowledge of the situation and the confidence of management must be willing to provide estimates as well as the assumptions behind the estimates. Internal experts may be found in the department in which the data originated—sales, marketing, procurement, engineering, labor relations or any number of other functions, even EHS. Most experts have their own methodologies for developing these values. So when their input is required, it is important to explain the full scope of what is needed and to provide as many specifics as possible.

If internal experts have a strong bias regarding the measure or are not available, external experts are sought. External experts should be selected based on their experience with the unit of measure. Fortunately, many experts are available who work directly with important measures, such as employee attitudes and customer satisfaction. They are often willing to provide estimates of the cost, or value, of these.

The credibility of the expert, whether internal or external, is a critical issue if the monetary value placed on a measure is to be reliable. Foremost among the factors behind an expert's credibility is the individual's experience with the process or measure at hand. Ideally, he or she should work with this measure routinely. Also, the person must be unbiased. Experts should be neutral in connection with the measure's value and should have no personal or professional interest in it.

In addition, the credentials of external experts—published works, degrees and other honors or awards—are important in validating their expertise. Many of these people are tapped often and their track records can and should be checked. If their estimate has been validated in more detailed studies and was found to be consistent, this can serve as a confirmation of their qualifications in providing such data.

9.4.4 External Databases

If standard values and historical costs are unavailable and if obtaining expert input is not feasible, it may be appropriate to use values developed through the research of others. This technique makes use of external databases that contain studies and research projects focusing on the cost of data items. Fortunately, many databases include cost studies of data items related to EHS projects and most are accessible on the Internet. Data are available on the costs of complaints, safety, hazardous and nonhazardous waste, customer satisfaction and more. The difficulty lies in finding a database with studies or research germane to the particular project. Ideally, the data should originate from a similar setting in the same industry, but that is not always possible. Sometimes, data on industries or organizations in general are sufficient, with adjustments possibly required to suit the project at hand.

9.4.5 Linkage with Other Measures

When standard values, records, experts and external studies are not available, a feasible alternative may be to find a relationship between the measure in question and some other measure that can be easily converted to a monetary value. This involves identifying existing relationships that show a strong correlation between one measure and another with a standard value.

A classic relationship is the correlation between job satisfaction and employee turnover. Studies repeatedly show that job satisfaction and employee loyalty have a strong correlation with reduction in employee turnover. Using standard data or external studies, the cost of turnover can be determined. Therefore, a change in job satisfaction can be immediately converted to a monetary value or at least an approximate value.

Finding a correlation between a customer satisfaction measure and another measure that can easily be converted to a monetary value is sometimes possible. A strong correlation often exists between customer satisfaction and revenue. Connecting these two variables allows the monetary value of customer satisfaction to be estimated.

In some situations, a chain of relationships may establish a connection between two or more variables. A measure that may be difficult to convert to a monetary value is linked to other measures that, in turn, are linked to measures to which values can be assigned. Ultimately, these measures are traced to a monetary value typically based on profits. Figure 9.1 shows the model used by Sears (Ulrich, 1998). The model connects job attitudes (collected directly from the employees) to customer service, which is directly related to revenue growth. The rectangles in the figure represent survey information and the ovals represent hard data. The shaded measurements are collected and distributed in the form of Sears' total-performance indicators.

As the model shows, a 5-point improvement in employee attitudes leads to a 1.3-point improvement in customer satisfaction. This, in turn, drives a 0.5 percent increase in revenue growth. If an employee's attitude at a local store improved by 5 points and the previous rate of revenue growth was 5 percent, the new rate of revenue growth would then be 5.5 percent.

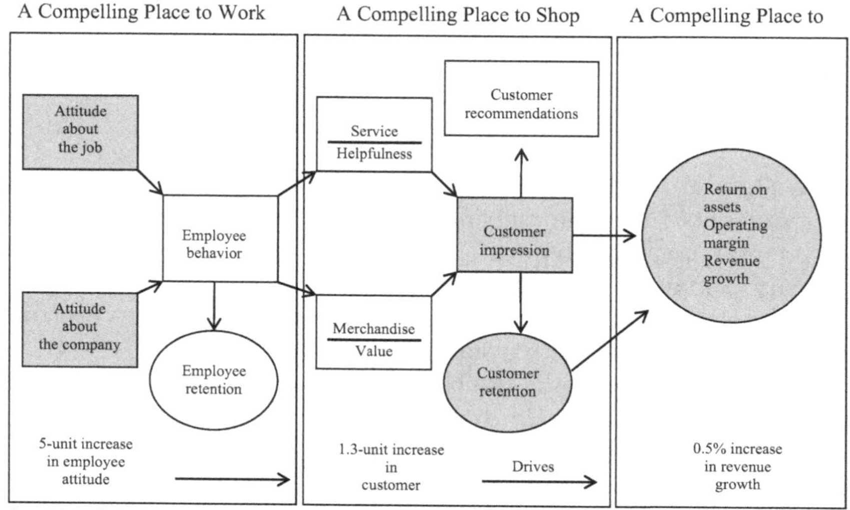

Figure 9.1 Relationship Between Attitudes and Profits.

9.4.6 Estimates

In some cases, estimating the value of improvement in a measure is the best choice. This estimate may come from participants involved in the project, the management team or the staff charged with project implementation. The key is selecting from whom to get the estimate depends on who has the most knowledge about the measure and who can provide the most credible estimate. The advantage of participant estimates is that the individuals who are most closely connected to the improvement are often able to provide the most reliable estimates of its value. As with isolating project effects, when estimates are used to convert measures to monetary values, adjustments are made to reduce the error in those estimates.

In some situations, participants in a project may be incapable of placing a value on the improvement. Their work may be so far removed from the ultimate value of the process that they cannot provide reliable estimates. In these cases, the team leaders, supervisors or managers of participants may be able to provide estimates. Thus, they may be asked to provide a value for a unit of improvement linked to the project. In other situations, managers are asked to review and approve participants' estimates and confirm, adjust or reject those values.

Senior management can often provide estimates of the value of data. In this approach, senior managers concerned with the project are asked to place a value on the improvement based on their perception of its worth. This approach is used when calculating the value is difficult or when other sources of estimation are unavailable or unreliable.

The final source of estimates is the project staff. Using all available information and experience, the staff members most familiar with the situation provide estimates of the value. Although the project staff may be qualified to provide accurate estimates, this approach is sometimes perceived as biased. It should therefore be used only when other approaches are unavailable or inappropriate.

9.5 Considerations When Selecting Data Conversion Methods

With so many techniques available, the challenge is selecting one or more strategies that are appropriate for the situation and available resources. Developing a table or list of values or techniques for the situation may be helpful. The guidelines that follow may aid in selecting a technique and finalizing the values.

9.5.1 Type of Data

Some strategies are designed specifically for hard data, whereas others are more appropriate for soft data. Thus, the type of data often dictates the strategy. Standard values are developed for most hard data items and company records and cost statements are used in the process. Soft data often involve the use of external databases, links with other measures and estimates. Experts are used to convert both types of data to monetary values.

9.5.2 Accuracy of Method

The techniques in this chapter are presented in order of accuracy. Standard values are always most accurate and, therefore, the most credible. But, as mentioned earlier, they are not always readily available. When standard values are not accessible, the following sequence of operational techniques should be tried:

- Historical costs from company records
- Internal and external experts
- External databases

- Links with other measures
- Estimates

Each technique should be considered in turn based on its feasibility and applicability to the situation. The technique associated with the highest accuracy is always preferred if the situation allows.

9.5.3 Source Availability

Sometimes the availability of a particular source of data determines the method selection. For example, experts may be readily accessible. In other situations, the convenience of a technique is a major factor in the selection. The Internet, for example, has made external database searches more convenient.

As with other processes, keeping the time investment for this phase to a minimum is important so that the total effort directed to the ROI study does not become excessive. Devoting too much time to the conversion process may dampen otherwise enthusiastic attitudes about the use of the methodology, plus drive up the costs of the evaluation.

9.5.4 Perspective

As noted in the third guiding principle in Table 4.4, the most credible data source must be used. The individual providing estimates must be knowledgeable of the processes and the issues surrounding the valuation of the data. For example, consider the estimation of a safety complaint in a manufacturing plant. Although a supervisor may have insight into what caused a particular safety issue, he or she may have a limited perspective. A high-level manager may be able to grasp the overall impact of the complaint and how it will affect other areas. Thus, a high-level manager would be a more credible source in this situation.

9.5.5 Need for Multiple Techniques

The availability of more than one technique for obtaining values for the data is often beneficial. When appropriate, multiple sources should be used to provide a basis for comparison for additional perspectives. The data must be integrated using a convenient decision rule, such as the lowest value. The conservative approach of using the lowest value was presented as the fourth guiding principle in Chapter 4, but this applies only when the sources have equal or similar credibility.

9.5.6 Credibility

The discussion of techniques in this chapter assumes that each data item collected and linked to a project can be converted to a monetary value. Highly subjective data, however, such as changes in employee attitudes or a reduction in the number of customer complaints, are sometimes difficult to convert. Although estimates can be developed using one or more strategies, such estimates may lack credibility with the target audience, which can render their use in analysis questionable.

The issue of credibility in combination with resources is illustrated in Figure 9.2. This is a logical way to decide whether to convert data to monetary values or leave them intangible. Essentially, in the absence of standard values, many other ways are available to capture the data or convert them to monetary values. However, there is a question to be answered: Can it be done with minimal resources? Some of the techniques mentioned in this chapter, such as linking measures to others that have been converted to money, cannot be performed with minimal use of resources. However, an estimate obtained from a group or from a few individuals is available with minimal use of resources. Then we move to the next question: Will the executive who is interested in the project buy into the monetary value assigned to the measure with minimum explanation? If so, then it is credible enough to be included in the analysis; if not, then move it to the intangibles. The intangible benefits of a project are also important, they just happen not to be included in the ROI equation. The last section of this chapter discusses the topic of intangibles.

9.5.7 Management Adjustment

In organizations where soft data are common and values are derived using imprecise methods, senior managers and administrators are sometimes offered the opportunity to review and approve the data. Because of the subjective nature of this process, management may factor or reduce the data to make the final results more credible.

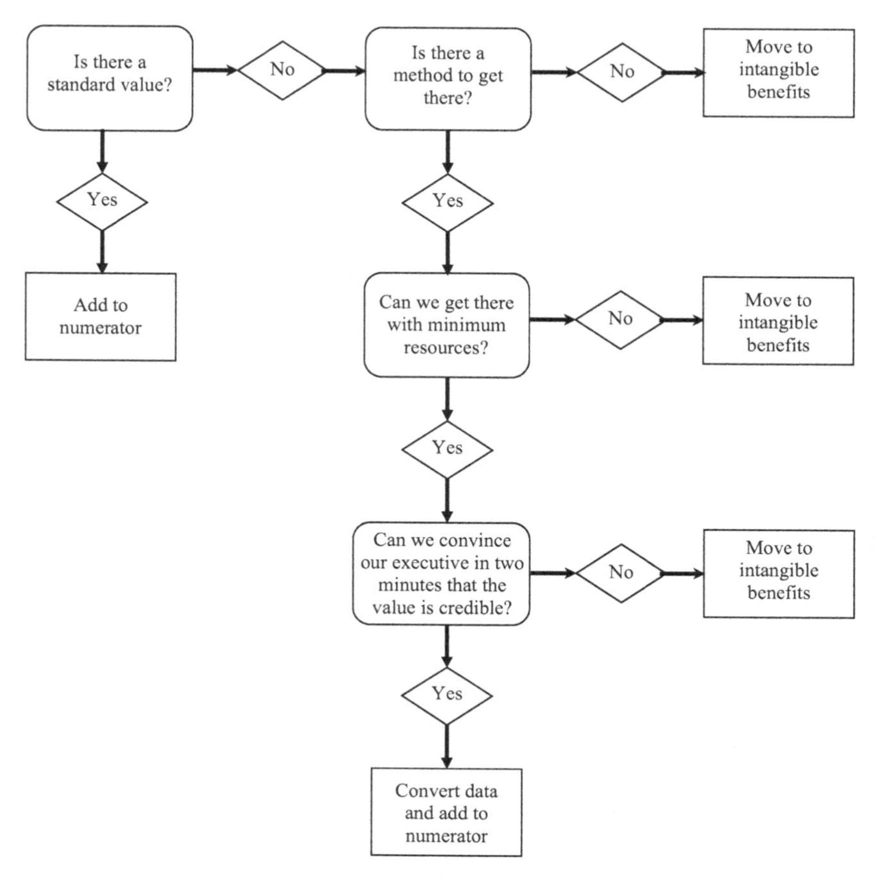

Figure 9.2 Four-Part Test: To Convert or Not to Convert?.

9.5.8 Short-Term Versus Long-Term Issue

When data are converted to monetary values, usually one year's worth of data is included in the analysis. This follows the ninth guiding principle in Table 4.4, which states that for short-term solutions, only the first year's benefits are used. The issue of whether a project is short-term or long-term depends on the time it takes to complete or implement the project and the life cycle of the project. If one group participating in the project and working through the process takes months to complete it, then it is probably not short-term. Some environment projects take years to implement even for one particular group. Workplace or equipment modification for safety purposes would normally be long-term solutions.

In general, it is appropriate to consider a project short-term when one individual takes one month or less to learn what needs to be done to make the project successful. When the lag between project implementation and the consequences is relatively brief, a short-term solution is appropriate.

When a project is long-term, no time limit for data inclusion is used, but the time value should be set before the project evaluation is undertaken. Time value input should be secured from all stakeholders, including the sponsor, champion, implementer, designer and evaluator. After some discussion, the estimates of the time factor should be conservative and perhaps reviewed by finance and accounting. When a project involves a long-term solution, forecasting will need to be used to estimate multiple years of value. No sponsor will wait several years to see how a project turns out.

9.5.9 Time Value of Money

Since investment in a project is made in one time period and the return is realized at a later time, some organizations adjust project benefits and costs to reflect the time value of money using discounted-cash-flow techniques.

Although this may not be an issue for every project, it should be considered for each project and some standard discount rate should be used. Consider the following example of how this is calculated. Assume that a project costs $100,000 and is expected to take two years for the full value of the investment to be realized. This is a long-term solution spanning two years. Using a discount rate of 6 percent, the cost for the project for the first year would be $100,000 × 106 percent = $106,000. For the second year it is $106,000 × 106 percent or $112,360. Thus, the project cost has been adjusted for a two-year value with a 6 percent discount rate. This assumes that the project sponsor could have invested the money in some other project and obtained at least a 6 percent return on that investment.

9.6 Intangible Benefits of EHS Projects

A variety of techniques are available to convert impact measures to money. But what happens when the technique costs too much or the monetary value is not deemed credible? In these cases, as noted earlier, the improvement in the measure is reported as an intangible benefit.

9.6.1 The Importance of Intangibles

Although intangible measures are not new, they are becoming increasingly important. Intangibles secure funding and drive the economy and organizations are built on them. In every direction we look, intangibles are becoming not only increasingly important, but also critical to organizations.

The success behind many well-known organizations often includes intangibles. A highly innovative company continues to develop new and improved products; a government agency reinvents itself. An organization shares knowledge with employees, providing a competitive advantage. A successful organization is able to develop strategic partners and alliances. These intangibles do not often appear in cost statements and other record keeping, but they are there and they make a huge difference. Table 9.2 lists intangible measures often measured and monitored in a variety of types of organizations.

From the 1950s forward, the world has moved from the Industrial Age into the Technology and Knowledge Age, which has resulted in more intangibles. During this time, a natural evolution of technology has occurred. During the Industrial Age, companies and individuals invested in tangible assets like plants and equipment. In the Technology and Knowledge Age, companies invest in intangibles ike brands or systems. The future holds more of the same, as intangibles continue to evolve into an important part of the overall economic system.

The good news is that more data, once regarded as intangible, are now being converted into monetary values. Because of this, classic intangibles are now accepted as tangible measures and their value is more easily understood. Consider, for example, customer satisfaction. Just a decade ago, few organizations had a clue as to the monetary value of customer satisfaction. Now more firms have taken the extra step to link customer satisfaction directly to revenues, profits and other measures. Companies are seeing the tremendous value that can be derived from intangibles. More data are being accumulated to show monetary values, moving some intangible measures into the tangible category.

Table 9.2 Common Intangibles

• Accountability
• Alliances
• Attention
• Awards
• Branding
• Capability
• Capacity
• Clarity
• Communication
• Corporate social responsibility
• Customer service (customer satisfaction)
• Employee attitudes
• Engagement
• Human life
• Image
• Environmental consciousness
• Intellectual capital
• Innovation and creativity
• Job satisfaction
• Leadership
• Loyalty
• Networking
• Organizational commitment
• Partnering
• Poverty
• Reputation
• Team effectiveness
• Timeliness
• Sustainability
• Work/life balance

Some EHS projects are implemented because of the intangibles or intrinsic benefits. For example, the need to engage communities, to increase social consciousness or to have greater collaboration, partnering, communication, teamwork or customer service will drive projects. Executives often pursue employee health projects because of image and reputation. From the outset, the intangibles are the important drivers and become the most important measures. Consequently, more executives include a string of intangibles on their scorecards, key operating reports, key performance indicators, dashboards and other routine reporting systems. In some cases, intangibles represent nearly half of all measures that are monitored.

The Federal Reserve Bank of Philadelphia recently estimated that investment in intangible assets amounts is over $1 trillion. Only 15 percent of the value of a contemporary organization can be tied to such tangible assets as buildings and equipment. Intangible assets have become the dominant investment in businesses. They are a growing economical force that can no longer be ignored and measuring their values poses challenges to managers and investors. Intangibles must be properly identified, selected, measured, reported and in some cases, converted to monetary values.

9.6.2 Measuring Intangible Benefits

In some projects, intangibles are more important than monetary measures. Consequently, these measures should be monitored and reported as part of the project evaluation. In practice, every project, regardless of its nature, scope and content, will produce intangible measures. The challenge is to identify them effectively and to report them appropriately.

Positions taken on intangibles usually come in the form of statements such as "You can't measure it." While it is true that intangibles are not things that can be counted, examined or seen in quantities like items produced on an assembly line, they can in fact be measured or they would not be a concern. A quantitative value can be assigned to or developed for any intangible. Consider human intelligence for example. Although human intelligence is vastly complex and abstract, with myriad facets and qualities, IQ scores are assigned to most people and most people seem to accept them as an accurate measurement. The software engineering institute of Carnegie-Mellon University assigns software organizations a score of 1 to 5 to represent their maturity in software engineering. This score has enormous implications for the organizations' business development capabilities, yet the measure goes practically unchallenged.

Several approaches are available for measuring intangibles. Intangibles that can be counted include customer complaints, employee complaints and conflicts. These can be recorded easily and they constitute one of the most acceptable types of measures. Unfortunately, many intangibles are based on attitudes and perceptions that must be measured. The key is in the development of the instrument. Measurement instruments are usually developed around scales of three, five and even ten points to represent levels of perception. The instruments to measure intangibles consist of three basic varieties.

9.6.2.1 *Five- or Ten-Point Scales*

The first instrument lists the intangible items and asks respondents to agree or disagree on a five-point scale (where the midpoint represents a neutral opinion). Instead of a five-point scale with a neutral position, a more efficient approach is to label each point in varying degrees of the issue. For example, a five-point scale can easily be developed to describe degrees of reputation, ranging from the worst rating—a horrible reputation (1) —to the best rating—an excellent reputation (5). The middle point, (3) — average reputation, is not neutral. Still other ratings are expressed as an assessment on a scale of one to ten, after respondents review a description of the intangible.

9.6.2.2 *Soft Measures Connected to Hard Measures*

Another instrument to measure intangibles connects them, when possible, to an item that is easier to measure or easier to value. As shown in Figure 9.3, most hard-to-measure items are linked to an easy-to-measure item. In the classic situation, a soft measure (typically the intangible) is connected to a hard measure (typically the tangible). Although this link can be developed through logical deductions and conclusions, having some empirical evidence through a correlation analysis (as shown in the figure) and

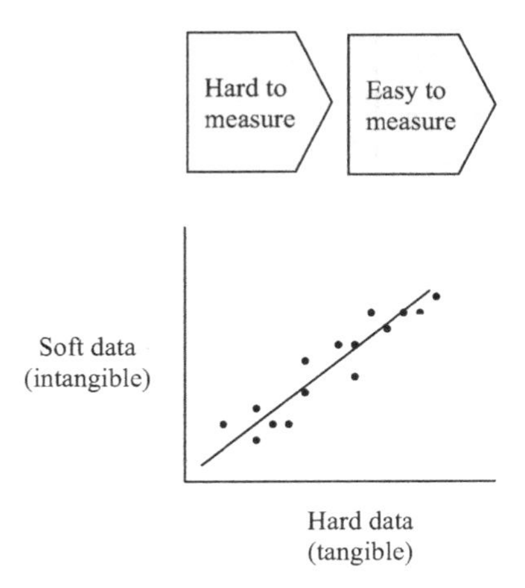

Figure 9.3 The Link between Hard-to-Measure and Easy-to-Measure Items.

developing a significant correlation between the items is the best approach. However, a detailed analysis would have to be conducted to ensure that a causal relationship exists. In other words, just because a correlation is apparent, it does not mean that one caused the other. Consequently, additional analysis, other empirical evidence and supporting data could pinpoint the actual causal effect.

9.6.2.3 *Indexes of Different Values*

A final instrument for measuring the intangible is the development of an index of different values. An index is a single score representing some complex factor that is constructed by aggregating the values of several different measures. Measures making up the index are sometimes weighted based on their importance to the abstract factor being measured. Index measures may be based strictly on hard data items; may combine hard and soft data items to reflect the performance of a business unit, function or project; or may be completely intangible, such as a customer satisfaction index.

9.6.3 Identifying and Collecting Intangibles

Intangible measures can be taken from different sources and at different times during the project life cycle, as depicted in Figure 9.4.

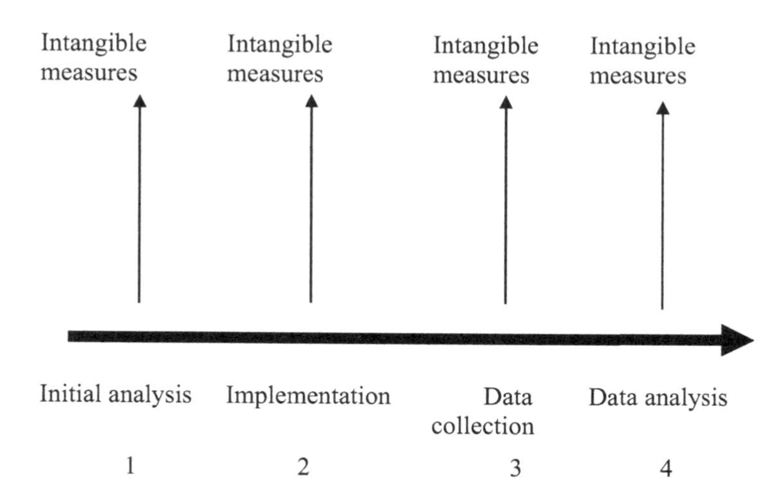

Figure 9.4 Identifying Intangible Measures During the Project Life Cycle.

They can be uncovered early in the process, during the needs assessment and their collection can be planned for as part of the overall data collection strategy.

A second opportunity to identify intangible benefits occurs during the planning process, when sponsors of the project agree on an evaluation plan. Key stakeholders can usually identify the intangible measures they expect to be influenced by the project.

A third opportunity to collect intangible measures presents itself during data collection. Although the measure may not be anticipated in the initial project design, it may surface on a questionnaire, in an interview or during a focus group. Questions are often asked about other improvements linked to a project and participants usually provide several intangible measures for which no plans are available to assign a value.

The fourth opportunity to identify intangible measures is during data analysis and reporting, while attempting to convert data to monetary values. If the conversion loses credibility, the measure is reported as an intangible benefit.

9.6.4 Analyzing Intangibles

For each intangible measure identified, some evidence of its connection to the project must be shown. However, in many cases, no specific analysis is planned beyond tabulation of responses. Early attempts to quantify intangible data sometimes result in aborting the entire process, with no further data analysis being conducted. In some cases, isolating the effects of the project may be undertaken using one or more of the methods outlined in Chapter 8. This step is necessary when project leaders need to know the specific amount of change in the intangible measure that is linked to the project.

Intangible data often reflect improvement. However, neither the precise amount of improvement nor the amount of improvement directly related to a project is always identified. Because the value of these data is not included in the ROI calculation, intangible measures are not normally used to justify another project or to justify continuing an existing project. A detailed analysis is not necessary. Intangible benefits are often viewed as additional evidence of the project's success and are presented as supportive qualitative data.

9.7 Final Thoughts

Showing the real money requires just that—money. Business impact data that have improved as a result of an EHS project must be converted to money. Standard values make this process easier, but easy is not always an option and other techniques must sometimes be used. However, if a measure cannot be converted with minimal resources or with assurance of credibility, the improvement in the measure should be reported as an intangible benefit. After the data are converted to monetary values, the next step is collecting the project costs and calculating the ROI. These processes are covered in the next chapter.

10

Calculating the ROI

Abstract

Chapter 10 focuses directly on the ROI calculation. Two components are required for the calculation: the monetary value and the total project cost. We discuss standards for the cost (direct and indirect) of EHS projects, which are needed for a credible calculation. We then present the different ways to calculate ROI. One method is a benefit-to-cost ratio, which shows the monetary benefits divided by the cost. Another method is the financial ROI, which is expressed as a percentage. The benefits minus cost divided by cost x 100 to result in a percentage. A final method is the payback period, which defines the number of months and years needed to pay back the cost of the project. We also explore misuse of ROI terms and presents cautions around presentation of ROI data.

Chapter 9 described how to convert business impact measures to money. These monetary benefits represent the financial output of an EHS project. However, until that output is compared to the project costs, the absolute contribution of investing in EHS is still unknown. This chapter explores the development of project costs. Often confused with the budget allocated to a project, project costs represent all that an organization has invested in a project. When the fully loaded costs are identified, a comparison is made to the benefits, resulting in the ROI calculation. This calculation is reported using a variety of metrics, including the benefit-cost ratio, ROI percentage and payback period.

Keywords: ROI calculation, monetary value, total project cost, indirect costs, direct costs, payback period

Jack Phillips, Patti Phillips, and Al Pulliam, Measuring ROI in Environment, Health, and Safety, (227–244) 2014 © Scrivener Publishing LLC

10.1 Why the Concern About Project Costs?

One of the main reasons for monitoring costs is to create budgets for projects. The initial costs of most projects are usually estimated during the proposal process and are often based on previous projects. The only way to have a clear understanding of costs so that they can be used to determine future projects and future budgets is to track them using different categories, identified later in this chapter.

Costs should be monitored in an ongoing effort to control expenditures and keep the project within budget. Monitoring cost activities not only reveals the status of expenditures but also gives visibility to expenditures and encourages the entire project team to spend wisely. Additionally, of course, monitoring costs on an ongoing basis is much easier, more accurate and more efficient than trying to reconstruct events to capture costs retrospectively. Developing accurate costs by category builds a database for understanding and predicting future costs.

Monitoring project costs is an essential step in developing the ROI calculation because it represents the denominator in the ROI formula. ROI has become a critical measure demanded by many stakeholders, including sponsors and senior executives. It is the ultimate level of evaluation, showing the actual payoff of the project expressed as a percentage and based on the same formula as the evaluation for other types of capital investment.

A brief example will highlight the importance of costs and ROI. A new green suggestion system was implemented in a large electric utility company. This new plan provided cash awards for employees when they submitted a suggestion that was implemented and resulted in cost savings associated with green projects (e.g., recyclable materials, trash, supplies). This project was undertaken to help lower the costs of this publicly owned utility as well as to engage employees in environmentally conscious activities. As the project was rolled out, project leaders captured the employees' reaction to ensure that they perceived the suggestion system as fair, equitable, motivating and challenging. At Level 2, they measured learning to make sure that the employees understood what types of waste to look for, how to document their suggestions and how and when the awards would be made. Application data (Level 3) considered the actual submission of the awards, indicating that employees were making appropriate and realistic suggestions. The company had the goal of a 10 percent participation rate in the process. Level 4 data corresponded to the actual monetary value of cost savings due to implementation of suggestions. In this case, $1.5 million in cost savings was achieved over a two-year period.

In many organizations, the evaluation would have stopped there. The project appeared to be a success since the goals were met at each of the four levels. However, the costs of the project for the same two-year period totaled $2 million. Costs included information sessions, motivational rallies, guest speakers, travel and lodging for field personnel and so forth. Thus, the utility company spent $2 million to save $1.5 million. This negative ROI would not have been recognized if the ultimate measure, the ROI, had not been developed. Incidentally, a negative ROI might be acceptable to some executives. After all, the intangible benefits reflected increased commitment, engagement in environmental activities, teamwork and cooperation. However, if the objective was a positive ROI, this system failed to achieve it, primarily because of excessive administrative costs. Full-time suggestion system administrators were situated in each of their twelve locations. After a review, it was decided that only two administrators were needed. The other administrators were transferred to other open jobs. This action reduced the costs going forward. The message: Just because the ROI is negative, does not mean the project is discontinued.

10.2 Fundamental Cost Issues

The first step in monitoring costs is to define and address issues relating to cost control. Several rules apply to tabulating costs. Consistency and standardization are necessary. In addition, all costs must be monitored, even if they are not needed. They must be realistic and reasonable, however, precision can be costly in and of itself, so estimates are okay to use. Finally, all costs should be disclosed. Consideration of these guidelines as well as the issues described in this section will help project owners define, monitor and manage all costs associated with their green projects.

10.2.1 Fully Loaded Costs

Because a conservative approach is used to calculate the ROI in an EHS project, costs should be fully loaded, which is the tenth guiding principle covered in Chapter 4. With this approach, all costs (direct and indirect) that can be identified and linked to a particular project are included. The philosophy is simple: For the denominator of the ROI equation, "when in doubt, put it in" (i.e., if there is any question as to whether a cost should be included, include it, even if the cost guidelines for the organization do not require it). When an ROI is calculated and reported to target audiences, the process should withstand even the closest scrutiny, when necessary, to ensure its credibility. The only way to meet this test is to include all costs. Of course, from a realistic viewpoint, if the controller or chief financial officer insists on not using certain costs, then leaving them out or reporting them in an alternative way is suggested, but always be certain as to why a cost is being omitted and ensure that you are consistent in your approach.

10.2.2 Costs Without Benefits

Because costs can easily be collected, they are presented to management in many ingenious ways, such as in terms of the total cost of the project, cost per day and cost per participant. While these may be helpful for efficiency comparisons, presenting them without identifying the corresponding benefits may be problematic. When most executives review project costs, a logical question is raised: What benefit was received from the project? This is a typical management reaction, particularly when costs are perceived to be high.

Unfortunately, many organizations have fallen into the trap of reporting costs without benefits. For example, in one organization, all costs associated with a major safety culture project were tabulated and reported to the senior management team. From an executive perspective, the total figure exceeded the perceived value of the project and the executive group's immediate reaction was to request a summary of monetary and nonmonetary benefits derived from the overall transformation. The conclusion was that few, if any, economic benefits could be connected to the project. Consequently, budgets for similar projects were drastically reduced in the future. While this may be an extreme example, it shows the danger of presenting only half the equation. Because of this, some organizations have developed a policy of not communicating cost data unless the benefits can be captured and presented along with the costs, even if the benefits are subjective and intangible. This helps maintain a balance between the two components.

10.2.3 Cost Guidelines

When multiple projects are being evaluated, it may be helpful to develop a philosophy and policy on costs in the form of guidelines for evaluators or others who monitor and report costs. Cost guidelines specifically detail which cost categories to include with projects and how to capture, analyze and report data. Guidelines include standards, unit cost guiding principles and generally accepted values. Cost guidelines can range from a one-page brief to a one-hundred-page document in a large, complex organization. The simpler approach is better.

When fully developed, cost guidelines should be reviewed and approved by the finance and accounting staff. The final document serves as the guiding force in collecting, monitoring and reporting costs. When the ROI is calculated and reported, costs are included in summary or table form and the cost guidelines are usually referenced in a footnote or attached as an appendix.

10.2.4 Sources of Costs

Four of the sources of costs are illustrated in Table 10.1. These sources include the project team expenses, vendor/supplier expenses, sponsor expenses and equipment, services and other expenses. The charges and expenses from the project team represent the major segment of costs and are usually transferred directly to the client for payment. These are often placed in subcategories under fees and expenses. A second major cost category relates to the vendors or suppliers who assist with the project. A variety of expenses, such as consulting or advisory fees, may fall in this category. A third major cost

Table 10.1 Sources of Project Costs.

Source of Costs	Cost Reporting Issues
Project team fees and expenses	• Costs are usually accurate. • Variable expenses are usually underestimated.
Vendor/supplier fees and expenses	• Costs are usually accurate. • Variable expenses are usually underestimated.
Sponsor expenses, direct and indirect	• Direct expenses are usually not fully loaded. • Indirect expenses are rarely included in costs.
Equipment, services and other expenses	• Costs are sometimes understated. • Expenses may lack accountability.

category is those expenses borne by the sponsor organization—both direct and indirect. In many projects, these costs are not identified but nevertheless are part of the project costs. The final cost category involves expenses not covered in the other three categories. These include payments for equipment and services needed for the project. Finance and accounting records should track and reflect the costs from these different sources and the process presented in this chapter can also help track these costs. An example will illustrate this. A plant has a problem with the proper handling of hazardous materials. The plant manager (the sponsor) contacts the Corporate Environment Health and Safety Team (project team) to address the problem. Training is conducted using purchased materials (vendor) and monitoring equipment is purchased (equipment).

10.2.5 Prorated Versus Direct Costs

Usually all costs related to a project are captured and charged to that project. However, some costs are prorated over a longer period. Equipment purchases, software development and acquisitions and the construction of facilities are all significant costs with a useful life that may extend beyond the project. Consequently, a portion of these costs should be prorated to the project. Under a conservative approach, the expected life of the project is fixed. Some organizations will assume a period of one year of operation for a simple project. Others may consider three to five years appropriate. If a question is raised about the specific time period to be used in this calculation, the finance and accounting staff should be consulted or appropriate guidelines should be developed and followed.

10.2.6 Employee Benefits Factor

Employee time is valuable and when time is required for a project, the costs for that time must be fully loaded, representing total compensation, including employee benefits. This means that the employee benefits factor should be included in the cost of employee time in the profit. This number is usually easily available in the organization and is used in other costing formulas. It represents the cost of all employee benefits expressed as a percentage of payroll. In some organizations, this value is as high as 50 to 60 percent. In others, it may be as low as 25 to 30 percent. The average in the United States is 38 percent ("Annual Employee Benefits Report," 2013).

Table 10.2 Project Cost Categories.

Cost Item	Prorated	Expensed
Initial analysis and assessment	✓	
Development of solutions	✓	
Acquisition of solutions	✓	
Implementation and application		
Salaries/benefits for project team time		✓
Salaries/benefits for coordination time		✓
Salaries/benefits for participant time		✓
Project materials		✓
Hardware/software	✓	
Travel/lodging/meals		✓
Use of facilities		✓
Capital expenditures	✓	
Maintenance and monitoring		✓
Administrative support and overhead	✓	
Evaluation and reporting		✓

10.3 Fully Loaded Cost Profile

Table 10.2 shows the recommended cost categories for a fully loaded, conservative approach to estimating project costs. Consistency in capturing all of these costs is essential and standardization adds credibility. Each category is described in this section.

10.3.1 Initial Analysis and Assessment

One of the most underestimated items is the cost of conducting the initial analysis and assessment that leads to the need for the EHS project. In a comprehensive project, this involves data collection, problem solving, assessment and analysis. In some projects, this cost is near zero because the project is implemented without an initial assessment of need. However, as more project sponsors place attention on needs assessment and analysis in the future, this item will become a significant cost.

10.3.2 Development of EHS Projects

Designing and developing an EHS initiative has significant costs, particularly for those projects intended to solve an expensive problem and/or drive substantial improvement in key measures. Development costs include time spent in both the design and development and the purchase of supplies, technology and other materials directly related to the project. As with needs assessment costs, design and development costs are usually charged to the project. However, if the project or initiative is being rolled in with other projects using the same resources, development costs can be prorated.

10.3.3 Acquisition

In lieu of development costs, some EHS project leaders choose to purchase a solution or parts of it. When this is the case, the acquisition cost becomes a line item in the project cost profile. The costs for these solutions include the purchase price, support materials and licensing agreements. Some projects have both acquisition costs and development costs. Acquisition costs can be prorated if the acquired solutions can be used in other projects.

10.3.4 Implementation

The largest cost segment in a project is associated with implementation and delivery. The time (salaries and benefits), travel and other expenses of those involved in any way in the project are included. These costs can be estimated using average or midpoint salary values for corresponding job classifications. When a project is targeted for an ROI calculation, participants can provide their salaries directly in a confidential manner. Project materials, such as field journals, instructions, reference guides, case studies, surveys, participant workbooks and job aids, should be included in the implementation costs, along with license fees, user fees and royalty payments. Supporting hardware, software, DVDs and remote access support systems should also be included.

The cost for the use of facilities needed for the project should be included. For external meetings, this is the direct charge for the conference center, hotel or motel. If the meetings are conducted in-house, the conference room represents a cost for the organization and the cost should be estimated and incorporated, even if it is uncommon to include facilities costs in other cost reporting. If a facility or building is constructed or purchased for the project, it is included as a capital expenditure, but it is prorated. The same is true for the purchase of major hardware and software when they are considered capital expenditures.

10.3.5 Maintenance and Monitoring

Maintenance and monitoring costs, which may be significant for some projects, involve routine expenses necessary to maintain and operate the project. These are ongoing expenses that allow the new project solution to continue.

10.3.6 Support and Overhead

The cost of support and overhead includes additional costs not directly charged to the project or any project cost not considered in the previous calculations. Typical items are the cost of administrative/clerical support, telecommunication expenses, office expenses, salaries of client managers and other fixed costs. Usually, this is provided in the form of an estimate allocated in some convenient way.

10.3.7 Evaluation and Reporting

Evaluation costs complete the fully loaded cost profile. Activities under evaluation costs include developing the evaluation strategy, designing instruments, collecting data, analyzing data, preparing a report and communicating the results. Cost categories include time, materials, purchased instruments, surveys and any consulting fees.

10.4 Cost Classifications

Project costs can be classified in two basic ways. One is with a description of the expenditures, such as labor, materials, supplies or travel. These are expense account classifications, which are standard with most accounting systems. The other way to classify costs is to use the categories in the project steps, such as initial analysis, development, implementation, maintenance, overhead and evaluation. An effective system monitors costs by account category according to the description of those accounts, but it also includes a method for accumulating costs for the process steps. Many systems stop short of this second task. Although classifying costs by expenditure description (e.g., labor, materials, supplies, travel) adequately states the total project costs, it does not allow for a useful comparison with other projects to understand which steps may incur excessive costs.

10.5 The ROI Calculation

The term *return on investment* is occasionally misused, sometimes intentionally. A broad definition for ROI is often given that includes any benefit from the project. ROI then becomes a vague concept in which even subjective data linked to a program are included. In this book, the term *return on investment* is defined more precisely and represents an actual value determined by comparing project costs to benefits of project implementation. The two most common measures are the benefit-cost ratio (BCR) and the ROI percentage. Both are presented in this section along with other approaches to calculate the return or payback.

Project benefits, making up the numerator of the ROI equation, are annualized. For short-term projects, the assumption is based on first-year only. Using annualized values is becoming an accepted practice for developing the ROI in many organizations. This approach is a conservative way to develop the ROI, since many short-term projects have added value in the second or third year, but to assume those future benefits will inflate the ROI based on highly presumptuous values. For long-term projects, however, longer time frames should be used. For example, in an ROI analysis of a project involving major software and technology purchases, a five-year time frame was used because the likely benefits would not surface until the project was implemented to its fullest. However, for short-term projects that take only a few weeks to implement (such as a recycling campaign), first-year values are appropriate.

In selecting the approach to measure ROI, the formula used and the assumptions made in arriving at the decision to use this formula should be communicated to the target audience. This helps prevent misunderstandings and confusion surrounding how the ROI value was developed. Although several approaches are described in this chapter, two stand out as preferred methods: the benefit-cost ratio and the basic ROI formula.

10.5.1 Benefit-Cost Ratio

One of the original methods for evaluating projects is the benefit-cost ratio. This method compares the monetary benefits of the project with the costs, using a simple ratio. The formula form is as follows:

$$\text{BCR} = \frac{\text{Project benefits}}{\text{Project costs}}$$

In simple terms, the BCR compares the annual economic benefits of the project with the costs of the project. A BCR of 1 means that the benefits equal the costs or that the project breaks even. This is usually written as 1:1. A BCR of 2, usually written as 2:1, indicates that for each dollar spent on the project, two dollars are returned in benefits.

The following example illustrates the use of the BCR. A safety audit program was implemented for a large construction materials company. Routine audits were conducted. The audits reduced injuries and OSHA fines. Data were monitored six months and one year after each audit. The months of improvements were isolated to the safety audit project. Accident data were converted to money and amended to avoid fines linked to the project. The first-year payoff for the company's combined efforts was $1,742,800. The total cost of implementing the safety audit project was $539,000. Thus, the BCR was

$$\text{BCR} = \frac{\$1{,}742{,}800}{\$539{,}000} = 3.23$$

For every dollar invested in the project, the company recovered $3.23 in benefits and it had a significant impact on stress, job satisfaction and image.

10.5.2 ROI

Perhaps the most appropriate formula for evaluating project investments is net project benefits divided by costs. This is the traditional financial ROI expressed as a percentage. This metric is related to the BCR, but expresses the return in terms of net gain. In formula form, ROI is the following:

$$\text{ROI} = \frac{\text{Net project benefits}}{\text{Project costs}} \times 100$$

Net project benefits are project benefits minus costs. Another way to calculate ROI is to subtract 1 from the BCR and multiply by 100 to get the ROI percentage. For example, a BCR of 2.45 translates to an ROI of 145 percent (1.45 x 100). This formula is essentially the same as the ROI for capital investments. For example, when a firm builds a new plant, the ROI is developed by dividing annual earnings by the investment. The annual earnings are comparable to net benefits (annual benefits minus the cost). The investment is comparable to the fully loaded project costs.

An ROI of 50 percent (BCR = 1.50) means that the costs were recovered and an additional 50 percent ($.50) of the costs were returned. A project ROI of 150 percent indicates that the costs have been recovered and an additional 1.5 ($1.50) times the costs are returned. Using the ROI formula to calculate the return on project investments essentially places these investments on a level playing field with other investments whose evaluation uses the same formula and similar concepts. Key management and financial executives who regularly work with investments and their own ROIs easily understand the ROI calculation.

Using the previous example of the safety audit project, the ROI becomes the following formula.

$$\text{ROI} = \frac{\$1{,}742{,}800 - \$539{,}000}{\$539{,}000} = 2.23 \times 100 = 223\%$$

This robust ROI is compared to the ROI objective described later in this chapter.

10.6 ROI Misuse

The ROI formula described here should be used consistently throughout an organization. Deviations from or misuse of the formula can create confusion, not only among users but also among finance and accounting staff. The CFO and the finance and accounting staff should become partners in the implementation of the ROI Methodology, using the same financial terms. Without the support, involvement and commitment of these individuals, the wide-scale use of ROI will be unlikely.

Table 10.3 shows some of the financial terms that are often misused in literature. Terms such as *return on intelligence* (or *information)*, abbreviated as ROI, do nothing but confuse most CFOs, who assume that ROI refers to the return on investment described here. Sometimes *return on expectations* (ROE), *return on anticipation* (ROA) and *return on client expectations* (ROCE) are used, which can also confuse CFOs who assume the abbreviations refer to return on equity,

Table 10.3 Misused Financial Terms.

Term	Misuse	CFO Definition
ROI	Return on information Return on involvement Return on intelligence	Return on investment
ROE	Return on expectation Return on engagement Return on environment	Return on equity
ROA	Return on anticipation	Return on assets
ROCE	Return on client expectation	Return on capital employed
ROS	Return on safety	?
ROP	Return on people	?
ROR	Return on resources	?
ROT	Return on technology	?
ROW	Return on web	?
ROM	Return on marketing	?
ROO	Return on objectives	?
ROQ	Return on quality	?

return on assets and return on capital employed, respectively. The use of these terms in the payback calculation of a project will also confuse and perhaps lose the support of the finance and accounting staff. Other terms, such as *return on people, return on resources, return on safety, return on technology and return on web,* are often used with almost no consistency regarding financial calculations. The bottom line: Don't confuse the CFO. Consider this person an ally and use the same terminology, processes and concepts when applying financial returns for projects.

10.7 ROI Targets

Specific expectations for ROI should be developed before an evaluation study is undertaken. Although no generally accepted standards exist, four strategies have been used to establish a minimum expected requirement or hurdle rate, for the ROI of a project or program. The first approach is to set the ROI using the same values used for investing in capital expenditures, such as equipment, facilities and new companies. For North America, Western Europe and most of the Asian Pacific area, including Australia and New Zealand, the cost of capital is low and the internal hurdle rate for ROI is usually in the 15 to 20 percent range. Thus, using this strategy, organizations would set the expected ROI for a project at the same value expected from other investments.

A second strategy is to use an ROI minimum target value that is above the percentage expected for other types of investments. The rationale is that the ROI process for projects and programs is still relatively new and often involves subjective input, including estimations. Because of this, a higher standard is required or suggested.

A third strategy is to set the ROI value at a break-even point. A 0 percent ROI represents breaking even; this is equivalent to a BCR of 1:1. This approach is used when the goal is to recapture the cost of the project only. This is the ROI objective for many public-sector organizations, where the primary value and benefit from the program come through the intangible measures, which are not converted to monetary values. Thus, an organization will use a break-even point for the ROI based on the reasoning that it is not attempting to make a profit from a particular project.

A fourth and often the recommended, strategy is to let the program sponsor set the minimum acceptable ROI value. In this scenario, the individual who initiates, approves, sponsors or supports the project establishes the acceptable ROI. Almost every project has a major sponsor and that person may be willing to specify an acceptable value. This links the expectations for financial return directly to the expectations of the sponsor.

10.8 Intangibles Revisited

Chapter 9 describes the importance of developing and reporting intangible benefits. These benefits deserve mentioning here as well. Although they are not converted to money, their importance cannot be ignored. When reporting the ROI in a profit, it is essential to balance that financial metric with other nonfinancial measures to tell the true story of the project's impact. So, when reporting ROI, report it in the context of other measures of performance. This includes the intangible benefits linked directly to the project as well as the lower levels of data that represent the chain of impact. More detail on how best to report results is presented in Chapter 11.

10.9 Other ROI Measures

In addition to the traditional ROI formula, several other measures are occasionally used under the general heading of return on investment. These measures are designed primarily for evaluating other financial measures but sometimes work their way into project evaluation.

10.9.1 Payback Period (Break-Even Analysis)

The payback period is commonly used for evaluating capital expenditures. With this approach, the annual cash proceeds (i.e., savings) produced by an investment are compared against the original cash outlay for the investment to determine the point at which cash proceeds equal the original investment. Measurement is usually in terms of years and months. For example, if the cost savings generated from a project are constant each year, the payback period is determined by dividing the original cash investment (including development costs and expenses) by the expected or actual annual savings.

To illustrate this calculation, assume that the initial cost of a project is $100,000 and the project has a three-year useful life. Annual savings from the project are expected to be $40,000. Thus, the payback period may be written as a formula:

$$\text{Payback period} = \frac{\text{Total investment}}{\$539{,}000} = \frac{\$100{,}000}{\$40{,}000} = 2.5 \text{ years}$$

The project will "pay back" the original investment in 2.5 years. Since the projected lifespan is three years, one can assume a benefit over and beyond the cost at the end of the three years.

The payback period method is simple to use but has the limitation of ignoring the time value of money. It has not enjoyed widespread use in the evaluation of project investments.

10.9.2 Discounted Cash Flow

Discounted cash flow is a method of evaluating investment opportunities in which certain values are assigned to the timing of the proceeds from the investment. The assumption behind this approach is that a dollar earned today is more valuable than a dollar earned a year from now, based on the accrued interest possible from investing the dollar.

There are several ways of using the discounted cash flow concept to evaluate a project investment. The most common approach uses the net present value of an investment. The savings each year are compared with the outflow of cash required by the investment. The expected annual savings are discounted based on a selected interest rate and the outflow of cash is adjusted by the same interest rate. If the present value of the savings exceeds the present value of the outlays, after the two have been adjusted by the common interest rate, the investment is usually considered acceptable by management. The discounted cash flow method has the advantage of ranking investments, but it requires calculations that can become difficult. Also, for the most part, it is subjective in terms of assumed future benefits as well as assumed future value of the dollar.

10.9.3 Internal Rate of Return

The internal rate of return (IRR) method determines the interest rate necessary to make the present value of the cash flow equal zero.

This represents the maximum rate of interest that could be paid if all project funds were borrowed and the organization was required to break even on the project. The IRR considers the time value of money and is unaffected by the scale of the project. It can be used to rank alternatives and to accept or reject decisions when a minimum rate of return is specified. A major weakness of the IRR method is that it assumes all returns are reinvested at the same internal rate of return. This can make an investment alternative with a high rate of return look even better than it really is and make a project with a low rate of return look even worse. In practice, the IRR is rarely used to evaluate project investments.

10.10 Final Thoughts

ROI, the final evaluation level, compares project benefits to project costs. From a practical standpoint, some costs may be optional and depend on the organization's guidelines and philosophy, but all costs should be included in some way, even if this goes beyond the requirements of the organization's policy. After the benefits are collected and converted to monetary values and the project costs are tabulated, the ROI calculation itself is easy. Plugging the values into the appropriate formula for the ROI or the benefit-cost ratio is the final step.

Of course, the ROI calculation and results at all other levels are useless data unless they are communicated in terms that resonate with all stakeholders. The next chapter addresses this issue.

11

Reporting Results

Abstract

An often overlooked step in the evaluation process is reporting study results to the appropriate audiences. If this step is not done efficiently and effectively, the value of the study will be diminished. First the audience must be defined. There can be many audiences, but four stand out as critical: the client for the project, managers of project participants, the participants themselves and the remainder of the EHS team. In a large organization, the group could include all employees and all shareholders. This chapter covers the forms of results, from a one-page summary to a full impact study. The information is distributed through a variety of techniques, from face-to-face meetings to video conferences to internal publications. The key is to communicate results in a manner that makes programs better and positions EHS as a critical function. We also discuss how to conduct an executive briefing to present EHS performance data to the C-suite in order to ensure commitment and future funding.

Applying a comprehensive evaluation process to EHS projects is important to show how that investments in these projects have benefited the organization, employees and the environment. But the presentation of that data is just as essential, since decisions are made based on the communication of results. A first step in the communications process is to understand how the data will be used. Should the results be used to modify the project, change the process, demonstrate the contribution, justify new projects, gain additional support or build goodwill? The answers will help you determine how best to present the data. Remember though, the worst course of action is to do nothing. Achieving results without communicating them is like planting seeds and failing to fertilize and cultivate the seedlings—the yield will be less than optimal. This chapter describes how to present the results of your green project evaluations to various audiences in the form of both oral and written reports.

Jack Phillips, Patti Phillips, and Al Pulliam, Measuring ROI in Environment, Health, and Safety, (245–270) 2014 © Scrivener Publishing LLC

Keywords: Reporting results, audience, client, managers, participants, summary, impact study, executive briefing

11.1 Why the Concern About Communicating Results?

Communicating results is critical to the success of an EHS project. The results achieved must be conveyed to stakeholders not just at project completion but also throughout the project implementation. Continuous communication maintains the flow of information so that adjustments can be made and all stakeholders are kept up to date on the status of the project.

Mark Twain once said, "Collecting data is like collecting garbage—pretty soon we will have to do something with it." Measuring project success and gathering evaluation data mean nothing unless the findings are communicated promptly to the appropriate audiences so that they are aware of the results and can take action in response if necessary. Communication is a critical need that should never be overlooked, even with underfunded in projects. Following are just a few of the important reasons for communicating results.

11.1.1 Make Improvements

Information is collected at different points during EHS project implementation and providing feedback to involved groups enables them to take action and make adjustments if necessary. Thus, the quality and timeliness of communication are critical to making improvements. Even after the project is completed, communication is necessary to make sure the target audience fully understands the results achieved and how the results may be enhanced in future projects or in the current project, if it is still operational. Communication is the key to making important adjustments at all phases of the project.

11.1.2 Explain the Contribution

The overall contribution of the EHS project, as determined from the six major types of measurement, is often unclear. The different target audiences will each need a thorough explanation of the results showing a clear connection between the project and the results. The communication strategy, including techniques, media and the overall process, will determine the extent to which each group understands the contribution. Communicating results, particularly in terms of business impact and ROI, can quickly overwhelm even the most sophisticated target audiences. Communication must be planned and implemented with the goal of making sure the respective audiences understand the full contribution.

11.1.3 Manage Sensitive Issues

Communication is one of those issues that can cause major problems. Because the results of a project may be closely linked to political issues within an organization, communicating the results can upset some individuals while it pleases others. If certain individuals do not receive the information or if it is delivered inconsistently between groups, problems can quickly surface. Not only must the information be understood, but issues relating to fairness, quality and political correctness make it crucial that the communication be constructed and delivered effectively to all key individuals.

11.1.4 Address Diverse Audience Needs

With so many potential target audiences requiring communication on the success of an EHS project, the communication must be tailored to each audience. A varied audience has varied needs. Planning and effort are necessary to ensure that each audience receives all the information it needs, in the proper format, at the proper time. A single report for presentation to all audiences is usually inappropriate. The scope, the format and even the content of the information will usually vary from one group to another. Thus, the target audience is the key to determining the appropriate method of communication.

11.2 Principles of Communicating Results

The skills one must possess to communicate results effectively are almost as sophisticated as those necessary for obtaining results. The style of the communication is as important as the substance. Regardless of the message, audience or medium, a few general principles apply. These are vital to the overall success of the communication effort and should serve as a checklist for the project team planning the dissemination of EHS project results.

11.2.1 Time Communication Strategically

In general, EHS project results should be communicated as soon as they become known. From a practical standpoint, however, it is sometimes best to delay the communication until a convenient time, such as the publication of the next company newsletter or the next general management meeting. Several questions are relevant to the timing decision. Is the audience ready for the results in view of other issues that may have developed? Is the audience expecting results? When will the delivery have the maximum impact on the audience? Do circumstances dictate a change in the timing of the communication?

11.2.2 Target Communication to Specific Audiences

As stated earlier, communication is usually more effective if it is designed for the specific group being addressed. The message should be tailored to the interests, needs and expectations of the target audience. The results of the project should reflect outcomes at all levels, including the six types of data presented in this book. Some of the data are developed earlier in the project and communicated during project implementation. Other data are collected after project implementation and communicated in a follow-up repost. The results, in their broadest sense, may incorporate early feedback in qualitative form all the way to ROI values expressed in varying quantitative terms.

11.2.3 Select Media Carefully

Certain media may be more appropriate for a particular group than others. For example, face-to-face meetings may be preferable to special bulletins. A memo distributed exclusively to top executives may be a more effective outlet than the company newsletter. The proper format of communication can determine the effectiveness of the process.

11.2.4 Keep a Communication Objective

For communication to be effective, fact must be separated from fiction and accurate statements distinguished from opinions. Disappointing news must be included along with the good news. Some audiences may approach the communication with skepticism, anticipating the presence of biased opinions. Boastful statements can turn off recipients and most of the content will be lost. Observable phenomena and credible statements carry much more weight than extreme or sensational claims, which may get an audience's attention but often detract from the importance of the results.

11.2.5 Use Communication Consistently

The timing and content of the communication should be consistent with past practices. A special presentation at an unusual time during the course of the project may provoke suspicion. Also, if a particular group, such as top management, regularly receives communication regarding outcomes, it should continue receiving communication even if the results are not positive. Omitting unfavorable results leaves the impression that only positive results will be reported.

11.2.6 Include Anecdotal Data from Respected Sources

Opinions are strongly influenced by other people, particularly those who are respected and trusted. This respect may be related to leadership ability, position, special skills or knowledge. Comments about project results, when solicited from individuals who are respected within the organization, can influence the effectiveness of the message. A comment from an individual who commands little respect and is regarded as a substandard performer can have a negative impact on the message.

11.2.7 Account for Audience Opinion

Opinions are difficult to change and a negative opinion toward a project or project team may not change with the mere presentation of facts. They need convincing from a credible person. A project team with a high level of credibility and respect may have a relatively easy time communicating results. Low credibility can create problems for teams that are trying to be persuasive. However, the presentation of facts alone may strengthen the opinions held by those who already support the project. The presentation of the results reinforces their position and provides them with a defense in discussions with others.

11.3 The Process for Communicating Results

The communication of EHS project results must be systematic, timely and well planned. A seven-step approach will help ensure your communication plan reaps the desired rewards. The seven steps are the following:

1. Analyze the need for communication
2. Plan the communication strategy
3. Select the target audience
4. Develop the report
5. Select the media
6. Present the results
7. Analyze reactions

By following these seven steps, which are described in more detail in this section, you will ensure that your communication strategy is methodical and consistent.

11.3.1 Analyze the Need for Communication

Because there may be various reasons for communicating results, depending on the specific EHS project, the setting and the unique needs of each party, a list of needs should be tailored to the organization and adjusted as necessary. Some of the most common reasons are mentioned below:

- Securing approval for the project and the allocation of time and money
- Gaining support for the project and its objectives
- Securing agreement on the issues, solutions and resources
- Enhancing the credibility of the project leader

- Reinforcing the processes used in the project
- Driving action for improvement in the project
- Preparing participants for the project
- Optimizing results throughout the project and the quality of future feedback
- Showing the complete results of the project
- Underscoring the importance of measuring results
- Explaining techniques used to measure results
- Motivating participants to become involved in the project
- Demonstrating accountability for expenditures
- Marketing future project

There may be other reasons for communicating results. Just as you analyze the needs of the organization to ensure that the sustainability initiative is in alignment, you must make the needs for the communication clear so the presentation of results is positioned for success.

11.3.2 Plan the Communication Strategy

Any activity must be carefully planned to achieve maximum results. This is a critical part of communicating the results of the project to ensure that each audience receives the proper information at the right time and that necessary actions are taken. Several issues are crucial in planning the communication of results:

- What will be communicated?
- When will the data be communicated?
- How will the information be communicated?
- Where will the information be communicated?
- Who will communicate the information?
- Who is the target audience?
- What are the specific actions required or desired?

The communication plan is usually developed as the project plan is developed. This plan details how specific information is to be developed and communicated to various groups and the expected actions. In addition, the communication strategy details how the overall results will be communicated, the time frame for communication and the appropriate groups to receive the information. The project leader, key managers and stakeholders need to agree on the detail of the strategy. If you don't know where you are going, you won't know if you get there–so plan your communication of results. For major projects, a communication plan is developed along with the project plan.

11.3.3 Select the Target Audience

Audiences range from top management to past participants and each audience has its own reasons for hearing about EHS project success. The following questions should be asked about each potential audience:

- Are they interested in the project?
- Do they really want to receive the information?
- Has a commitment been made to include them in the communications?
- Is the timing right for this audience?
- Are they familiar with the project?
- How do they prefer to have results communicated?
- Do they know the project leader? The project team?
- Are they likely to find the results threatening?
- Which medium will be most convincing to this group?

For each target audience, three steps are necessary. First, to the greatest extent possible, the project leader should get to know and understand the target audience. Next, the project leader should find out what information is needed and why. Each group will have its own required amount of information; some will want detailed information while others will prefer a brief overview. Rely on the input from others to determine the audience's needs. Finally, the project leader should take into account audience bias. Some audiences will immediately support the results, others may oppose them and still others will be neutral. The staff should be empathetic and try to understand the basis for the differing views. Given this understanding, communications can be tailored to each group. This is critical when the potential exists for the audience to react negatively to the results.

Determining which groups will receive a particular item of communication requires careful thought, because problems can arise when a group receives inappropriate information or is overlooked altogether. A sound basis for audience selection is to analyze the reason for the communication, as discussed earlier. Table 11.1 identifies common target audiences and the basis for audience selection. Several audiences stand out as critical. Perhaps the most important audience is the project's sponsor. This individual or group initiates the project, reviews data, usually selects the project leader and weighs the final assessment of the effectiveness of the project. Another important target audience is top management. This group is responsible for allocating resources to the project and needs information to help them justify expenditures and gauge the effectiveness of the efforts.

Table 11.1 Common Target Audiences.

Primary Target Audience	Reason for Communication
Sponsor, top executives	To secure approval for the project
Immediate managers	To gain support for the project
Participants	To secure agreement with the issues
Top executives, EHS leaders	To enhance the credibility of the project
Immediate managers	To reinforce the processes
EHS project team	To drive action for improvement
Immediate managers	To prepare participants for the project
Participants	To improve the results and quality of future feedback
Stakeholders	To show the complete results of the project
Sponsor, EHS project team	To underscore the importance of measuring results
Sponsor, project support staff	To explain the techniques used to measure results
All employees	To create the desire for a participant to be involved
All employees Prospective sponsors	To demonstrate accountability for expenditures To market future projects

Communication of results to EHS project participants is often overlooked, with the assumption that when the project is completed, they do not need to be informed of its success. However, participants need feedback on the overall success of the effort. Some individuals may not have been as successful as others in achieving the desired results. Communicating the results creates additional pressure to implement the project effectively and improve results in the future. For those achieving excellent results, the communication will serve as reinforcement.

Communicating with the participants' immediate managers is also essential. In many cases, these managers must encourage participants to implement the project. Also, they are keys in supporting and reinforcing the objectives of the project. An appropriate ROI strengthens the commitment to projects and enhances the credibility of the project team.

The EHS project team must receive information about project results. Whether for small projects in which team members receive a project update or for larger projects, in which a complete team is involved, those who design, develop, facilitate and implement the project require information on the project's effectiveness. Evaluation data are necessary so that adjustments can be made if the project is not as effective as it was projected to be.

11.3.4 Develop the Report

The impact study report details the purpose of the program and evaluation. It describes the methodology used and reports results at each level in detail. The report closes out with recommendations about the project as well as lessons learned through the process. The type of formal evaluation report to be issued to the various audiences depends on the degree of detail needed to achieve the purpose of the communication. Brief summaries of project results with appropriate charts may be sufficient for some communication efforts. In other situations, particularly those involving major projects that require extensive funding, a detailed evaluation report is crucial. A complete and comprehensive impact study report is always necessary at least for the project team. This report can then be used as the basis for more streamlined information aimed at specific audiences and using various media. One possible format for an impact study report is presented in Table 11.2.

Table 11.2 Format of an Impact Study Report.

- General information
  - Background
  - Objectives of study
- Methodology for the study
  - Levels of evaluation
  - ROI Methodology
  - Data collection procedures
    - Data collection methods
    - Data sources
    - Data collection timing
  - Data analysis procedures
    - Techniques to isolate effects of projects
    - Techniques to convert data to money
    - Cost categories
- Results: General information
  - Response profile
  - Success with objectives
- Results: Reaction
  - Data sources
  - Data summary
  - Key issues
- Results: Learning
  - Data sources
  - Data summary
  - Key Issues
- Results: Application and Implementation
  - Data Sources
  - Data Summary
  - Barriers and enablers to application
- Results: Impact
  - General comments
  - Linkage with EHS impact measures
  - Key issues
- Results: ROI
  - Monetary value of impact measure
  - Project costs
  - ROI
- Results: Intangible Measures
- Conclusions and Recommendations
  - Conclusions
  - Recommendations
- Exhibits

While the impact study report is an effective, professional way to present ROI data, several cautions are in order. Since this report documents the success of a project involving a large group of employees, credit for the success must go completely to the participants and their immediate leaders. Their performance generated the success. Also, it is important to avoid boasting about results. Grand claims of overwhelming success can quickly turn off an audience and interfere with the delivery of the desired message.

The evaluation methodology should be clearly explained, along with the assumptions made in the analysis. The reader should easily see how the values were developed and how specific steps were followed to make the process more conservative, credible and accurate. Detailed statistical analyses should be placed in an appendix.

11.3.5 Select the Media

Many options are available for the dissemination of project results. In addition to the impact study report, commonly used media are meetings, interim and progress reports organization publications and case studies. Table 11.3 lists a variety of options to develop the content and the message.

11.3.5.1 Meetings

If used properly, meetings are fertile ground for the communication of project results. All organizations hold a variety of meetings and some may provide the proper context to convey project results. Staff meetings are held to review progress, discuss current problems and

Table 11.3 Options for Communicating Results.

Detailed Reports	Brief Reports	Electronic Reporting	Mass Publications
Impact study	Executive summary	Website	Announcements
Case study (internal)	Slide overview	E-mail	Bulletins
Case study (external)	One-page summary	Blog	Newsletters
Major articles	Brochure	Video	Brief articles

distribute information. These meetings can be an excellent forum for discussing project results. Project results can be sent to executives for use in a staff meeting or a member of the project team can attend the meeting to make the presentation.

Regular meetings with management groups are a common practice. Typically, discussions will focus on items that might be of help to work units. The discussion of a project and its results can be integrated into the regular meeting format. A few organizations have initiated the use of periodic meetings for all key stakeholders, in which the project leader reviews progress and discusses next steps. A few highlights from interim project results can be helpful in building interest, commitment and support for the project.

11.3.5.2 Interim and Progress Reports

A highly visible way to communicate results, although usually limited to large projects, is the use of interim and routine memos and reports. Published or disseminated by e-mail on a periodic basis, they are designed to inform management about the status of the project, to communicate interim results of the project and to spur needed changes and improvements.

A second reason for the interim report is to enlist additional support and commitment from the management group and to keep the project intact. This report is produced by the project team and distributed to a select group of stakeholders in the organization. The report may vary considerably in format and scope and may include a schedule of planned steps or activities, a brief summary of reaction evaluations, initial results achieved from the project and various spotlights recognizing team members or participants. Other topics may also be appropriate. When produced in a professional manner, the interim report can boost management support and commitment.

11.3.5.3 Routine Communication Channels

To reach a wide audience, the project leader can use internal, routine publications. Whether a newsletter, magazine, newspaper or electronic file, these media usually reach all employees or stakeholders. The content can have a significant impact if communicated appropriately. The scope should be limited to general-interest articles, announcements and interviews. Project results communicated must be important enough to arouse general interest.

11.3.5.4 *E-Mail and Electronic Media*

Internal and external Internet pages, company-wide intranets and e-mails are excellent vehicles for releasing results, promoting ideas and informing employees and other target groups of project results. E-mail, in particular, provides a virtually instantaneous means of communicating results to and soliciting responses from large groups of people. For major projects, some organizations create blogs to present results and solicit reactions, feedback and suggestions.

11.3.5.5 *Project Brochures and Pamphlets*

A brochure might be appropriate for a project conducted on a continuing basis or for which the audience is large and continuously changing. The brochure should be attractive and present a complete description of the project, with a major section devoted to results obtained with previous participants, if available. Measurable results and reactions from participants or even direct quotes from individuals, can add spice to what may otherwise be perceived as a dull brochure.

11.3.5.6 *Case Studies*

Case studies represent an effective way to communicate the results of an EHS project. A typical case study describes the situation, provides appropriate background information (including the events that led to the project), presents the techniques and strategies used to develop the study and highlights the key issues in the project. Case studies tell an interesting story of how the project was implemented and how the evaluation was developed, including the problems and concerns identified along the way.

11.3.6 Present the Results

With a plan in place and appropriate media at hand, it is now time to deliver the results. Through the development of content appropriate for the media, the purpose of the communication can be achieved. Perhaps one of the most challenging and stressful types of communication involves presenting an impact study to the senior management team. This group also serves as the client for a project and to convince this highly skeptical and critical group that outstanding results have been achieved (assuming that they have) in a reasonable time frame, addressing the salient points and making sure the managers understand the process can be difficult. Two potential reactions can create problems. First, if the results are impressive, making the managers accept the data may be tough. On the other extreme, if the data are negative, making sure that managers do not overreact to the results and look for someone to blame is important.

Arrange a face-to-face meeting with senior team members to review the results. If they are unfamiliar with the ROI Methodology, this meeting is necessary to make sure they understand the process. The good news is that they will probably attend the meeting because they have never seen ROI data developed for this type of project. The bad news is that it takes precious executive time, usually about an hour, for this presentation. After the meeting, an executive summary may suffice. At this point, the senior members will understand the process, so a shortened version may be appropriate. When a particular audience is familiar with the process, a brief version may be developed, including a one- to two-page summary with charts and graphs showing the six types of measures.

The results should not be disseminated before the initial presentation or even during the session, but should be saved until the end of the session. This will allow enough time to present the process and collect reactions to it before the target audience sees the ROI calculation. Present the ROI Methodology step by step, showing how the data were collected, when they were collected, who provided them, how the effect of the project was isolated from other influences and how data were converted to monetary values. Present the various assumptions, adjustments and conservative approaches along with the total cost of the project, so that the target audience will begin to buy into the process of developing the ROI.

When the data are actually presented, give the results one level at a time, starting with Level 1, moving through Level 5 and ending with the intangibles. This allows the audience to observe the reaction, learning, application and implementation, impact and ROI procedures. After some discussion of the meaning of the ROI, present the intangible measures. Allocate time for each level as appropriate for the audience. This helps to defuse potential emotional reactions to a positive or negative ROI.

Show the consequences of additional accuracy if this is an issue. The tradeoff for more accuracy and validity often is more expense.

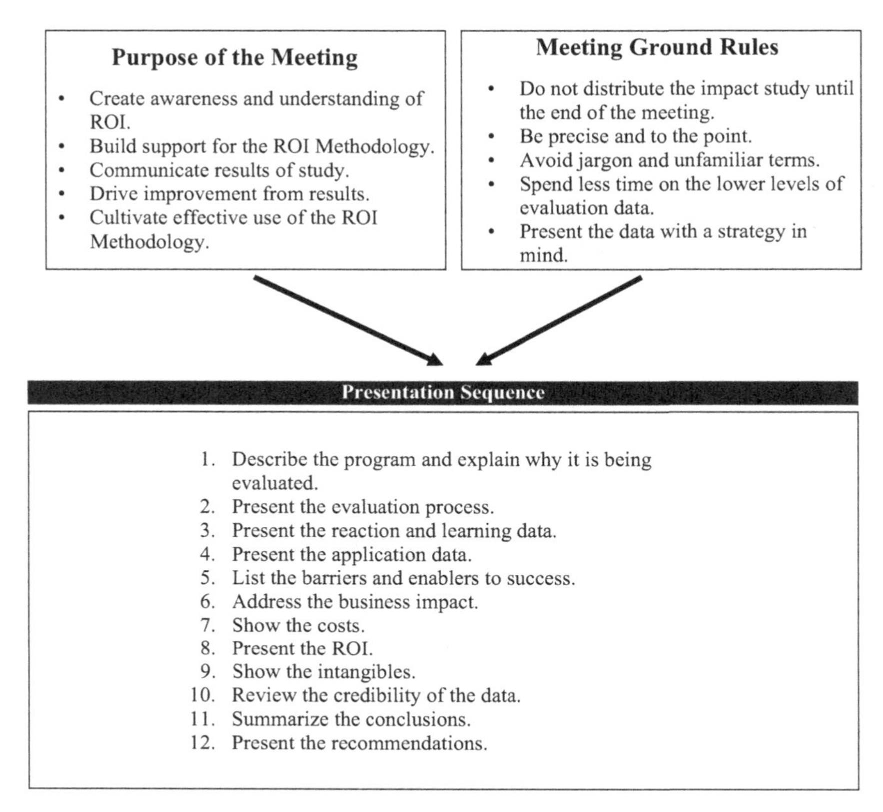

Figure 11.1 Presenting the Impact/ROI Study to Executive Sponsors.

Address this issue when necessary, agreeing to add more data if they are required. Collect concerns, reactions and issues involving the process and make adjustments accordingly for the next presentation.

Collectively, these steps will help in the preparation and presentation of one of the most important meetings in the ROI Methodology. Figure 11.1 shows the recommended approach to an important meeting with the sponsor.

11.3.7 Analyze Reactions

The best indicator of how effectively the results of an EHS project have been communicated is the level of commitment and support from the managers, executives and sponsors. The allocation of requested resources and voiced commitment from top management are strong evidence of the management's positive perception of the results. In addition to this macro-level reaction, a few techniques can also be helpful in measuring the effectiveness of the communication effort.

11.3.7.1 Monitor Reactions

When results are communicated, monitor the reactions of the target audiences. These reactions may include nonverbal gestures oral remarks, written comments or indirect actions that reveal how the communication was received. Usually, when results are presented in a meeting, the presenter will have some indication of how they were received by the group and the interest and attitudes of the audience can be quickly evaluated. Comments about the results, formal or informal, should be noted and tabulated.

11.3.7.2 Discuss Reactions

Project team meetings provide an excellent arena for discussing the reaction to communicated results. Comments can come from many sources depending on the particular target audience. When major project results are communicated, a feedback questionnaire may be administered to the entire audience or a sample of the audience to determine the extent to which the audience understood and/or believed the information presented. This is practical only when the effectiveness of the communication will have a significant impact on future actions by the project team.

11.4 The EHS Scorecard

Reporting results of an EHS project or sustainability initiative requires a strategy. A detailed report provides a historical look at the project along with the evaluation results. This history serves as backup for questions that arise as well as the basis for future evaluations. Executive summaries, human interest stories and presentations around key measures are important types of reports that engage audiences at all levels of the organization. Sometimes, however, it is important to communicate project success in a simple format that includes only the critical data. This format, often referred to as a *scorecard,* summarizes the key measures and shows how the project fared in achieving objectives beyond the activity surrounding it.

11.4.1 The Importance of a Scorecard

As mentioned throughout this book organizations are moving away from a focus on activity to one on results, which are defined based on a five-level framework. Progressive organizations and communities place great emphasis on results-focused initiatives, including those initiatives that support a greater good, yet if the projects appear to be more costly than beneficial, they are often scraped and either a new approach is taken or nothing is done at all.

A scorecard provides clear evidence of the contribution made by an EHS project. While all projects require resource inputs, contribution is defined in terms of the following:

- The reaction of employees to the project
- The knowledge and skills needed to implement the project
- The new behaviors and actions being taken after project implementation
- The business impact resulting as a consequence of the project
- The financial return on investment
- The intangible benefits, which are not converted to money

The green scorecard primarily provides information to the sponsor group, including top executives. However, it also provides useful measures for the project team. The scorecard provides a direct linkage between the investment in EHS initiatives and the organizational strategy.

11.4.2 Macro-Level Scorecards

Most evaluation processes concentrate on micro-level activities, evaluating one project at a time. The ROI Methodology presented in this book does the same. The final output of any one of these evaluations is a detailed report with all the peripheral summary reports and presentations, including a micro-level summary or scorecard for that specific project. But for most organizations, multiple EHS initiatives are under way, ranging from the very simple to the very complex. As mentioned earlier and discussed in more detail in the next chapter, most EHS projects should not be evaluated at the ROI level, which is intended for the expensive, high-profile projects. Yet there is a need to connect projects not evaluated to ROI to strategy as well. This is accomplished by integrating the data from all evaluations in a meaningful way to show the overall contribution of EHS investments. In essence, this process takes a micro-level activity (evaluation of a specific EHS initiative) and presents a macro-level view (evaluation of all EHS initiatives).

Figure 11.2 illustrates the concept. Though each program is evaluated on the micro-level, only a few selected measures in each of the micro-evaluations are captured for the macro-level evaluation. It takes the most critical, important and executive-friendly measures to go on the EHS scorecard.

11.4.3 Measures to Include in the Scorecard

Figure 11.3 shows an outline of a comprehensive EHS scorecard. As shown, eight categories of data are included.

11.4.3.1 Inputs and Indicators

The traditional approach to measuring project success is to report the inputs or activities. These measures are important in that they represent the organization's commitment to green and sustainability projects, efficiencies and trends in processes. They do not, however, represent results of the project. The number of inputs is vast, so it is important to include in the scorecard measures important to top managers. Ideally, the management group should approve

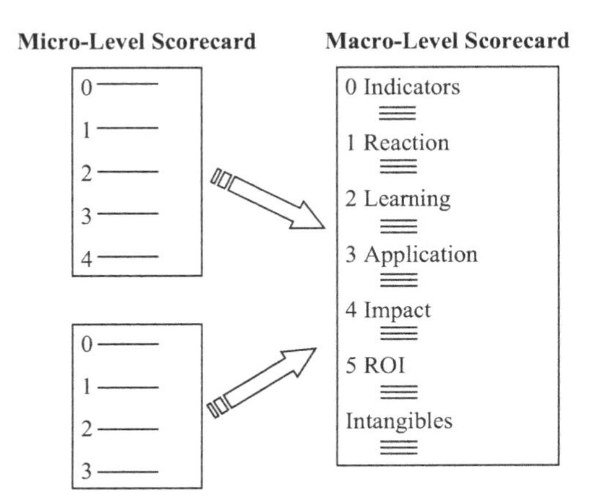

Figure 11.2 Micro-Level Versus Macro-Level Scorecard.

0. **Indicators**
    1. Number of Employees Involved
    2. Total Hours of Involvement
    3. Hours Per Employee
    4. Project Investment as a Percent of Payroll
    5. Cost Per Participant

I. **Reaction**
    1. Percent of Projects Evaluated at This Level
    2. Ratings on 4 Items vs. Target
    3. Percent with Action Plans
    4. Percent with ROI Forecast

II. **Learning**
    1. Percent of Projects Evaluated at This Level
    2. Types of Measurements
    3. Self-Assessment Ratings on 3 Items vs. Targets

III. **Application and Implementation**
    1. Percent of Projects Evaluated at This Level
    2. Ratings on 3 Items vs. Targets
    3. Percent of Action Plans Complete
    4. Barriers (List of Top 10)
    5. Enablers (List of Top 10)
    6. Management Support Profile

IV. **Impact**
    1. Percentage of Projects Evaluated at This Level
    2. Linkage with Measures (List of Top 10)
    3. Linkage with EHS Measures
    4. Types of Measurement Techniques
    5. Types of Methods to Isolate the Effects of Projects
    6. Investment Perception

V. **ROI**
    1. Percent of Projects Evaluated at This Level
    2. ROI Summary for Each Study
    3. Methods of Converting Data to Monetary Values
    4. Fully Loaded Cost per Participant

Intangibles
    1. List of Intangibles (Top 10)
    2. How Intangibles Were Captured

Awards and Recognition
    1. Industry Awards
    2. EHS Awards

Figure 11.3 Sample Outline of Reporting for an EHS Function.

the indicators and the inputs reported should stimulate interest with executives. A few possible measures to report in this category include the following:

- The number of employees participating in EHS projects
- The number of hours employees engage in EHS projects
- Various statistics, such as demographics and completion rates
- Financial investment in EHS projects (total costs, cost per employee, etc.)
- Cost recovery, if there is a financial benefit for implementing EHS projects (tax breaks, grants, etc.)

11.4.3.2 *Reaction*

There are many measures that can be taken at this level of evaluation. The first measure to report is the percentage of projects evaluated at this level. Typically, this will be 100 percent of the projects. This measure provides senior management with a picture of the investment in the evaluation of green projects. When reporting Level 1 results, it is important to capture other measures that provide some indication that success will be achieved through the EHS project. Some measures to consider are listed below:

- Relevance to the job
- Amount of new information
- Recommendation to others
- Importance of the information
- Intention to apply knowledge, skill or information

These measures have been shown to have a statistical correlation with measures of application (Level 3).

Another potential measure in this category is the percentage of project participants with actions plans. This can become an important measure, since the action plans will drive specific steps participants will take to make the project successful.

11.4.3.3 *Learning*

This level of evaluation, as described earlier, indicates the extent to which project participants are able to proceed with the project given the information they have received or the knowledge and skill they have acquired. Every project has a learning component. How and the extent to which that component is measured depends on the level of learning required to make the project successful. Simple projects, like changing from traditional lighting to LED lighting, do not require a lot of new information. But a major safety improvement initiative targeting a variety of types of safety measures will require knowledge and sometimes skills, action and impact. For these types of projects, evaluation at Level 2 (learning) will occur. Given these issues, the first measure to report in this category is the percentage of projects evaluated at this level.

Because some learning components will require comprehensive Level 2 evaluation and others will require less formal methods, it is sometimes difficult to compare results at this level across all programs. So, to make it simple to roll up Level 2 measures across projects, it may be useful to develop a set of common measures to determine the extent to which participants acquired the knowledge they need to make the project successful. Measures may include self-reports of learning change, such as those listed below:

- Acquisition of knowledge, skill or information
- Enhanced awareness of need for project
- Change in attitude toward EHS efforts
- Ability to apply knowledge, skill or information
- Confidence to use knowledge, skill or information

Using this type of measurement scheme, it is important that the same scale (e.g., a 1–5 Likert scale) be used for every Level 2 evaluation.

11.4.3.4 Application and Implementation

To measure the actual change in behavior and progress with application, measurements must be taken after a project is launched and implemented. The first measure reported on the scorecard at Level 3 is percentage of projects evaluated at this level. Again, by reporting this measure, leaders will have an indication of the investment being made in evaluation. Follow-up measures for all projects may include these measures:

- The extent of use of knowledge, skill or information
- The frequency of use of knowledge, skill or information
- The effectiveness of use of knowledge, skill or information

Because application and implementation of EHS projects vary based on the focus of the EHS project, it is important to capture measures that can be summarized and integrated across all EHS projects. These three measures make that possible. Essentially, these items can be collected for every project to show the success in changing behaviors or applying concepts associated with the EHS project.

Another important set of measures to report in the macro-level green scorecard identifies barriers and enablers. Recognizing and reporting these inhibitors to and supporters of project success are important because they explain why progress is or is not being made. Adding them to the macro-level scorecard is an easy way to show senior managers how the system is supporting EHS projects.

11.4.3.5 *Impact*

Success at this level is reported for only a few EHS projects. So again, the first measure to report here is the percentage of projects evaluated at this level. In addition, any linkage of the project to EHS impact measures is also reported. Typically, for macro-level scorecard reporting, this linkage is measured through the use of questionnaires, with the top two high scores reported as a composite score. By listing the top ten measures that improved due to a project, it is easy to show how closely aligned EHS projects are to strategy.

Another important measure reported at this level is the method used to isolate the effects of the project. As presented in Chapter 8, a variety of methods are available. Some organizations are moving to more research and analytical methods, while others are using subjective estimations to save time and costs. Either way, reporting the techniques employed in the evaluation of green projects shows senior managers that efforts are under way to make the clearest possible connection between projects and results.

11.4.3.6 *Return on Investment*

The ultimate level of evaluation is the actual return on investment. The first measure to report in this category of results is the percentage of projects evaluated at this level. Because the ROI, BCR and payback period are self-explanatory to senior leaders and administrators, little additional information is needed on the scorecard. Still, it is sometimes helpful to report data conversion methods used so the audience understands how the monetary value is developed.

11.4.3.7 *Intangible Benefits*

It is important to include the intangible benefits on the macro-level scorecard. These measures may also be included in the impact category, but it is important to highlight them here. The top ten intangibles are sometimes reported, providing an opportunity to check alignment with organizational strategy.

11.4.3.8 Awards

The extent to which an organization or community is recognized by professional associations, the community and other entities is often important to senior leaders. These types of recognitions make employees feel proud of their organization, help with recruiting and often help economic development. Include awards on the green scorecard so senior managers can see that the awards being pursued (and won) are in alignment with the strategy of the organization.

11.4.4 Getting Started

EHS project success is reported in a variety of ways. Individual project evaluations are described in detailed reports, summaries, newsletters and human interest stories. Key results of the individual projects are also summarized in micro-level scorecards. But because a vast number of projects are often initiated, it is important to integrate results into a macro-level scorecard, which presents the success of all green initiatives and provides evidence of a link between activity and strategy.

To get started developing a macro-level scorecard, follow these simple guidelines:

1. *Start the scorecard development with the ultimate desired outcomes in mind.* In Chapter 5, we described a process to align green projects with strategy. While implementing the alignment process, focus on the payoff needs, business needs, performance needs and learning needs. Use these data as a basis for your scorecard measures.
2. *Start simple.* The initial scorecard can be brief, showing results of only the key measures. As the scorecard evolves, additional measures can be incorporated.
3. *Limit the focus on activity.* While you will certainly want to capture some input data, limit the focus on activity and gives stakeholders the data they want: results!

11.5 Final Thoughts

The final step in the ROI Methodology, reporting results, is a crucial step in the overall evaluation process. If this step is not executed adequately, the full impact of the results will not be recognized and the study may amount to a waste of time. The principles and steps shown in this chapter for communicating EHS project results can serve as a guide for any significant communication effort. Various target audiences should be identified, with an emphasis on the executive group because of its importance and the suggested format for a detailed evaluation report can be followed. An EHS scorecard is an excellent way to present the success of all EHS projects. The final issue in developing a successful approach to accounting for green investments is developing a strategy for implementing and sustaining the process. This issue is covered in the final chapter, which follows.

12

Implementing and Sustaining ROI

Abstract

This final chapter examines the challenge of making this type of analysis routine and consistent. Building on many successful examples of the use of the ROI Methodology for more than a decade, the chapter details the building blocks to overcome resistance to this analysis. Initially, there will be resistance within the organization and even within the EHS team. However, there is little resistance at the top of the organization, where funding for EHS is initiated—they want to see this type of data. Over a dozen techniques and approaches are available to ensure that proper capability is developed, procedures are in place, responsibilities are assigned, projects are conducted and information is routinely distributed to appropriate organizational audiences both internally and externally. This includes our recommended review of the process from time to time to make adjustments along the way.

To be effective, even the best designed process or model must be integrated into the organization. Often, resistance to the ROI Methodology arises. Some of this resistance is based on fear and misunderstanding; some is based on actual barriers and obstacles. Although the ROI Methodology presented in this book is a step-by-step, methodical and simplistic procedure, it can fail if it is not integrated properly. This integration includes full acceptance and support by those who must make it work within the organization. This chapter focuses on some of the most effective means of overcoming resistance to implementing the ROI process.

Keywords: Resistance to analysis, funding, fear, misunderstanding, integration, ROI methodology implementation

---

Jack Phillips, Patti Phillips, and Al Pulliam, Measuring ROI in Environment, Health, and Safety, (271–290) 2014 © Scrivener Publishing LLC

12.1 The Concern About Implementing and Sustaining ROI

With any new process or change, there is resistance, which may be especially great when implementing a process as complex as ROI. To implement ROI and sustain it as an important accountability tool, the resistance must be minimized or removed. Four key factors serve as the basis for developing a detailed plan to sustaining the use of ROI:

- Resistance
- Implementation
- Consistency
- Efficiency

12.1.1 Resistance is Inevitable

Resistance to change is a constant. Sometimes there are good reasons for resistance, but often it exists for the wrong reasons. The important point is to sort out both kinds of resistance and try to dispel any myths. When legitimate barriers are the basis for resistance, minimizing or removing them altogether is the challenge.

12.1.2 Implementation is Key

As with any process, effective implementation is the key to its success. This occurs when the new technique, tool or process is integrated into the routine framework. Without effective implementation, even the best process will fail. A process that is never removed from the shelf will never be understood, supported or improved. Clear steps must be in place for designing a comprehensive implementation process that will overcome resistance.

12.1.3 Consistency is Necessary

Consistency is an important consideration as the ROI Methodology is implemented. With consistency come accuracy and reliability and accountability. The only way to make sure consistency is achieved is to follow clearly defined processes, procedures and standards each time the ROI Methodology is used. Proper, effective implementation will ensure that this occurs.

12.1.4 Efficiency is Essential

Cost control and efficiency are important considerations in any major undertaking and the ROI Methodology is no exception. During implementation, tasks must be completed efficiently and effectively. Doing so will help ensure that process costs are kept to a minimum, that time is used economically and that the process remains affordable.

12.2 Implementing the Process: Overcoming Resistance

Resistance shows up in the form of comments, remarks, actions or behaviors. Table 12.1 lists representative comments that indicate open resistance to the ROI process. Each comment signals an issue that must be resolved or addressed in some way. A few are based on realistic barriers, whereas others are based on myths that must be dispelled. Sometimes resistance to the process reflects underlying concerns. For example, the individuals involved may fear losing control of their processes and others may feel vulnerable to whatever action may follow if the process is not successful. Still others may be concerned about any process that brings change or requires the additional effort of learning.

EHS project team members may resist the ROI process and openly make comments similar to those listed in Table 12.1. It may take heavy persuasion and evidence of tangible benefits to convince team members that it is in their best interest to make the project a success. Although most clients do want to see the results of the project, they may have concerns about the information they are asked to provide and about whether their personal performance is being judged while the project is undergoing evaluation. Participants may express the same fears listed in the table.

The challenge is to implement the process systematically and consistently so that it becomes normal business behavior and a routine and standard process that is built into projects. The implementation necessary to overcome resistance covers a variety of areas. Figure 12.1 shows actions outlined in this chapter that are presented as building blocks to help overcome resistance. They are all necessary to build the proper base or framework to dispel myths

Table 12.1 Typical Objections to Use of ROI Methodology.

Open Resistance
1. It costs too much.
2. It takes too much time.
3. Who is asking for this?
4. This is not in my job description.
5. I did not have input on this.
6. I do not understand this.
7. What happens when the results are negative?
8. How can we be consistent with this?
9. The ROI looks too subjective.
10. Our managers will not support this.
11. ROI is too narrowly focused.
12. This is not practical.

and remove or minimize barriers. The remainder of this chapter presents specific strategies and techniques devoted to each of the nine building blocks identified in Figure 12.1. They apply equally to the project team and the client organization. In some situations, a particular strategy will work best with the project team. In certain cases, all strategies may be appropriate for both groups.

12.3 Assessing the Climate

As a first step toward implementation, some organizations assess the current climate for achieving results. One way to do this is to develop a survey to determine the current perspectives of the management team and other stakeholders. Another way is to conduct interviews with key stakeholders to determine their willingness to follow the project through to ROI. With an awareness of the current status, the project leaders can plan for significant changes and pinpoint particular issues that need support as the ROI process is implemented.

12.4 Developing Roles and Responsibilities

Defining and detailing specific roles and responsibilities for different groups and individuals addresses many of the resistance factors and helps pave a smooth path for implementation.

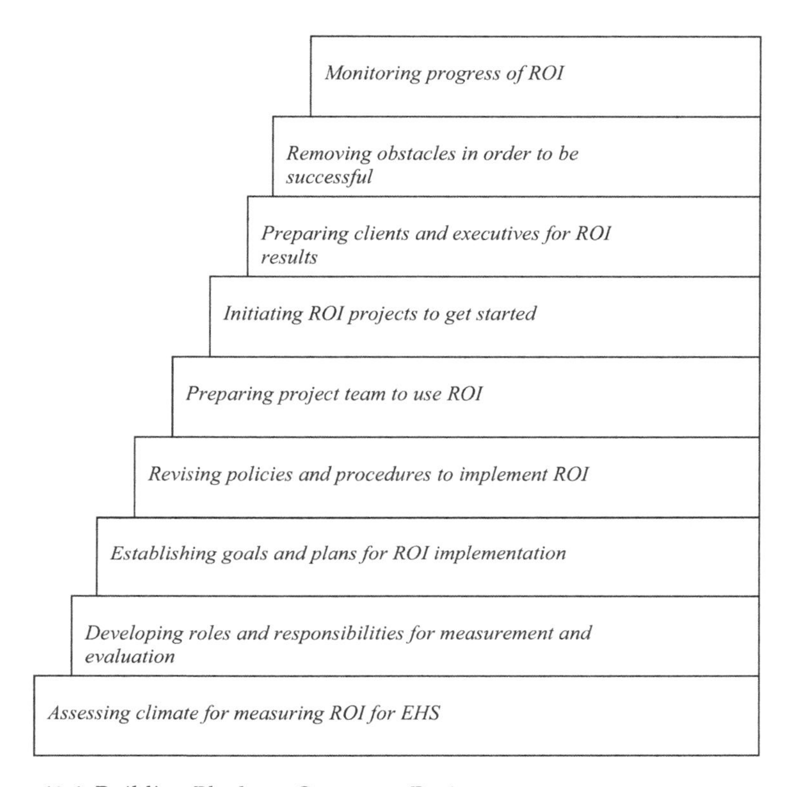

Figure 12.1 Building Blocks to Overcome Resistance.

12.4.1 Identifying a Champion

As an early step in the process, one or more individuals should be designated as the internal leader or champion for the ROI Methodology. As in most change efforts, someone must take responsibility for ensuring that the process is implemented successfully. This leader serves as a champion for ROI and is usually the one who understands the process best and sees vast potential for its contribution. More important, this leader is willing to teach others and will work to sustain sponsorship.

12.4.2 Developing the ROI Leader

The ROI leader is usually a member of the project team who has the responsibility for evaluation. This person holds a full-time position in larger project teams or a part-time position in smaller teams. Sponsor organizations may also have an ROI leader who pursues the ROI Methodology from the sponsor's perspective. The typical job title for a full-time ROI leader is manager of measurement and evaluation. Some organizations assign this responsibility to a team and empower it to lead the ROI effort.

In preparation for this assignment, individuals usually receive special training that builds specific skills and knowledge of the ROI process. The role of the implementation leader is broad and serves a variety of specialized duties. In some organizations, the implementation leader can take on many roles, ranging from problem solver to communicator to cheerleader. Leading the ROI process is a difficult and challenging assignment that requires unique skills. Some programs are designed to certify individuals who will be assuming leadership roles in the implementation of the ROI Methodology. This certification is built around ten specific skill sets linked to successful ROI implementation, focusing on the critical areas of collecting data, isolating the effects of the project, converting data to monetary value, presenting evaluation data and building capability. This process is comprehensive but may be necessary to build the skills necessary for taking on this challenging assignment.

12.4.3 Establishing a Task Force

Making the ROI Methodology work well may require the use of a task force, which usually comprises a group of individuals from different parts of the project or client team who are wiling to develop the ROI Methodology and implement it in the organization. The selection of the task force may involve asking for volunteers or participation may be mandatory depending on specific job responsibilities. The task force should represent the cross section necessary for accomplishing stated goals. Task forces have the additional advantage of bringing more people into the process and developing more ownership of and support for the ROI Methodology. The task force must be large enough to cover the key areas but not so large that it becomes too cumbersome to function. Six to twelve members is a good size.

12.4.4 Assigning Responsibilities

Determining specific responsibilities is critical because confusion can arise when individuals are unclear about their specific assignments in the use of the ROI Methodology. Responsibilities apply to two areas. The first is the measurement and evaluation responsibility of the entire project team. Everyone involved in projects must have some responsibility for measurement and evaluation, which includes providing input on designing instruments, planning specific evaluations, analyzing data and interpreting the results. Typical responsibilities include the following:

- Ensuring that the initial analysis for the project includes specific business impact measures
- Developing specific application and business impact objectives for the project
- Keeping participants focused on application and impact objectives
- Communicating rationale and reasons for evaluation
- Assisting in follow-up activities to capture application and business impact data
- Providing assistance for data collection, data analysis and reporting

Although involving each member of the project team in all of these activities may not be appropriate, each individual should have at least one responsibility as part of his or her routine job duties. This assignment of responsibility keeps the ROI Methodology from becoming disjointed and separated during EHS projects. More important, it brings accountability to those directly involved in project implementation.

Another issue involves technical support. Depending on the size of the project team, establishing an individual or a group of technical experts to provide assistance with the ROI Methodology may be helpful. When the group is established, the project team must understand that the experts have been assigned not for the purpose of relieving the team of its evaluation responsibilities, but to supplement its ROI efforts with technical expertise. Technical experts are typically the individuals who participated in the certification and training process to build special skills. Responsibilities of the technical support group include six elements:

1. Designing data collection instruments
2. Providing assistance for developing an evaluation strategy
3. Analyzing data, including specialized statistical analyses
4. Interpreting results and making specific recommendations

5. Developing an evaluation report or case study to communicate overall results
6. Providing technical support in all phases of the ROI Methodology

Although the project team must be assigned specific responsibilities during an evaluation, requiring others to serve in support functions to help with data collection is not unusual. These responsibilities are defined when a particular evaluation strategy plan is developed and approved.

12.5 Establishing Goals and Plans

Establishing goals, targets and objectives is critical to the implementation, particularly when several projects are planned. The establishment of goals can include detailed planning documents for the overall process and for individual ROI projects.

Establishing specific targets for evaluation levels is an important goal-setting strategy. Not every green initiative should be evaluated to ROI. Knowing in advance to which level a project will be evaluated helps in planning which measures will be needed and how detailed the evaluation must be. Table 12.2 presents examples of targets set for evaluation at each level when different types of projects exist.

Another important part is establishing a timetable for the complete integration of the ROI Methodology. This document becomes a master plan for completion of the different elements presented earlier. Beginning with forming a team and concluding with meeting the targets previously described, this schedule presents a project plan for transitioning from the current situation to the desired future situation. Items on the schedule include developing specific ROI projects, building staff skills, developing policy and teaching managers the process. Figure 12.2 is an example of an implementation plan. The more detailed the document, the more useful it becomes. The project plan is a living, long-range document that should be reviewed frequently and adjusted as necessary. More important, those engaged in work on the ROI Methodology should always be familiar with the implementation plan.

Table 12.2 Evaluation Targets in a Large Organization with Many EHS Projects.

Level	Target
Level 1, Reaction	100%
Level 2, Learning	80%
Level 3, Application and Implementation	40%
Level 4, Business Impact	25%
Level 5, ROI	10%

	Month																			
	J	F	M	A	M	J	J	A	S	O	N	D	J	F	M	A	M	J	J	A
Team Formed	■																			
Responsibilities Defined		■																		
Policy Developed	■	■																		
Targets Set	■		■																	
Workshops Developed				■																
ROI Project (A) Complete																■				
ROI Project (B) Complete																	■			
ROI Project (C) Complete																		■		
ROI Project (D) Complete																			■	
Project Teams Trained								■												
Managers Trained																				■
Support Tools Developed					■															
Guidelines Developed										■										

Figure 12.2 Implementation Plan for a Large Organization with Many EHS Projects.

12.6 Revising or Developing Policies and Guidelines

Another building block to overcoming resistance to the ROI process is revising or developing the organization's policy on project measurement and evaluation. The policy statement contains information developed specifically for the measurement and evaluation process, using input from the project team and key managers or stakeholders. Sometimes, policy issues are addressed during internal workshops designed to build measurement and evaluation skills. The policy statement addresses critical matters that will influence the effectiveness of the measurement and evaluation process. These may include adopting the framework presented in this book, requiring objectives at all levels for some or all projects and defining responsibilities for the project team.

Policy statements are important because they provide guidance and direction for the staff and others who work closely with the ROI Methodology, keeping the process clearly focused and enabling the group to establish goals for evaluation. Policy statements also provide an opportunity to communicate basic requirements and fundamentals of performance and accountability. More than anything else, they serve as learning tools to teach others, especially when they are developed in a collaborative way. If policy statements are developed in isolation, staff and management will be denied a sense of ownership, which makes them neither effective nor useful.

Guidelines for measurement and evaluation are important for showing how to use the tools and techniques, guide the design process, provide consistency in the ROI process, ensure that appropriate methods are used and place the proper emphasis on each of the areas. The guidelines are more technical than policy statements and often include detailed procedures that show how the process is undertaken and developed. They often include specific forms, instruments and tools necessary to facilitate the process.

12.7 Preparing the Project Team

Project team members may resist the ROI Methodology. They often see evaluation as an unnecessary intrusion into their responsibilities that absorbs precious time and stifles creative freedom. The cartoon character Pogo perhaps characterized it best when he said, "We have met the enemy and he is us." Several issues must be addressed when preparing the project team for ROI implementation.

12.7.1 Involving the Project Team

For each key issue or major decision involving ROI implementation, the project team should be involved in the process. As policy statements are prepared and evaluation guidelines developed, team input is essential. Resistance is more difficult if the team helped design and develop the ROI process. Convene meetings, brainstorming sessions and task forces to involve the team in every phase of developing the framework and supporting documents for ROI.

12.7.2 Using ROI as a Learning and Project Improvement Tool

One reason the project team may resist the ROI process is that the effectiveness of the project will be fully exposed, putting the reputation of the team on the line. They may have a fear of failure. To overcome this, the ROI Methodology should be clearly positioned as process improvement, not as an individual performance evaluation tool (at least not during the early years of use). Team members will not be interested in developing a process that may reflect unfavorably on their performance.

Evaluators can learn as much from unsuccessful projects as they can from successful projects. If the project is not working, it is best to find out quickly so that issues can be understood firsthand, not from others. If a project is ineffective and it is not producing the desired results, the failure will eventually be known to the sponsors and the management group (if they are not aware of it already). A lack of results will make managers less supportive of immediate and future projects. However, if the project's weaknesses are identified and adjustments are quickly made, not only can more effective projects be developed, but the credibility of and respect for project implementation will be enhanced.

12.7.3 Teaching the Team

The project team and project evaluator often have minimal skills in measurement and evaluation, so they will need to develop some expertise. Consequently, the project team leader must learn the ROI Methodology and its systematic steps. The evaluator must learn to develop an evaluation strategy and specific plan, to collect and analyze data from the evaluation and to interpret results from data analysis. A one- to two-day workshop can help evaluators to build the skills and knowledge needed to understand the process and to appreciate what it can do for project success and for the sponsoring organization. Teach-the-team workshops can be valuable tools in ensuring successful implementation of ROI Methodology.

12.8 Initiating ROI Projects

The first tangible evidence of the value of using the ROI Methodology arises from the initiation of the first project for which an ROI calculation is planned. Identifying appropriate projects and keeping then on track is critical to successful ROI sustainability.

12.8.1 Selecting the Initial Project

Projects that qualify for comprehensive, detailed ROI analysis are those that have the following qualities:

1. Are important to strategic objectives
2. Involve large groups of participants
3. Will be linked to major operational problems and opportunities upon completion
4. Are expensive
5. Are time-consuming
6. Have high visibility
7. Have the interest of management in performing their evaluation

Using these or similar criteria, the project leader must select the appropriate projects to consider for ROI evaluation. Ideally, sponsors should agree with or approve the criteria.

12.8.2 Developing the Planning Documents

As described earlier, perhaps the two most useful ROI documents are the data collection plan and the ROI analysis plan. The data collection plan shows what data will be collected, the methods used, the sources, the timing and the assignment of responsibilities. The ROI analysis plan shows how specific analyses will be conducted, including how to isolate the effects of the project and how to convert data to monetary values. Each evaluator should know how to develop these plans.

12.8.3 Reporting Progress

As the projects are developed and the ROI implementation gets under way, status meetings should be conducted to report progress and discuss critical issues with appropriate team members. These meetings keep the project team focused on the critical issues, generate the best ideas for addressing problems and barriers and build a knowledge base for better evaluation of future projects. Sometimes, an external consultant, perhaps an expert in the ROI process, facilitates these meetings. In other cases, the project leader may facilitate. In essence, the meetings serve three major purposes: reporting progress, learning and planning.

12.8.4 Establishing Discussion Groups

Because the ROI Methodology is considered difficult to understand and apply, establishing discussion groups to teach the process may be helpful. These groups can supplement formal workshops and other learning activities and they are often flexible in format. An external ROI consultant or the project leader usually facilitates these groups. In each session, a new topic is presented for discussion, which should extend to how the topic applies to the organization. The process can be adjusted for different topics as new group needs arise, driving the issues. Ideally, participants in group discussions will have an opportunity to apply, explore or research the topics between sessions. Group assignments such as reviewing a case study or reading an article are appropriate between sessions to further the development of knowledge and skills associated with the process.

12.9 Preparing Sponsors and Management Team

Perhaps no group is more important to the use of the ROI Methodology than the management team that must allocate resources for the project and support its implementation. In addition, the management team often provides input to and assistance for the ROI Methodology. The preparation, training and development of the management team should be carefully planned and executed.

One effective approach for preparing executives and managers for ROI is to conduct a briefing. Varying in duration from one hour to half a day, a practical briefing such as this can provide critical information and enhance support for ROI use. Managers leave these briefings with greater appreciation of the use of ROI and its potential impact on projects and with a clearer understanding of their role in the ROI process. More important, they often renew their commitment to react to and use the data collected by the ROI Methodology.

A strong, dynamic relationship between the project team and key managers is essential for successful implementation of the ROI Methodology. A productive partnership is needed that requires each party to understand the concerns, problems and opportunities of the other. The development of such a beneficial relationship is a long-term process that must be deliberately planned for and initiated by key project team members. The decision to commit resources and support to a project may be based on the effectiveness of this relationship.

12.10 Removing Obstacles

As the ROI Methodology is implemented, there will inevitably be obstacles to progress. The obstacles are usually based on the concerns discussed in this chapter, which may be valid or based on unrealistic fears or misunderstandings.

12.10.1 Dispelling Myths

As part of the implementation, attempts should be made to dispel the myths and remove or minimize the barriers or obstacles. Much of the controversy regarding ROI stems from misunderstandings about what the process can and cannot do and how it can or should be implemented in an organization. After years of experience with ROI and having noted reactions during hundreds of projects and workshops, we have recognized many misunderstandings about ROI, including the following:

- ROI is too complex for most users.
- ROI is expensive and consumes too many critical resources.
- If senior management does not require ROI, there is no need to pursue it.
- ROI is a passing fad.
- ROI is only one type of data.
- ROI is not future-oriented; it only reflects past performance.
- ROI is rarely used by organizations.
- The ROI Methodology cannot be easily replicated.
- ROI is not a credible process; it is too subjective.
- ROI cannot be used with soft projects.

- Isolating the influence of other factors is not always possible.
- ROI is appropriate only for large organizations.
- No standards exist for the ROI Methodology.

12.10.2 Delivering Bad News

One of the obstacles most difficult to overcome is receiving inadequate, insufficient or disappointing news. Addressing a bad-news situation is an issue for most EHS project leaders and other stakeholders involved in a project. Table 12.3 presents the guidelines to follow when addressing bad news. As the table makes clear, the time to think about bad news is early in the process, using it to recognize that things need to change and that the situation can improve. The team and others may need to be convinced that good news can be found in a bad-news situation.

12.10.3 Using the Data

Too often EHS projects are evaluated and data are collected, but nothing is done with the data. This creates a major obstacle to successful implementation because once the project has concluded, the team has a tendency to move on to the next project or issue and get

Table 12.3 How to Address Bad News.

- Never fail to recognize the power to learn from and improve with a negative study.
- Look for red flags along the way.
- Lower outcome expectations with key stakeholders along the way.
- Look for data everywhere.
- Never alter the standards.
- Remain objective throughout the process.
- Prepare the team for the bad news.
- Consider different scenarios.
- Find out what went wrong.
- Adjust the story line to "Now we have data that show how to make this program more successful," which puts a positive spin on data that are less than positive.
- Drive improvement.

Table 12.4 How Data Should Be Used.

Use of Evaluation Data	Appropriate Level of Data				
	1	2	3	4	5
Adjust project or program design	✓	✓			
Improve implementation			✓	✓	
Influence application and impact			✓	✓	
Improve management support for the project			✓	✓	
Improve stakeholder satisfaction			✓	✓	✓
Recognize and reward participants		✓	✓	✓	
Justify or enhance budget				✓	✓
Reduce costs		✓	✓	✓	✓
Market projects or programs in the future	✓		✓	✓	✓

on with other priorities. Table 12.4 shows how the different levels of data can be used to improve projects.

Failure to use the data may mean that the entire evaluation was a waste of time and resources. As the table illustrates, data can become action items for the team to ensure that changes and adjustments are made.

12.11 Monitoring Progress

A final building block to overcoming resistance is monitoring the overall progress made and communicating that progress. Although often overlooked, an effective progress report can help keep the implementation on target and let others know what the ROI Methodology is accomplishing for project leaders and the client.

The initial schedule for implementation of ROI is based on key events or milestones and routine progress reports should be developed to communicate their status. Reports are usually developed at six-month intervals, but they may be more frequent for short-term projects. Two target audiences—the project team and senior managers—are critical for progress reporting. All project team members should be kept informed of the progress and senior managers should know the extent to which ROI is being implemented and how it is working within the organization.

12.12 Final Thoughts

This chapter explored the implementation of the ROI Methodology and ways to sustain its use. If the ROI process is not approached in a systematic, logical and planned way, it will fail to be integrated as part of the evaluation and accountability of green projects and sustainability initiatives. This chapter presented the different elements that must be considered and issues that must be addressed to ensure that implementation is smooth and uneventful, which is the most effective means of overcoming resistance to ROI. The result provides a complete integration of ROI as a mainstream component of green projects.

References

Alden, Jay. 2006. "Measuring the 'Unmeasurable.'" *Performance Improvement*, May/June.

Anderson, M., Bensch, I. and Pigg. S. 2007. "Measuring ROI in the Better Buildings, Better Business Conference Energy Center of Wisconsin." In *Proving the Value of Meetings and Events*, edited by J. Phillips, M. Myhill and J. McDonough. Dallas: MPI Foundation and Birmingham: ROI Institute, Inc.

Anderson, Ray, with White, Robin. 2009. *Confessions of a Radical Industrialist: Profits, People Purpose—Doing Business by Respecting the Earth.* New York: St. Martin's Press.

"Annual Employee Benefits Report." 2006. *Nation's Business*, January.

Baliey, B. and Dandrade, R. 1995. "Employee Satisfaction + Customer Satisfaction = Sustained Profitability." *Center for Quality Management Journal* 4, no. 3 (Employee Involvement Special Issue, Fall).

Berns, Maurice, Townend andrew, Khayat, Zayna, Balagopal, Balu, Reeves, Martin, Hopkins, Michael and Kruschwitz, Nina. 2009. "The Business of Sustainability: What it Means to Managers Now." *MIT Sloan Management Review*, Fall.

Boulton, Richard E. S., Libert, Barry D. and Samek, Steve M. 2000. *Cracking the Value Code*. New York: HarperBusiness.

Bray, Illona, *Healthy Employees, Healthy Business: Easy, Affordable Ways to Promote Workplace Wellness*, NOLO Publishing: 2010.

Brewer, Clint. 2009. "Green Business." *Media Planet*, July.

Campanella, Jack, ed. 1999. *Principles of Quality Costs*, 3d ed. Milwaukee: American Society for Quality.

Clinton, Bill. 2009. "Creating Value in an Economic Crisis." *Harvard Business Review*, September.

Cokins, Gary. 1996. *Activity-Based Cost Management: Making it Work—A Manager's Guide to Implementing and Sustaining an Effective ABC System*. New York: McGraw-Hill.

Crain, W. Mark and Crain, Nicole V., *The Impact of Regulatory Costs on Small Firms*. Easton: Lafayette College. September 2010.

Esty, Daniel C. and Winston andrew S. 2006. *Green to Gold: How Smart Companies Use Environmental Strategy to Innovate, Create Value and Build Competitive Advantage*. London: Yale University Press.

Frangos, Cassandra A. 2004. "Aligning Learning with Strategy." *Chief Learning Officer*, March.

Friedman, Thomas. 2008. *Hot, Flat and Crowded: Why We Need a Green Revolution— and How It Can Renew America*. New York: Farrar, Straus and Giroux.

"Getting Warmer: A Special Report on Climate Change and the Carbon Economy." 2009. *The Economist*, December.

Friend, Mark A. and Kohn, James P., *Fundamentals of Occupational Safety and Health*. Fifth Edition Government Institutes: 2010.

Graham, Morris, Bishop, Ken and Birdsong, Ron. 1994. "Self-Directed Work Teams." *In Action: Measuring Return on Investment*, vol.1., edited by Jack J. Phillips. Alexandria, VA: American Society for Training and Development.

Hopkins, Michael. 2009. "8 Reasons Sustainability Will Change Management (That You Never Thought of)." *MIT Sloan Management Review*, Fall.

Hurd, Mark and Nyberg, Lars. 2004. *The Value Factor: How Global Leaders Use Information for Growth and Competitive Advantage*. New York: Bloomberg Press.

Jones, Van. 2008. *The Green-Collar Economy: How One Solution Can Fix Our Two Biggest Problems*. New York: HarperOne.

Kaplan, Robert and Norton, David. 1996. *The Balanced Scorecard: Translating Strategy into Action*. Boston: Harvard Business Press.

Kerr, Steve, 1995. "On the Folly of Rewarding A, While Hoping for B." *Academy of Management Journal* 18.

Langdon, Danny G., Whiteside, Kathleen S. and McKenna, Monica M., eds. 1999. *Intervention Resource Guide: 50 Performance Improvement Tools*. San Francisco: Jossey-Bass, Pfeiffer.

Langreth, Robert and Herper, Matthew. 2010. "The Planet Versus Monsanto." *Forbes*, January.

MacKay, David J. 2009. *Sustainable Energy—Without the Hot Air.* Cambridge, UK: UIT

Mayo andrew. 2003. "Measuring Human Capital." *The Institute of Chartered Accountants*, June.

Morse, Gardiner. 2009. "On the Horizon, Six Sources of Limitless Energy." *Harvard Business Review*, September.

Myrow, Rachel. 2007. "UPS Takes Left Turn Out of Deliveries." *www.npr.org/templates/story/story.php?storyId=7000908.*

Owen, David. 2009. *Green Metropolis: Why Living Smaller, Living Closer and Driving Less Are the Keys to Sustainability*. New York: Riverhead Books.

Parrs, C. and Weinberg, I. 2009. "Green Marketing: Communicating with the Green Consumer." In *Inside the Minds: Greening Your Company.* Boston: ASPATORE.

Phillips, J. J. 1997. Return on investment in performance improvement programs. Houston: Gulf Publishing

Phillips, Jack J. and Aaron, Bruce C. 2008. *Isolation of Results: Defining the Impact of the Program.* San Francisco: Pfeiffer.

Phillips, Jack J. and Phillips, Patricia Pulliam. 2004. "Return to Sender: Improving Response Rates for Questionnaires and Surveys." *Performance Improvement Journal*, August.

———. 2007. *Show Me the Money: How to Determine ROI in People, Places, Projects and Programs*. San Francisco: Berrett Koehler.

———. 2010. *The Consultant's Guide to Results Driven Proposals: How to Write Proposals that Forecast Impact and ROI.* New York: McGraw-Hill.

Rust, Roland T., Zahorik, Anthony J. and Keiningham, Timothy L. 1994. *Return on Quality: Measuring the Financial Impact of Your Company's Quest for Quality*. Chicago: Probus.

Schendler, Auden. 2009. *Getting Green Done: Hard Truths from the Front Lines of the Sustainability Revolution*. New York: PublicAffairs.

Stringer, Leah. 2009. *The Green Workplace: Sustainable Strategies that Benefit Employees, the Environment and the Bottom Line.* New York: Palgrave Macmillan.

Surowieki, James. 2004. *The Wisdom of Crowds: Why the Many Are Smarter than the Few and How Collective Wisdom Shapes Business, Economics, Societies and Nations*. New York: Doubleday.

The New England Chapter of the System Safety Society, *System Safety: A Science and Technology Primer,* April: 2002.

Tsai, J. 2010. "Mail Model of the Year." *Customer Relationship Management* 14, no. 4.

Ulrich, Dave, ed. 1998. *Delivering Results*. Boston: Harvard Business School Press.

Wal-Mart (2010). Wal-Mart 2010 Annual Report. Retrieved from http://cdn.walmartstores.com/sites/AnnualReport/2010/PDF/01_WMT%202010_Financials.pdf

Werbach, Adam. 2009. *Strategy for Sustainability: A Business Manifesto.* Boston: Harvard Business Press.

Winston andrew. 2009. *Green Recovery: Get Lean, Get Smart and Emerge from the Downturn on Top.* Boston: Harvard Business Press.

13

Measuring ROI in Safety Management for Project Leaders

Global Engineering and Construction Co.

Jack J. Phillips, Ph.D.

Abstract
This case study shows the power of a safety management program for project safety leaders on construction sites. These are large construction sites and the safety project leader is a full-time safety and health professional. Responding to disappointing safety performance, a thorough needs analysis was conducted, yielding a variety of actions that needed to be taken through the project safety leaders. This program involved a two-day workshop with action plans to drive business performance measures. Each participant selected three measures to improve; using the content of the program and the action planning process provided the detailed steps necessary to achieve that. The results are very impressive, underscore the benefit of having an action plan built into the program and show the power of a focus on results in the program.

Keywords: Needs analysis, safety performance, action plan, improvement measures, focus on results

13.1 Background

Global Engineering and Construction Company (GEC) designs and builds large commercial projects such as chemical plants, paper mills and municipal water systems. The company employs 35,000 full-time associates. In addition, for each project another 200

Jack Phillips, Patti Phillips, and Al Pulliam, Measuring ROI in Environment, Health, and Safety, (291–318) 2014 © Scrivener Publishing LLC

to 1,500 contract workers are involved during peak construction phases. During a typical year, contract workers account for another 100,000 at construction sites. Safety is always a critical matter at GEC and usually commands much management attention.

From a corporate perspective, safety is managed by a Safety and Health Team of specialists and managers who report to the director of environment, health and safety (EHS). Each project has at least one person responsible for safety who functions as a project safety leader.

13.1.1 The Need

Over the previous two years, the safety performance had deteriorated or remanded flat at unacceptable levels. Because of this disappointing and sometimes erratic safety performance, the chief operating officer (COO) asked the EHS director to explore the causes of the unacceptable performance and to offer a remedy. The department reviewed the safety records, safety procedures and safety administration, searching for common threads of causality. Questionnaires were sent to all of the project safety leaders at each site. Finally, a selective group of safety leaders were interviewed, attempting to pinpoint what could be done to improve safety. From this initial needs assessment, the following conclusions were made:

1. There is still a lack of knowledge about the different tools and techniques available for the project safety leaders to use to improve safety performance.
2. There was clear evidence that project safety leaders are not operating on a proactive basis, but merely reacting to events and issues as they happen.
3. Routine safety meetings need more content, better planning and improved coordination.
4. Project safety leaders need to use available tools for investigation, causation analysis and corrective action.

With this in mind, the EHS Team recommended a two-day workshop for all of the project safety leaders. This workshop would focus on the gaps defined in the needs assessment and would provide the motivation, knowledge, skills and tools to improve safety performance.

The program was designed for project safety leaders, who usually had the title of safety manager, safety engineer or safety superintendent. The program focused on safety leadership, safety planning, safety inspections and audits, safety meetings, accident investigation, safety policies and procedures, safety standards and safety problem solving.

The objectives for the program are listed in Table 13.1.

These topics are fully explored in a two-day safety management training program conducted regularly. Safety leaders (i.e., participants) were expected to improve the safety performance of their individual construction projects. The safety performance measures used in the company are reviewed and discussed in the workshop.

This particular program will be expensive, because it would be necessary for all the project safety leaders to travel and two days

Table 13.1 Objectives for Safety Management Program.

Level	Measurement Focus
1. Reaction	Obtain favorable reaction to program and materials on • Need for program • Relevance to project • Importance to project success Identify planned actions
2. Learning	After attending this program, participants should be able to: • Establish safety audits • Provide feedback and motivate employees • Investigate accidents • Solve safety problems • Follow safety procedures and standards • Counsel problem employees • Conduct safety meetings
3. Application and Implementation	• Use knowledge, skills and tools routinely in appropriate situations • Complete all steps of action plan
4. Business Impact	• Improve at least three safety and health measures
5. Return on Investment	20%

of their work would be missed while they are participating in the program. The COO wants to make sure that this is the right solution and that it represents a good investment. He is asking for success measures, showing how safety performance has improved. Ideally, he would like to see the ROI for conducting this particular program.

13.1.2 Business Alignment

The program facilitator asked participants to provide limited needs assessment data before attending the program. Each participant was asked to review the safety performance of his or her project and identify at least three safety measures that, if improved, should enhance safety performance. Each measure selected should be important and have the possibility of being improved using the topics in the safety management program. These business impact measures could be disabling injury rate, accident severity rate, first aid treatments, OSHA citations, OSHA penalties, property accidents, hazardous material incidents, or near misses. Each participant may have different measures. The important point in this step is to avoid selecting measures that cannot be enhanced through the efforts of the team and the content covered in the program.

As participants register for the program, they are reminded of the requirement to complete an action plan to improve the three measures. This requirement is presented as an integral part of the program and not as an add-on data collection tool. Action planning is necessary for participants to see the actual improvements generated from the program as well as the entire group of participants.

13.2 Why Evaluate this Program?

Although the COO had suggested the ROI calculation, the EHS director was convinced that this program would add value and he wanted to convince top executives that investments in safety and health had high payoffs. The ROI is the way to do it. The Safety Team decided at the outset that improvement data would be collected from this program and presented to the C-suite. The team built the evaluation into the program with the action plan. This decision was based on three issues.

- This program is designed to add value at the construction project level and the outcome is expressed in project level measures that are well known and respected by the Management Team. The evaluation should show the actual monetary value of improvement.
- The application data enables the team to make improvements and adjustments.
- The data also helps the team gain respect for the program from the operating executives and project managers.

13.3 The ROI Process

The Safety and Health Team staff used a comprehensive evaluation process to develop the ROI. The ROI Methodology generates six types of data:

- Reaction
- Learning
- Application/Implementation
- Business Impact
- ROI
- Intangible Measures

To determine the contribution the program makes to the changes in business impact measures, a technique to isolate the effects of the program is included in the process. Figure 13.1 shows the ROI process model used, which had its beginning with detailed objectives for reaction, learning, application and impact. Data collection plans and an ROI analysis plan are developed before data collection actually begins. Four levels of data are collected, which represents the first four types of data listed above. The process includes a method to isolate the effects of a program and techniques to convert data to monetary value. The ROI is calculated when comparing the monetary benefits to the cost of the program. The intangible measures, the sixth type of data, are those impact measures not usually converted to monetary value such as job satisfaction and image. This comprehensive model allows the organization to follow a consistent standardized approach each time it is applied to evaluate safety programs.

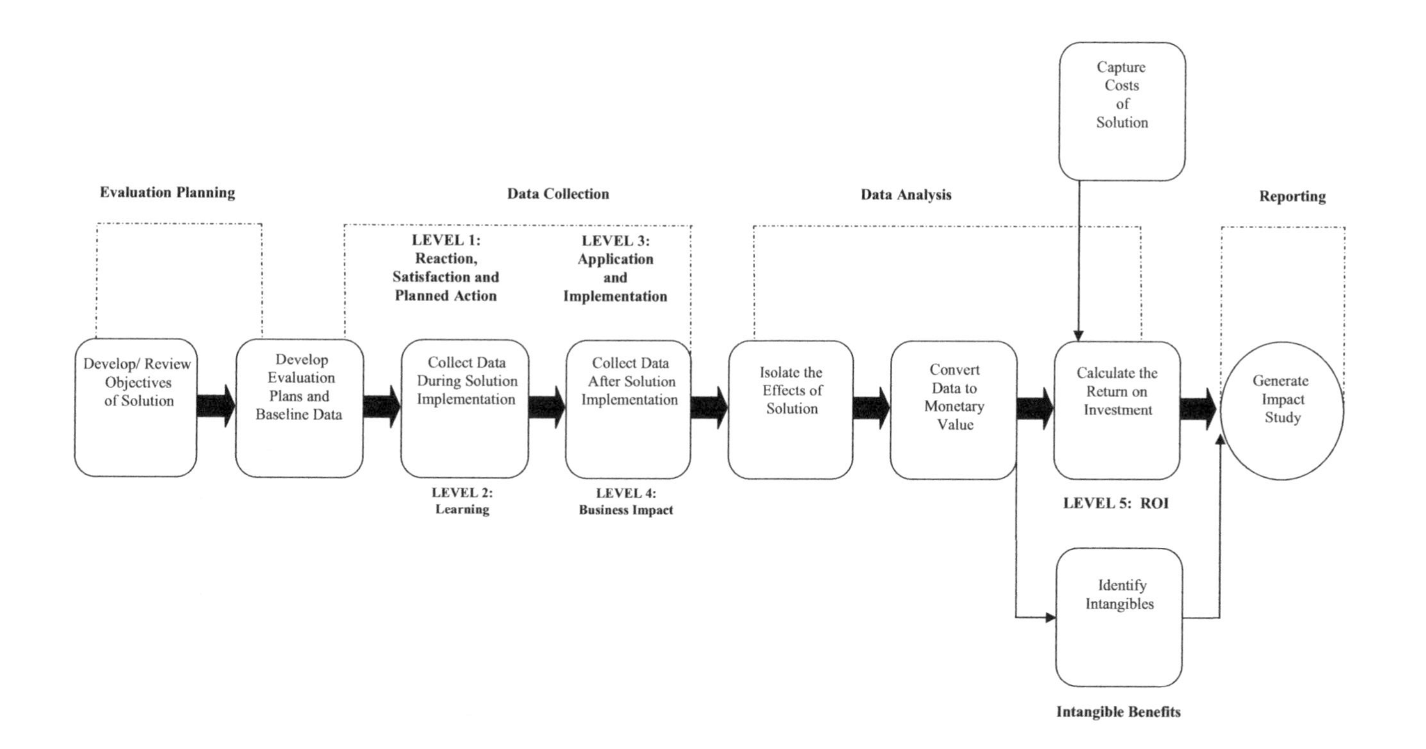

Figure 13.1 The ROI Methodology.

13.4 Planning for Evaluation

Planning for the evaluation is critical to save time and improve the quality and quantity of data collection. It also provides an opportunity to clarify expectations and responsibilities and shows the client group – in this case, the Senior Operating Team – exactly how this program is evaluated. Two documents are created: the data collection plan and the ROI analysis plan.

13.5 Data Collection Plan

Figure 13.2 shows the data collection plan for this program. Program objectives are detailed along the five levels of evaluation, which represent the first five types of data collected for programs. As the figure illustrates, the typical reaction and learning data are collected at the end of the program by the facilitator. Learning objectives focus on six major areas of the program.

Through application objectives, participants focus on two primary broad areas. The first is to use the knowledge, skills and tools routinely in appropriate situations and the second is to complete all steps on their action plans. A follow-up questionnaire was selected to measure the use of knowledge, skills and tools. This was planned for two months after the program. For the second area, action plan data are provided to show the actual improvement in the safety measures planned.

Business impact objectives vary with the individual as each project safety leader identifies at least three safety and health measures needing improvement. These are detailed on the action plan and serve as the basic principal document for the Safety and Health Team to tabulate the overall improvement. The ROI objective is 20 percent, which was higher than the ROI target for capital expenditures at GEC.

13.6 ROI Analysis Plan

The ROI analysis plan, which appears in Figure 13.3, shows how data are analyzed and reported. Safety performance data are listed and they form the basis for the rest of the analysis. The method

Level	Objective(s)	Measures/Data	Data Collection Method	Data Sources	Timing	Responsibilities
1	**Reaction** ▪ Obtain favorable reaction to program and materials on – Need for program – Relevance to project – Importance to project success ▪ Identify planned actions	▪ Average rating of 4.0 out of 5.0 on feedback items ▪ 100% submit planned actions	▪ Standard questionnaire	▪ Participant	▪ End of program	▪ Facilitator
2	**Learning** After attending this session, participants should be able to: ▪ Establish safety audits ▪ Provide feedback and motivate employees ▪ Investigate accidents ▪ Solve safety problems ▪ Follow procedures and standards ▪ Counsel problem employees ▪ Conduct safety meetings	▪ Achieve an average of 4 on a 5 point scale	▪ Questionnaire	▪ Participant	▪ End of program	▪ Facilitator
3	**Application/Implementation** ▪ Use knowledge, skills and tools in appropriate situations ▪ Complete all steps of action plan	▪ Ratings on questions (4 of 5) ▪ The number of steps completed on action plan	▪ Questionnaire ▪ Action plan	▪ Participant ▪ Participant	▪ Two months after program ▪ Three months after program	▪ Safety and Health Team
4	**Business Impact** ▪ Improve three safety and health measures	▪ Varies	▪ Action plan	▪ Participant	▪ Six months after program	▪ Safety and Health Team
5	ROI ▪ 20%	▪ Comments: Several techniques will be used to secure commitment to provide data on the questionnaire and action plan.				

Figure 13.2 Data collection plan.

Program Safety Management Program **Responsibility:** ____________ **Date:** ______

Data Items (Usually Level4)	Methods for Isolating the Effects of the Program/ Process	Methods of Converting Data to Monetary Values	Cost Categories	Intangible Benefits	Communication Targets for Finale Report	Other Influences/ Issues During Application	Comments
▪ Three safety and health measures identified by project safety leader	▪ Participant estimation	▪ Standard values ▪ Expert input (Safety Team) ▪ Participant estimation	▪ Needs assessment ▪ Program development ▪ Program materials ▪ Travel & lodging ▪ Facilitation & coordination ▪ Participant salaries plus benefits while in the program ▪ Extra project expenses related to program ▪ Evaluation	▪ Job Engagement ▪ Job satisfaction ▪ Stress ▪ Image ▪ Brand	▪ Construction project General Manager ▪ Participants ▪ Director, Environment, Health and Safety ▪ Corporate Safety and Health team ▪ Operating executives ▪ Senior VPHuman resources		

Client Signature: ____________ Date: ____________

Figure 13.3 ROI analysis plan.

for isolating the effects of the program was estimations from the safety project leader. The method to convert data to monetary values relied on three techniques: standard values (when they are available), expert input, or participant's estimates. The good news is that most of the costs of safety measures were readily available in the safety function. Cost categories represent a fully loaded profile of program costs, including direct and indirect costs, anticipated intangibles are detailed and the communication audiences for the results are outlined. The ROI analysis plan represents the approach to process business impact data to develop the ROI analysis and to capture the intangible data. Collectively, these two planning documents outline the approach for evaluating this program.

13.7 Action Planning: A Key to ROI Analysis

Figure 13.4 shows the sequence of activities as the action planning process is introduced to participants and reinforced throughout the program. The requirement for the action plan is communicated prior to the program along with the request for needs assessment information.

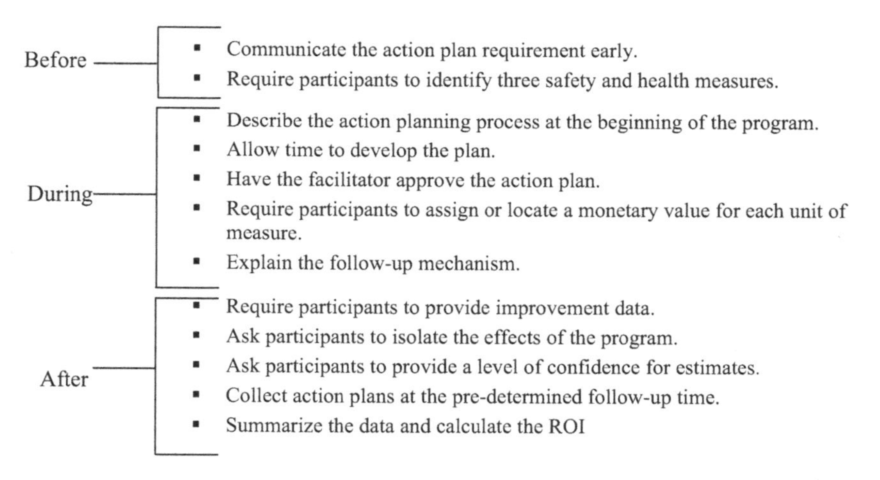

Figure 13.4 Sequence of activities for action planning.

13.7.1 Teaching and Explaining the Plan

On the first day of the program, the facilitator describes the action planning process in a fifteen-minute discussion. This discussion includes three parts:

- The guidelines for developing action plans are presented. The SMART (Specific, Measurable, Achievable, Realistic and Time-Based) requirements are used.
- Five blank forms are provided (3 must be completed).
- Examples are presented to illustrate what a complete action plan should look like.

This discussion reinforces the need for action plans and the importance of the tool for participants.

13.7.2 Developing the Plan

At the end of the second day, the booklets are completed in a session lasting about ninety minutes. Participants work in teams to complete all three action plans, which took twenty to thirty minutes each. Figure 13.5 shows a blank action plan. During the session, participants completed the top portion of the action plan, Figure 13.5, the left column on which they listed the action steps and parts A, B and C on the right column. The remainder of the form is completed in a three-month follow up. A facilitator monitored the session and several operations executives were present. The involvement of operations executives not only keeps participants focused on the task, it usually leaves executives impressed with the program and the quality of the action planning process.

The action plan could focus on any specific steps as long as they are consistent with the content in the program and are related to the safety and health improvement measures. The most important part of developing the plan is to convert the measure to a monetary value (B and C). Three approaches are offered to the participants. First, standard values are used if they are available. A standard value is a value already known to project safety leaders. Fortunately, standard values are available for most of the EHS measures. Safety and health previously had assigned a cost to particular measures to use in controlling costs and to develop an appreciation for the impact

Name:	________	Facilitator Signature:	________	Follow-Up Date	________
Objective:		Evaluation Period:	________	To	________
Improvement Measure:	________	Current Performance		Target Performance	________

Action Steps	Analysis
1. ________	A. What is the unit of measure? ________
2. ________	B. What is the value (cost) of one unit? $________
3. ________	C. How did you arrive at this value? ________
4. ________	D. How much did the measure change during the evaluation period? (monthly value) ________
5. ________	E. What other factors could have caused the improvement? 1. ________ 2. ________ 3. ________
6. ________	F. What percent of this change was actually caused by this program? ________%
7. ________	G. What level of confidence do you place on the above information? (100%=certainty and 0%=no confidence) ________%
8. ________	H. How many months to project completion? ________
9. ________	**Other Benefits and consequences**
Comments:	

Figure 13.5 Action plan Form.

of these measures. Second, if a standard value is not available, the participants are encouraged to use expert input such as from a corporate safety and health team member who may know the value of a particular item. The program facilitator encouraged participants to call the expert and include the value in the action plan. Third, if a standard value or expert input is not available, participants are asked to estimate the cost or value using knowledge and resources available to them. It was important to require this value to be developed during the program.

13.8 ROI Forecast with Reaction Data

At the end of the two-day meeting, participants completed a questionnaire to evaluate the safety management program. A few questions were asked to provide a forecast. Participants were asked to provide a one-year estimated monetary value of their planned actions. In addition, participants explained the basis for estimates and placed a confidence level on their estimates. Table 13.2 presents this data. Nineteen of the twenty-five participants supplied data. The estimated cost of the program, including participants' salaries for the time devoted to the project, was $120,000.

The monetary values of the planned improvements were extremely high, reflecting the participants' optimism and enthusiasm at the end of an impressive program from which specific actions were planned. As a first step in the analysis, extreme data items were omitted (one of the guiding principles of the ROI Methodology). Data such as "millions," "unlimited," and "$4 million" were discarded and each remaining value was multiplied by the confidence value and totaled. This adjustment is one way of reducing highly subjective estimates.

The resulting tabulations yielded a total improvement of $990,125 (rounded to $990,000). The projected ROI, which was based on the feedback questionnaire, is

$$\text{ROI} = (\$990{,}000 - \$120{,}000)/\$120{,}000 \times 100 = 725\%$$

Although these projected values are subjective, the results were generated by project safety leaders (participants) who should be aware of what they could accomplish. The follow-up study will determine the true results delivered by the group.

Table 13.2 Level 1 data for ROI forecast calculations.

Participant No.	Estimated Value ($)	Basis	Confidence Level	Adjusted Value ($)
1	80,000	Reduction in accidents	90%	72,000
2	91,200	OSHA reportable injuries, OSHA Fines	80%	72,960
3	55,000	Accident reduction	90%	49,500
4	10,000	First-aid visits/visits to doctor/DIR	70%	7,000
5	150,000	Reduction in lost-time injuries, OSHA Fines	95%	142,500
6	Millions	Total accident cost	100%	—
7	74,800	Workers' compensation, Injury	80%	59,840
8	7,500	OSHA citations, Accidents	75%	5,625
9	50,000	Reduction in accidents	75%	37,500
10	36,000	Workers' compensation (lost time)	80%	28,800
11	150,000	Reduction in total accident costs	90%	135,000
12	22,000	OSHA fines/accidents	70%	15,400
13	140,000	Accident reductions	80%	112,000
14	4 Million	Total cost of safety	95%	—
15	65,000	Total workers' compensation	50%	32,500
16	Unlimited	Accidents	100%	—
17	20,000	Accidents	95%	19,000
18	45,000	Injuries	90%	40,500
19	200,000	Lost-time injuries	80%	160,000
				Total: $990,125

Collecting these types of data focuses increased attention on project outcomes. This issue becomes clear to participants as they anticipate results and convert them to monetary values.

This simple exercise is productive because of the important message it sends to participants. Participants will understand that specific action is expected, which produces results for the project. The data collection helps participants plan the implementation of what they are learning.

Because a follow-up evaluation of the program is planned, the post-project results will be compared to the ROI forecast. Comparisons of forecast and follow-up data are helpful. If there is a defined relationship between the two, the less expensive forecast may be substituted for the more expensive follow up in the future.

13.9 Improving Response Rates

Data were collected at Level 1 and 2 (reaction and learning) at the end of the two-day workshop. As expected, the facilitator was able to secure a 100 percent response rate directly from the participants. Not everyone completed the forecast of results, with only nineteen of the twenty-five providing data for the ROI forecast described earlier. A follow-up questionnaire, which was completed two months after the program, had an impressive response, with twenty-two out of the twenty-five providing data.

This was achieved by taking on the following techniques:

1. A questionnaire was reviewed at the workshop with an expectation that the data would be provided in two months.
2. The questionnaire was positioned as a tool for them to see the progress they were making.
3. The questionnaire was designed for ease of responding, with the expectation that it would take only about twenty minutes of their time.
4. The chief operating officer signed the memo to participants, asking for the data and encouraging them to reflect over what they actually are doing as a result of this program. The participants were promised the summary results of the questionnaire and were

promised that actions would be taken to improve the program as a result of their comments.
5. Two follow-up reminders were provided, one was an email and the other was a phone call directly from the facilitator.
6. A new book on the importance of safety was provided as an incentive for responding. This was an exchange, the questionnaire for the book.

Action plans were collected three months after the program; this provided an opportunity for the participants to show the impact of their work. Because of their commitment and ownership of the data, a response rate of 92 percent was achieved. To achieve this response, several techniques were used, similar to those used with the questionnaire.

In summary, the data collection was extremely effective with high levels of commitment and participation by the individuals.

13.10 Results

The Safety and Health Team reported results in all six data categories developed by the ROI Methodology, beginning with reaction and moving through ROI and the intangibles. Here are the results in each category with additional explanation about how some of the data was processed.

13.10.1 Reaction and Learning

Reaction data, collected at the end of the program using a standard questionnaire, focused on issues such as relevance of and intention to use the content. The delivery and facilitation also are evaluated. Table 13.3 shows a summary of the reaction data with ratings. Learning improvement was measured at the end of the program using self-assessment. Table 13.4 shows the summary of the learning results. Although these measures are subjective, they provide an indication of improvements in learning.

13.10.2 Application and Implementation

To determine the extent to which the knowledge, skills and tools are actually being used and to check progress of the action plan, a

Table 13-3 Reaction Measurements
1= Unsatisfactory 5 = Exceptional.

Topic	Rating
Need for the program	4.3
Relevance to construction project	4.5
Importance to project success	4.5
Delivery of the program	4.2
Facilitation of the program	4.2
Planned actions developed	100%

Table 13.4 Learning Measurements. 1 = Cannot do this
5 = Can do this extremely well.

Topic	Rating
Establish safety audits	4.2
Provide feedback/motivation To employees	4.0
Investigate accidents	4.9
Follow safety procedures and standards	4.2
Counsel problem employees	3.9
Conduct safety meetings	4.8

questionnaire was distributed two months following participation in the program. This two-page, user-friendly questionnaire evaluated the success of the program at the application level. Table 13.5 provides a summary of the results. As shown in Table 13.5, participants reported progress in each of the areas and success using the content, in addition to the projects involving action plans. Also, the safety leaders indicated linkage of this program with other safety measures beyond the three measures selected for action planning. Typical barriers of implementation that they reported included lack of time, understaffing, changing culture and pressures to get work done. Typical enablers were support from the project general manager, which was highly ranked. This follow-up questionnaire allowed project safety leaders an opportunity to summarize briefly the progress with the action plan.

Table 13.5 Application Results 1= *Unsuccessful* 5= *Very Successful.*

Success With:	Rating
1. Conducting safety audits	4.1
2. Providing feedback to employees	3.9
3. Investigating accidents	4.8
4. Solving safety problems	4.9
5. Follow safety procedures and standards	4.7
6. Counseling problem employees	4.2
7. Conducting safety meetings	4.6

In essence, this served as a reminder to continue with the plan as well as a process check to see if there were issues that should be explored.

13.10.3 Business Impact

Project safety leaders provided safety improvement data specific to their construction projects. Although the action plan contains some Level 3 application data (the left side of the form in Figure 13-6), the primary value of the action plan was business impact data obtained from the documents.

In the three-month follow up, the participants were required to furnish the following seven items:

1. The actual change in the measure on a monthly basis is included in part D of the action plan. This value is used to develop an annual (first year) improvement.
2. The only feasible way to isolate the effects of the program is to obtain an estimate directly from the participants. As they monitor the measures and observe their improvement, the project safety leaders probably know the other influences driving a particular measure, which are detailed on the action plan. As an initial step in the analysis, the project safety leaders listed the other factors that could have caused the improvement. This is part D.

3. Realizing that the other factors could have influenced the improvement, project safety leaders were asked to estimate the percent of improvement resulting from the application of the content from the safety management program (the action steps on the action plan). Each project safety leader was asked to be as accurate as possible with the estimate and express it as a percentage (part E on the action plan).
4. Recognizing that the above value is an estimate, the project safety leaders were asked to indicate the level of confidence in their allocation of the contribution to this program. This is included in part F on the action plan, using 100 percent for certainty and 0 percent for no confidence. This reflects the degree of error in the allocation.
5. Project safety leaders were asked to provide an estimate of the months to project completion. This allowed for the calculation of the duration of the benefits.
6. The participants were asked to provide input on intangible measures observed or monitored during the three months that were directly linked to this program.
7. Participants were asked to provide additional comments including explanations, if necessary.

Figure 13.6 shows an example of a completed action plan. The example focuses directly on first aid visits from participant number five. This participant was averaging 22 per month and had a goal to reduce it to 10. Specific action steps are indicated on the left side of the form. The average cost of a first aid visit is $300 per visit, an amount that represents a standard value. The actual change on a monthly basis is 11 visits, slightly below the target. Three other factors contributed to the improvement. The participant estimated that 60 percent of the change is directly attributable to this program and that he is 80 percent confident in this estimate. The confidence estimate frames a range of error for the 60 percent allocation, allowing for a possible 20 percent (plus or minus) adjustment in the estimate. To be conservative, the estimate is adjusted to the low side, bringing the contribution rate of this program to a reduction of 48 percent:

$$60\% \times 80\% = 48\%$$

Name:	*Roger Gerson*	Facilitator Signature:		Follow-Up Date	*1 July*
Objective:	*Reduce first aid treatments*	Evaluation Period:	*January*	to	*April*
Improvement Measure:	*Firstaid visits*	Current Performance	*22 / Month*	Target Performance	*10 / Month*

Action Steps	Date	**Analysis**
1. *Review first aid records for each employee – look for trends and patterns.*	*20 Feb*	A. What is the unit of measure? *One first aid visit* What is the value (cost) of one unit? $ 300 How did you arrive at this value? *Standard Value*
2. *Meet with team to discuss reasons for first aid visits – using problem solving skills*	*28 Feb*	
3. *Counsel with "problem employees" to correct habits and explore opportunities for improvement*	*As needed*	B. How much did the measure change during the evaluation period? (monthly value) 11 What other factors could have caused the improvement? 1. Required OSHAtraining 2. Project leadership (General Manager) 3. Safety first program for all employees
4. *Conduct a brief meeting with an employee returning to work after a visit to first aid*	*As needed*	
5. *Provide recognition to employees who have perfect accident records.*	*As needed*	
6. *Follow-up with each discussion and discuss improvement or lack of improvement and plan other action.*	*31 March*	C. What percent of this change was actually caused by this program? 60% What level of confidence do you place on the above information? (100%=Certainty and 0% - No Confidence) 80%
7. *Monitor improvement and provide recognition when appropriate.*	*Routinely*	D. How many months to project completion? 18
Other Benefits: *Greater Productivity* Comments: The action plan kept me on track with this problem		OPTIONAL: Calculate the value (BxDx12xFx9) 11 x 300 x 12 x 60% x 80% = $19,008

Figure 13.6 Action Plan.

The actual improvement value for this example can be calculated as follows:

$$11 \text{ visits} \times \$300 \text{ per visit} \times 12 \text{ months} = \$39{,}600$$

The number of months to project completion is 18, making it appropriate to use the one-year rule for benefits. In the last three months of a project, most of the employees have left the job. Consequently, a project has to have at least 15 months remaining to use one year of data. Otherwise, an adjustment is made. For example, a project with 14 months remaining would use 11 months of benefits, instead of one year.

Table 13.6 shows the annual improvement values on the first measure only for the first 25 participants in this group. Similar tables are generated for the second and third measures. The values are adjusted by the contribution estimate and the confidence estimate. In the number 5 example, the $39,600 is adjusted by 60 percent and 80 percent to yield $19,008. This same adjustment is made for each of the values, with a total first-year adjusted value for the first measure of $320, 309. The same process is followed for the second and third measures for the group, yielding totals of $162, 310 and $57, 320, respectively. The total benefit is $539,939.

13.10.4 Program Cost

Table 13.6 details the program costs reflecting a fully loaded cost profile. The estimated cost of the needs assessment ($5,000) is prorated over the life of the program, which will be with three sessions. The estimated program development cost ($7,500) is prorated over the life of the program, as well. The program materials and facilitators are direct costs and the program includes a book on safety management. Travel and lodging are estimates of an average for each participant. Facilitation and coordination costs were estimated as well. Time away from work represents lost opportunity and is calculated by multiplying two days times daily salary costs adjusted for 40 percent employee benefits factor. The average hourly rate for these leaders is about $50. When adjusted for benefits, the rate is $70, which is $560 per day, or $1,120 per participant for the two

Table 13.6 Business Impact Data.

Participant	Annualized Improvement ($ Values)	Other Measures	Other Factors	Contribution Estimate from Safety project leaders	Confidence Estimate	Adjusted $ Value
1	5,000	Medical treatment	2	40%	90%	23,400
2	5,500	Property damage	4	25%	70%	963
3	32,800	Disabling injuries	2	70%	60%	13,776
4	21,800	First aid	1	80%	80%	13,952
5	39,600	First Aid	3	60%	80%	19,008
6	19,800	Disabling injuries	2	70%	90%	12,474
7	25,000	OSHA citations	3	30%	70%	5,250
8	23,000	Property damage	4	30%	40%	2,760
9	34,500	Medical treatment	1	75%	800%	20,700
10	50,000	Near miss	0	100%	100%	50,000
11	75,000	Disabling injury rate	2	45%	75%	23,313
12	42,350	Medical treatment	3	50%	75%	15,881
13	40,000	OSHA fine	4	25%	80%	8,000
					Total (this page) = $209,477	

Participant	Annualized Improvement ($ Value)	Other Measures	Other Factors	Contribution Estimate From Safety Project Leaders	Confidence Estimate	Adjusted $ Value
14	59,000	Disabling injuries	3	40%	85%	20,060
15	75,000	Disabling injuries	2	20%	90%	13,500
16	missing					
17	24,900	Hazmat violations	2	40%	70%	6,972
18	25,000	Property damage	5	20%	80%	4,000
19	missing					
20	39,000	OSHA citations	2	60%	95%	22,230
21	13,500	OSHA citations	2	70%	90%	850
22	15,000	First aid	0	100%	90%	13,500
23	1,000,000	Near miss	0	100%	100%	1,000,000 (leave this out of calculation)*
24	54,000	Hazardous materials	3	60%	70%	22,680
25	22,000	Property damage	3	40%	80%	7,040

* Extreme data – omitted from analysis

Total (this page) = $110,832
Total first measure = $320,309

Total Annual Benefit for Second Measure is $162,310
Total Annual Benefit for Third Measure is $57,320

days. For 25 participants, the total is $28,000, the second-largest cost item after travel. The cost for the evaluation was estimated. These total costs of $106,087 represent a very conservative approach to cost accumulation.

13.10.5 ROI Analysis

The total monetary benefits are calculated by adding the values of the three measures, totaling $539,939. This leaves a benefits-to-cost ratio (BCR) and ROI as follows.

$$\text{BCR} = \frac{539{,}939}{128{,}067} = 4.22$$

$$\text{ROI} = \frac{(539{,}939 - 128{,}067)}{128{,}067} \times 100 = 322\%$$

When this actual ROI is compared to the forecasted ROI, a significant difference surfaces. The return is 54 percent less than the forecast, which is expected because of the optimism experienced at the end of the workshop.

Table 13.7 Program Cost Summary.

Needs Assessment (Prorated over 3 Sessions)	$1,667
Program Development (Prorated over 3 Sessions)	2,500
Program Materials – 25 @ $100	2,500
Travel & Lodging – 25 @ $2000	37,500
Facilitation & Coordination	50,000
Facilities & Refreshments – 2 days @ $700	1,400
Participants Salaries Plus Benefits	28,000
ROI Evaluation	4,500
Total	$128,067

13.10.6 Credibility of Data

This ROI value of over 300 percent greatly exceeds the 20 percent target value. The ROI value was considered to be credible, although it is extremely high. Its credibility rests on these principles on which the study was based.

1. The data comes directly from the participants.
2. The data could be audited to see if the changes were actually taking place.
3. To be conservative, only the first year of improvements is used. With the changes reported in the action plans, there should be some second and third year value that has been omitted from the calculation.
4. The monetary improvement has been discounted for the effect of other influences. In essence, the participants take credit only for the part of the improvement related to the program.
5. This estimate of contribution to the program is adjusted for the error of the estimate, which represents a discount, adding to the conservative approach.
6. The costs are fully loaded to include both direct and indirect costs.
7. The business impact does not include value obtained from using the skills to address other problems or to influence other measures. Only the values from three measures taken from the action planning projects were used in the analysis.

The ROI process develops convincing data connected directly to project construction costs. From the viewpoint of the chief financial officer, the data can be audited and monitored. It should be reflected as actual improvement at the project site.

13.10.7 Intangible Data

As a final part of the complete profile of data, the intangible benefits were itemized. The participants provided input on intangible measures at two time frames. The follow-up questionnaire provided an opportunity for participants to indicate intangible measures they perceived to represent a benefit directly linked to this program. Also, the action plan had an opportunity for participants to add additional intangible benefits. Collectively, each of the following benefits was listed by at least five individuals.

- Improved productivity
- Improved teamwork
- Improved work quality
- Improved job satisfaction
- Improved job engagement
- Enhanced image
- Reduced stress

To some executives, these intangible measures are just as important as the monetary payoff.

13.10.8 The Payoff: Balanced Data

This program drives six types of data items: satisfaction, learning, application, business impact, ROI and intangible benefits. Collectively, these six types of data provide a balanced, credible viewpoint of the success of the program.

13.11 Communication Strategy

Table 13.8 shows the strategy for communicating results from the study. All key stakeholders received the information. The communications were credible and convincing. The information to project general managers helped build confidence in the program. The CEO and CFO were pleased with the results. The data provided to employees, shareholders and future participants was motivating and helped to bring more focus on safety.

13.12 Lessons Learned

It was critical to build evaluation into the program, positioning the action plan as an application tool instead of a data collection tool. This approach helped secure commitment and ownership for the process. It also shifted much of the responsibility for evaluation

Table 13.8 Communication Strategy.

Timing	Communication Method	Target Audience
Within one month of follow-up	Executive briefing	Regional executives CEO, CFO
Within one month of follow-up	Live briefing	Corporate and regional operation executives
Within one month of follow-up	Detailed impact study (125 pages)	Program participants; Safety and health staff • Responsible for this program in some way • Involved in evaluation
Within one month of follow-up	Report of results (1 page)	Project general managers
Within two months	Article in Project News	All employees
As Needed	Report of results (1 page)	Future participants in similar safety programs
End of Year	Paragraph in Annual Report	Shareholders

to the participants as they collected important data, isolated the effects of the program on the data and converted the data to monetary values, the three most critical steps in the ROI process. The costs are easy to capture and the report is easily generated and sent to the various target audiences.

This approach has the additional advantage of evaluating programs where a variety of measures are influenced. The improvements are integrated after they are converted to monetary value. Thus, the common value among measures is the monetary value representing the value of the improvement.

13.13 Discussion Questions

1. Is this approach credible? Explain
2. Is the ROI value realistic?

3. Were the differences in the ROI forecast and the RPI actual expected? Explain.
4. How should the results be presented to the senior team?
5. How can the action planning process be positioned as an application tool?
6. What type of programs would be appropriate for this approach?

14

Measuring ROI in a Modular/Reusable Safety Railing System in Commercial, Multi-Family & Residential Construction

Southeastern Framing Corporation

Al Pulliam

Abstract

Southeastern Framing Corporation (SFC) is a wood-framed structures design/build organization. The company was faced with increased regulatory obligations due to a change in the Occupational Safety and Health Administration (OSHA) enforcement policy. Southeastern Framing needed solutions to fall-protection safety issues while maintaining a competitive position in the marketplace. Company management believed that its current methods needed improvement to protect employees and comply with current regulations. It was ultimately decided to invest in a modular and reusable method of providing guardrail protection, resulting in a 341 percent ROI.

Keywords: OSHA policy change, regulatory obligations, fall protection standard, ROI analysis plan, standard compliance

---

Jack Phillips, Patti Phillips, and Al Pulliam, Measuring ROI in Environment, Health, and Safety, (319–338) 2014 © Scrivener Publishing LLC

14.1 Background

SFC is a design/build wood-framed structures contractor. Over the years, SFC succeeded in growing its business in a fragmented niche of the construction business. During an eight-year period, the business grew from a two-man operation to a $15 million-per-year business with approximately fifteen full-time employees and as many as 100 employees on any particular job.

As a construction company, SFC is subject to the safety and health regulations found in 29 CFR 1926, commonly known as the OSHA Construction Standard (most industries are subject to 29 CFR 1910, the General Industry Standard). A key part of the OSHA Construction Standard that impacted SFC is Subpart M, the Fall Protection Standard. Commercial and industrial contractors have been complying with the Fall Protection Standard since its development back in the early 1990s. However, OSHA Instruction STD 03-11-001 provided interim enforcement guidelines for certain residential construction activities that provided, to a large extent, more relaxed standards for all residential construction. This included multi-family condominium and apartment construction, which is a major portion of SFC's business. The guidance had been in place since June 1999. Under certain conditions, this guidance allowed residential contractors to utilize site-specific safe work practices rather than an engineered solution to protect workers from a fall above six feet. Examples of an engineered solution include guardrail systems, safety net systems and personal fall arrest systems. For residential contractors, the ability to use work practice and administrative controls provided a much cheaper —and less effective—means of providing worker protection under the standard.

Effective June 11, 2011, OSHA issued a new instruction, STD 03-11-002, which rescinded STD 03-11-001. The new directive is much more rigorous in its application of engineering controls as opposed to work practices and administrative controls. The new standard essentially requires that contractors in the residential building industry be held to the same standards with which industrial and commercial contractors are required to comply. Strict compliance with the standard has added significant cost to certain portions of the residential construction business.

14.2 Problem Definition

To understand the problem SFC faced, one must have some understanding of the industry and the competitive market in which SFC operates. The residential sector of the wood-frame construction industry is highly fragmented and dominated by small companies consisting of only a few employees. Often these companies operate with no insurance or workers' compensation. Larger framing contractors subcontract work to the small companies and let them roll into their insurance and workers' compensation. The smaller contractor is typically paid a unit price rate for his work. This acts to legitimize the independence of the smaller contractor and allows the larger contractor some ability to distance himself from the smaller contractor. As part of a standard subcontract, the larger contractor requires that the smaller contractor be responsible for complying with all OSHA safety regulations. Further, the subcontract will require that the smaller contractor hold harmless the issuing contractor for any fines, penalties or other negative impacts that may result from an OSHA inspection.

On the surface it seems reasonable to put the responsibility on the subcontractor actually performing the work. The residential construction industry is unique. This is particularly true for the wood structural framing portion of residential construction. First, the vetting process for many of these small framing subcontractors is weak at best. In fact, it is so weak that a lone man (or woman) with a hammer and a nail pouch can walk in off the street and claim to be a framing contractor. While that case is extreme, it is common for many of the framing subcontractors on a job to be very small, with three or four employees. Often these smaller operators do not have the background and many times the willingness to comply with all OSHA regulations. Compliance with OSHA regulations does add cost to a project. Operating margins for these small contractors is slim, and even if they wished to comply with every regulation, it is often financially impractical.

OSHA regulations do have certain provisions for multi-employer sites. These provisions should result in regulation and enforcement actions against the hiring contractor, if certain criteria are met. Most times, however, the contractor whose employees are actually exposed to the hazard receives the brunt of any enforcement action.

As previously mentioned, many of these contractors are very small. Given a meaningful OSHA enforcement action, they will pack up their tools, file for bankruptcy and restart their business under another name. The bottom line is that, given the structure and dynamics of the residential framing business, an unlevel playing field has developed. Contractors who wish to adopt a policy of ensuring compliance with all regulations—the fall protection standard, specifically —are faced with a competitive dilemma.

Before the change in OSHA's guidance, SFC was able to operate within the regulatory structure and be competitive. Even though faced with competition that operated under OSHA's radar, SFC had established a solid reputation for good work and was committed to meeting all its regulatory requirements. This allowed SFC to grow its business and prosper. Once the new directive was implemented, SFC was faced with significant cost increases that, given the market, it could not just pass on to its customers. These cost increases came in two forms. The first was the cost to competitively build and construct compliant fall protection. The second cost related to this issue was avoiding any penalties potentially levied by OSHA.

In August 2011, shortly after the directive was issued, OSHA inspected an SFC jobsite. SFC was found to be non-compliant with the new directive and was ultimately issued a Notice of Violation and subsequent penalties. After mitigation efforts and an informal conference with OSHA, a penalty of $19,000 ultimately was levied against SFC.

14.3 Project Background

To comply with the new OSHA directive, SFC rigorously enforced rules that all fall hazards be protected against or appropriate fall arrest systems be employed. On its multi-family jobs, many of the protection requirements were met with railing that was hand-built out of standard 2″x4″ lumber. This process was somewhat random and inefficient. Additionally, SFC had experienced some puncture wound injuries while workers were "stick building" fall protection framing. SFC felt that a more standardized, cost-efficient method of constructing this railing would improve its business.

In the fall protection marketplace, many manufacturers were offering modular systems that claimed all types of cost savings for contractors. Based on a variety of factors, SFC settled on a particular system that used a combination of 2″x4″ material with reusable fastening components. The savings described by the vendor were savings in lumber and labor. This system was attractive to SFC because it did not require any special tools not otherwise used or owned by contractors in the rough framing business. Other than the fastening components, no special material was required. Not only did SFC employees need no new power tools or have to work with different materials, but this system did not impact lay down space on a job or at the home office. Upon implementation of the system, SFC intuitively felt that there was a positive return but wished to thoroughly evaluate all costs and returns.

14.4 Business Alignment

In all competitive businesses, knowing the true cost and benefit of a program or method is critical to effective management and success. The residential framing business is no different. Jobs are bid on and awarded based on costs and the contractor's ability to perform the work. For SFC, knowing these costs was critical for two reasons. First, on smaller jobs where SFC performed all the work, the company needed to know the cost of protecting its own employees. Second, on many larger jobs, a framing contractor will have the assignment of installing fall protection to be used by all the trades. A meaningful evaluation of the costs and benefits of this new system also could result in additional business development.

14.5 Evaluation Methodology

14.5.1 Data Collection Plan

In implementing this change in fall protection, SFC recognized that there were many additional costs associated with the change not considered in the vendor claims of features, benefits and value. While the system did work as advertised, SFC wished to know the total value this change may have brought to the company. SFC felt that using the Phillips ROI Methodology would provide a systematic means of drilling down to all costs and revealing all returns in order to make an informed judgment on the value of using a new method of fall protection.

The ROI Methodology evaluates five different types of data focusing on result rather than inputs. Table 14.1 presents the data collection plan for the SFC project.

Table 14.1 Data Collection Plan.

	SFC Broad Objectives	Measures	Data Collection Method	Data Sources	Timing	Responsibilities
1	**Reaction and Planned Action** Carpenter satisfaction with new system	75% of full time carpenters are satisfied that the new system will work.	Informal survey	Carpenters	During 1st job	Foreman
2	**Learning and Confidence** Demonstrate competence In system installation	100% of carpenters can install new system according to the manufactures design.	Superintendent observations	Field Superintendents	During 1st job	Superintendent
3	**Application and Implementation** Correct installation	100% compliance with safety checklist of top and mid-rail height, fastening systems and installation locations.	Unannounced safety manager audits	Safety Manager	During first 3 jobs	Safety Manager

	SFC Broad Objectives	Measures	Data Collection Method	Data Sources	Timing	Responsibilities
4	**Business Impact** Reduce costs due to new system Comply with company policy & OSHA regulations Reduce workers' compensation cases	Cost per lineal foot of rail. Cost of OSHA penalties 100% compliance with applicable regulations. Cost of injuries	Review material and labor costs from pre & post-change jobs. Safety Manager audits of OSHA Subpart M regulations. Review Records	Accounting & Job Audit Records	Review after 3rd job	Estimator & Safety Manager
5	**ROI**	Target: 25%				

Level 1 data measure *Reaction*. For SFC the reaction data captures the full-time employee's general feeling that the new system will work. It was felt that without some degree of buy-in from this core group, any new system would be compromised. Non-full-time employees hired for specific jobs were not considered in this measurement, because SFC dictates materials and methods to those employees and their acceptance of the system was not as meaningful as core employees.

Level 2 data measure *Learning*. SFC wanted to ensure that all carpenters knew how to install the systems according to the manufacturer's specifications.

Level 3 data measure *Application*. For this level of data, SFC added to its jobsite safety audit form key indicators of correct application of the system to be measured by unannounced safety audits.

The level 4 data measure changes in *Business Impact*. For this project, the changes would be the reductions in the line item cost of safety railing per lineal foot, the cost avoidance of non-compliance and improvement in safe work practices.

Level 5, *ROI*, compares the benefits of a program with its total cost. The target return for this program was 25 percent.

14.5.2 ROI Analysis Plan

Table 14.2 presents the ROI Analysis Plan. Determining reduced cost per lineal foot of safety rail is relatively straightforward. On previous multi-family projects, line item data on materials used for safety were unavailable. It was assumed that all lumber used in the safety line item was for railing. This was felt to be a valid assumption since there was limited use of lumber as a pure safety item on this type of project. All other items related to personal protective equipment, other fall protection devices and general safety are not lumber and easily segregated. For the labor portion of the analysis, the subsidiary ledger of individual multi-family specifically identified safety railing as a job task and was easily identified.

The cost of eliminating or reducing OSHA penalties was less straightforward but nonetheless achievable. This calculation was based on a history of jobs, OSHA inspections and penalties issued.

The total cost of injuries was calculated using the direct cost of the injury to estimate the total indirect costs. Based on the total cost of the injury and the profitability of the business, the revenue necessary to cover the cost of the injury can then be calculated.

Table 14.2 ROI Analysis Plan.

Data Items	**Methods for Isolating the Effects of the Program/ Process**	**Methods of Converting Data to Monetary Values**	**Cost Categories**	**Intangible Benefits**	**Communications Targets for Final Report**	**Other Influences/ Issues During Application**	**Comments**
Reduce Costs OSHA citations Workers' Compensation Claims	Compare actual costs of safety rail installation from previous jobs with new system after 3 jobs. Estimates Estimates	Estimation of all hard and soft costs of injures. Estimation of OSHA penalties assuming no change. Standard values	Training Time Purchase of new system Consideration Time Evaluation costs	Improved competitiveness. New business development.	Owner of Business Management Core Hourly employees.	Need for maintaining accurate cost accounting records.	

14.6 Evaluation Results

14.6.1 Level 1: Reaction and Planned Action

After a brief orientation and demonstration by the manufacturer's representative, the SFC jobsite supervisor took an informal survey of the full-time employees. The survey was captured on a simple form with five questions, as shown in Figure 14.1.

At this particular jobsite, not all 15 of the full-time employees were present, due to the size of the project. There were nine carpenters on the job. After the demonstration and some on-the-job training by the manufacturer's representative, nine of the nine carpenters felt that the system would result in lumber savings. Six of the nine felt that the system would be easier to install.

14.6.2 Level 2: Learning and Confidence

The target measure for demonstrating competence in the system was the observation that 100 percent of the carpenters had the

Question	Yes Answers	No Answers	Total
Do you believe than the new system of fall protection will result in less lumber usage?	9	0	9
Do you believe that the new system will be faster to install?	6	3	9
Do you believe that the new system will provide effective fall protection?	9	0	9
Do you feel you were trained adequately in the new system?	9	0	9
Do you feel that the new system will be a better system than we are currently using?	9	0	9
Totals	42	3	45
Percentage	93.3%	6.7%	100%

Figure 14.1 SFC Fall Protection System Survey Results.

knowledge and ability to install the new system as designed. Typically a two-man crew was assigned to install fall protection railing "in front" of the work. This basically means that, as each floor level was installed, a two-man crew would begin installation of protective railing in areas where a worker would be exposed to a fall greater than six feet. This protective railing would remain in place until walls, railing or another permanent structure was in place. As work progressed past the need for fall protection, this same crew would remove the railing, store the connection pieces and save the reusable lumber. On larger jobs, there may be numerous two-man crews performing this work. During the second and third week of the first job using the new system, SFC's field superintendent had the job foreman rotate carpenters through the fall protection crews in order to observe their competency installing the new fall protection system. After two weeks, the superintendent had seen each carpenter work on the system for approximately two days. Based on his observations, each carpenter had demonstrated the capability of correct installation.

14.6.3 Level 3: Application and Implementation

For the application and implementation evaluation, SFC wanted 100 percent compliance. A safety checklist was designed for this evaluation. The checklist, shown in Figure 14.2, consisted of a questionnaire addressing top rail height, mid-rail height, toe board installation, quality of installation (no jagged surfaces), adherence to the manufacturer's design and protection of all applicable areas.

Safety audits were conducted on the first three jobs where the new fall protection railing system was used. During the first job, three unannounced safety audits were conducted. In total there were thirty-two installation areas, using 427 lineal feet of the new system. All the installations were exactly as designed and there were no other findings.

During the course of the second job, only one unannounced inspection was performed, since the job was much smaller and was completed much more quickly. It was determined during that inspection that six sections (81 lineal feet) of railing were constructed correctly based on top and mid-rail heights and use of the manufacturer's fastening system. It was noted, however, that two locations were not protected by the guardrail system. Even though these two locations were not readily observable, by the strictest read of the OSHA standard the fall hazard did exist, and there

Question	Answer Yes or No (Any no answers require non-conformances to be listed)	
Are all top rails between 39 & 45 inches from the walking/working surface? **List Non-conformances:**	Yes	No
Do all sections have mid-rails approximately half-way between the top rail and working surface? **List Non-conformances:**	Yes	No
Do all sections have toe_boards (or other means of protecting workers below)? **List Non-conformances:**	Yes	No
Are the systems installed such that there are no rough or jagged surfaces that could cause punctures or lacerations? **List Non-conformances:**	Yes	No
Are the systems installed per the manufacturer's instructions? **List Non-conformances:**	Yes	No
Do all areas applicable to the new systems have fall protection installed? **List Non-conformances:**	Yes	No

Figure 14.2 Modular Fall Protection Checklist.

were employees of another trade contractor working in the area. In a worst-case scenario, SFC could have been exposed to a possible citation as a "controlling" employer, since by contract SFC was providing guardrailing for the project.

Only one unannounced safety audit was conducted on the third job. As with the second job, the project was not lengthy and only one audit was required based on the scope of the project. During this inspection, fourteen sections of railing (220 linear feet) were inspected. All the rails met the checklist requirements and no non-compliant locations were noted.

Based on all the data gathered during the safety inspections, SFC considered whether each section of railing was constructed appropriately as a unit rather than counting the top rail, mid-rail and fastening systems as individual metrics. The result was fifty-two correctly constructed guardrail units and two locations where the fall protection should have been installed and was not. While one could debate how to use the data for the two locations where no protection was installed, SFC felt that part of the program was to ensure compliance. Even though material compliance was achieved, the measure did not result in 100 percent compliance with the checklist, as designed. The measure for the level 3 data was 96 percent.

14.6.4 Level 4: Business Impact

Safety spending data were collected from all jobs performed in 2011. The total amount spent on lumber for safety purposes was $129,700. The estimated total lineal feet of guardrail installed was 53,000. The resulting material cost per lineal foot for 2011 was $2.45. For the three-job study in 2012, the total cost of lumber used for safety purposes was $21,300. The total lineal feet of guardrail installed was 8,000. The resulting material cost for the study period was $2.66 per lineal foot. At first glance, it appears that the cost of lumber per lineal foot of guardrail installed was increasing. In 2011, however, lumber prices rose considerably. SFC's average cost in 2011 was $301 per 1,000 board feet. In 2012, the cost had risen to $373 per 1,000 board feet. In order to have meaningful information, SFC used 2011 as its baseline pricing year. After adjusting for the 24 percent increase in lumber prices, the adjusted material cost for installing the new style guardrail was $2.02 per lineal foot. The resultant decrease in lumber costs normalized against the baseline was 17. 9 percent

Labor costs coded to safety (fall protection only) in 2011 were $246,900. Based on the 53,000 lineal feet of guardrail installed in 2011, the labor cost per lineal foot was $4.66. Over the study period, the labor costs coded to safety were $25,100. Based on the 8,000 lineal feet of guardrail installed during the study period, the resulting cost was $3.13 per lineal foot. The result on a lineal foot basis was a savings of 32. 8 percent

In summary, the total labor and material costs for 2011 to install fall protection guardrail was $7.11 per lineal foot ($2.45 for material, $4.66 for labor). In the study period, the material (lumber only)

and labor costs were $4.78 ($2.02 for material, $3.13 for labor). Over all, the resulting business impact for labor and materials on a lineal foot basis was a savings of 27.5 percent.

For the OSHA penalty calculation, SFC took the $19,000 penalty that was directly related to fall protection and amortized it across the same 53,000 lineal feet of railing constructed in 2011. This method assumed that, over time, if SFC continued to operate in the exact same manner, the same penalties would reoccur. While SFC recognized there are many variables in calculating an OSHA penalty, it was felt that, for purposes of this study, this was a good barometer. The resulting cost was $0.39 per lineal foot of railing constructed. SFC did not want to assume that all OSHA penalties would forever go away. To be conservative in its estimate, SFC assumed that the savings from penalty avoidance would be reduced by approximately 50 percent. SFC management settled on $0.20 per lineal foot as an acceptable cost of penalties for purposes of this analysis.

For the cost of the injuries, SFC began with a method of calculating the total cost of an injury as a function of the direct cost. During 2011, SFC had three minor, yet OSHA-recordable, injuries while installing stick-built fall protection. The total direct cost of the injuries was $2,970. Direct costs included the emergency room cost and other medical bills. Based on information from a variety of sources, SFC used a factor of four to calculate the indirect costs, which would include administrative costs in dealing with the injury, increased workers' compensation insurance costs and all other indirect costs associated with an injury. The total indirect cost was estimated at $11,880. This brings the total costs of injuries associated with installing fall protection for the year 2011 to $14, 850. As with the OSHA penalty, SFC amortized this over the 53,000 lineal feet of guardrail installed to arrive at a 2011 cost of $0.28 per lineal foot. As with the OSHA penalty calculation, SFC did not wish to assume that every injury would be prevented for all time, so again it settled on a 50 percent reduction in the injury cost. This was believed to be reasonable because, during the study period with the new system, workers had clear work instructions that standardized the process and there were no incidents with injury during the study. The assumed cost of injury for the future was set at $0.14 per foot.

The final business impact to be measured was compliance with the OSHA standard. Based on the safety audits conducted in the study period, a 96 percent compliance rate was achieved. While this did not meet the business impact target of 100 percent, SFC felt that, considering the nature of the observed non-compliance during the study, the failure to meet this target was not impacted by the implementation of the new product. This is a management or training issue that should be isolated and addressed separately. Collectively, the field supervision and the hourly workforce felt that compliance was easier with the new system. For the purposes of this analysis, the impact was deemed positive.

14.6.5 Level 5: ROI

Table 14.3 lists the data necessary to determine if the change to a new system of fall protection guardrail provided the targeted ROI on an annualized basis.

In 2011, it cost SFC $412,340 to install 53,000 lineal feet of safety railing using the old method. The total cost per foot was $7.78, which includes labor and material, OSHA penalty and injury costs. Assuming that SCF does the same amount of business on an annual basis, when adjusting for lumber prices, the total cost of labor and materials, along with the cost of OSHA penalties and injury, would be $290,970. This is a cost reduction of $121,370—a significant benefit to the company. Cost of purchasing the new system was $23,000. It was estimated that the fasteners and connectors either would be lost or destroyed at a rate of approximately 10 percent annually, resulting in a replacement part cost of $2,300 per year. Finally, there was some cost for training the 15 full-time employees. The cost was minimal, since much of the training was on the job. Training costs included two hours of training for 15 full-time employees paid an hourly rate of $20 (fully loaded); facilitator cost of $100 per hour; materials cost at $10 per person, including facilitator; and the cost of a small training room of $75 per hour. Total cost for formal learning was $1,110. Evaluation costs of approximately $1,100 also were considered even though this type of cost analysis was routinely conducted on this type of project. Total cost of the solution was $27,510. The BCR was 4.41 with an ROI of 341 percent.

14.6.6 Intangible Data

Upon completion of the data collection part of this evaluation, both management and hourly employees provided input on the benefits the new system might bring to the organization. These discussions were held informally and collected by the safety manager. The key

Table 14.3 Benefits and Costs.

Cost Savings Due to Solution			
Description	**2011**	**2012**	**Percent Change**
Material Cost ($/lf)	$ 2.45	$ 2.02	–17.9%
Labor Cost ($/lf)	$ 4.66	$ 3.13	–32.8%
OSHA Penalty ($/lf)	$ 0.39	$ 0.20	–48.7%
Injury Costs ($/lf)	$ 0.28	$ 0.14	–50.0%
Total Cost($/lf)	$ 7.78	$ 5.49	–27.6%
Annual Feet Installed (normalized to baseline year)	53,000	53,000	N/A
Annual Material, Labor, Penalty and Injury Costs	**$412,340**	**$290,970**	**–29.5%**
			$121,370

Cost of Solution			
Cost of New System		$23,000	
One Year of Replacement Parts		$ 2,300	
Formal Training of Hourly Employees Participant time: $600 Facilitator time: $200 Materials: $160 Facilities: $150		$ 1,110	
Evaluation Cost		**$ 1,100**	
Total Solution Cost		**$27,510**	
BCR $\frac{\$121,370}{\$27,510} = 4.41$			
ROI $\frac{\$121,370 - \$27,510}{\$27,510} = 341\%$			

intangible was the ability to bid more competitively and secure more jobs. Another intangible was the potential of an enhanced image of the company. Not only did the new guardrail system prove to be cost effective, but it also has a professional and high-quality look compared to the framing done only with 2″x4″ lumber. In addition to the observed productivity increases, the hourly employees felt that the system was easier to install, thus making their work more productive. There were no negative intangible-related comments.

14.7 Communication Strategy

SFC is a relatively small business compared to many of those using the Phillips ROI Methodology. It is closely held and employs a small staff. Every penny counts. The communication strategy for this ROI study was straightforward. There are three audiences within SFC: the owner, the managers and the hourly employees. All information from this project was summarized and provided to each of the three audiences. Since the project did result in an extraordinarily high return on investment, the communication strategy was to address in one document and one meeting results for each stakeholder group. For the owner, the ROI of 341 percent was well beyond the target 25 percent, but not entirely surprising. Any time a costly problem can be resolved with an inexpensive solution, there is a high return. Nevertheless, he was pleased and satisfied that his resources were being used appropriately. For the owner and managers, the business impact of being able to reduce cost and maintain competitiveness was met. Additionally, the ever-looming regulatory enforcement action was mitigated. For the full-time hourly employees, the communication strategy was to ensure they were aware that their efforts made a difference in the safety of their jobsites and the long-term sustainability of the SFC organization.

14.8 Lessons Learned

One lesson learned stems from the fact that when implementing the new system, the first project received a significant amount of attention focused on ensuring that all areas needing fall protection were addressed. In the second project, there was a bit less oversight and ultimately two areas of potential non-compliance with the regulations were discovered. The lesson is that no matter how easy or inexpensive the system, all potential fall hazards must be identified before they can be addressed. While this certainly seems obvious, if 100 percent compliance is the goal, then the same significant effort and oversight as demonstrated in the first study project must occur each and every day. More globally, the lesson is that, regardless of new capital programs, new tools or technical enhancements, management systems must be robust and sustainable.

14.9 Questions For Discussion

1. Is this study credible? Explain.
2. How could this study be improved?
3. Can the ROI approach work in smaller organizations with limited staff and resources?
4. How could this approach be presented to other small business owners?
5. What is the value in this type of rigor in analysis for small business owners? For large corporations?

15

Measuring ROI in an Ergonomics-Based Risk Management Intervention

National Furniture Company

George (Sonny) Blackwell

Abstract
A furniture manufacturer faced with mounting claims costs needed to quickly develop and implement solutions to reduce claims costs without interrupting productivity. Traditional compliance and training-based interventions had proven unsuccessful. With assistance from its insurance broker, National Furniture Company formed a small team of production employees to which the risk management consultant provided training in ergonomic risk factor recognition and a method to develop and estimate the effectiveness of engineering modifications to the assembly process. The team ultimately generated more than thirty ideas, many of which were tested or implemented. Those changes resulted in a significant reduction in the number of employee injuries and workers' compensation claim costs and 452 percent return on investment.

Keywords: Claims cost reduction, risk management, ergonomics

15.1 Background

National Furniture Company (NFC) produces upholstered furniture,including sofas, loveseats and motion recliners. The company grew for approximately twenty years and employs around 800. It had been privately owned since it was founded but was recently purchased by an investment group that chose to leave local management in place. NFC continued to use a

Jack Phillips, Patti Phillips, and Al Pulliam, Measuring ROI in Environment, Health, and Safety, (339–360) 2014 © Scrivener Publishing LLC

piece-rate compensation program under which employees were compensated based on the number of units produced rather than the number of hours worked. Work organization is typical for upholstered furniture production including receiving, cut-and-sew, wood working shop, frame building, upholstery and assembly and shipping.

Safety, while not disregarded, had never been a priority and accident costs were absorbed as a business cost. During the past several years, claims costs had escalated to a point that the workers' compensation experience modifier had increased to approximately 1.26, meaning that National Furniture Company experienced approximately 26 percent higher claim costs than actuarially expected and therefore paid approximately 26 percent more for workers' compensation insurance than a comparable business with average claims experience. Workers' compensation premiums had risen as a result to 3.15 percent of payroll. These metrics were unacceptable in a marketplace that includes significant foreign competition.

In addition to the cost of insurance, the claims experience significantly reduced the number of insurance carriers willing to cover the exposure. Various underwriters signaled that unless performance improved, National Furniture Company was at risk of being placed in the residual insurance market, which would result in an additional 30 percent penalty, potentially making the company noncompetitive and economically non-viable. The combination of new management, limited options and high costs created a stressful environment with openness to changes that may result in improved claims performance, lower costs and more options for risk finance.

15.1.1 Project Formation

National Furniture Company's previous insurance providers had approached loss control through traditional compliance-oriented means that included inspections and employee training with little or no measurable impact. Some improvements, based largely on ideas derived from "Voluntary Ergonomics Guidelines for the Furniture Manufacturing Industry," had been implemented, but management considered many too expensive or impractical. An alternative approach that would develop cost-effective remedies had to be developed. Management was very clear that whatever changes were made could not impede production rates.

The developing magnitude of the situation required management to consider new approaches that could lead to improved results. The insurance agent provided the services of a consulting risk manager to devise a plan of action. The first step was to evaluate several years of historical loss activity based on carrier "loss runs" and Occupational Safety and Health Administration form 300 records. Through this analysis it was determined that the majority of claims costs resulted from a few specific jobs and it became apparent that the majority of intervention efforts would need to focus on making changes to the approximately twenty assembly and upholstery production lines.

A plant tour was conducted after the claims analysis, which is when it became apparent that there would be opportunities to modify the work process to reduce ergonomic risk factors. Employees worked at a tremendous pace due to the piece-rate compensation structure and there were many examples of employees engaged in stressful manual material handling, awkward work postures, deferred maintenance and employee-modified work stations; all indicators of musculoskeletal stress.

It was agreed that the insurance agency risk management consultant would provide training in risk factor recognition to a small number of production and supervisory personnel. Those employees would form an ergonomics improvement team to develop plant-specific remedies to reduce risk factors contributing to the musculoskeletal injuries. Most of the employees had several years of experience working at National Furniture Company or similar operations, however none had training in recognizing ergonomic risk factors or selecting cost-effective remedies. The risk management consultant also was expected to provide a system for estimating and ranking the effectiveness of ideas developed by the team. Those ideas showing the most promise would be tested in the plant and, if effective, would then be implemented throughout the plant.

Risk factor recognition training involved seventeen employees and required approximately three hours. The team later met several times to identify problems, generate ideas and make recommendations to management.

15.1.2 Project Purpose

The intervention was necessary to reduce the drag on profitability caused by accident costs. Without significant improvement in this aspect of performance, NFC faced mounting claim and insurance costs that could have had several undesirable consequences, including higher labor and production costs. The purpose of the intervention was to improve financial performance by lowering claim costs and by creating additional risk finance options.

15.1.3 Project Objectives

The intervention had three ultimate objectives. The first was to quickly, significantly and demonstrably improve workers' compensation claim performance to prevent NFC's workers' compensation program from being placed into the "assigned risk" market and ultimately to create competition among insurance carriers willing to underwrite the coverage more aggressively. The second objective was to observe a measurable reduction in employee injuries and workers' compensation claim costs that ultimately would lead to an experience modifier of 1.0 or better. The experience modification factor is a premium adjustment technique used by underwriters intended to reflect the company's actual claim experience over time. The company's expected and actual claim costs are compared against a group of peer organizations. This factor is multiplied by a governing rate to determine the actual premium. Organizations with better-than-average claims experience receive a credit, while those with higher-than-expected claims are penalized. Organizations have significant control over their experience modifiers through their accident prevention and claims management processes, but the measure is a lagging indicator that can take up to four years to fully react to significant improvements in risk management.

The third objective involved the requirement for the intervention to have no adverse effect on production efficiencies.

15.1.4 Purpose of Evaluation

This study was primarily for the purpose of demonstrating the effectiveness of a focused risk management intervention and the value of the service the insurance agent provided to NFC.

A secondary purpose was to determine the applicability of the Phillips ROI Methodology in a hybrid intervention that involved both training and engineering components and to identify additional issues that may need to be considered when using the methodology with risk management programs.

Risk management, safety and loss control practitioners lament the fact that during difficult economic times, their programs are sometimes the earliest targeted for elimination because decision makers may perceive safety and health programs as costs rather than value contributors. An additional purpose of the evaluation was to explore and develop a means for risk management, loss control and safety professionals to transcend a common perception that risk management is simply a necessary cost of business and demonstrate that carefully selected and well-designed programs align with and contribute toward strategic organizational goals.

15.2 Evaluation Methodology

The "Hierarchy of Controls" has existed in some form for decades, since its introduction by H.W. Heinrich in 1931. The hierarchy recently was adopted as part of the ANSI Z10-P Standard for Occupational Health and Safety Management Systems. The hierarchy encourages risk managers to prioritize controls based on "feasibility" and infers, but does not specifically mention ROI. Following the hierarchy, risk managers are encouraged to first consider eliminating hazards and then other types of intervention in order of anticipated effectiveness based on years of anecdotal and quantified success.

1. Eliminate the hazard
2. Substitute a less hazardous option
3. Engineering controls
4. Warning signs, signals and labels
5. Administrative Controls, including training
6. Personal protective equipment

The traditional approaches to measuring the success of risk management interventions include the frequency of recordable lost time accidents, total cost of risk and workers' compensation experience modifier. Risk managers often struggle with converting these terms into a language that decision makers appreciate. Each requires some technical understanding of measurement unit definitions, making it awkward to communicate program value, or lack thereof, to decision makers unaccustomed to these terms.

Some risk managers promote the concept of total cost of risk, that being the cost of insurance in addition to the indirect costs of accidents. Indirect costs have been estimated to be several times the direct value of the claim and a 4:1 ratio is widely claimed. These indirect costs primarily include the impact on production efficiency caused by down time resulting from accidents. This 4:1 ratio has been used by risk managers for decades to magnify the total cost of industrial and construction-related accidents and to help justify the need for occupational safety and health programs. While it would be difficult to argue that indirect costs do not exist, it is difficult for some to accept a single, consistent indirect-to-direct cost ratio for all accidents in all business segments. Risk management professionals need a conservative, defendable and robust method more consistent with generally accepted accounting practices to demonstrate value.

This project was chosen for ROI analysis because it represented an opportunity to apply the Phillips ROI Methodology to a unique project that included both training and engineering components, offered readily obtainable cost and Level 4 impact data and represented a potential opportunity for the insurance broker to express a quantified business value proposition to the owner-client.

15.2.1 Categories or Levels of Data

The Phillips ROI framework evaluates six types of data and some may be disturbed to learn that traditional activity measures are not among them. Many traditional metrics such as billable hours, training hours, number of surveys or inspections more accurately measure input or cost than result. Risk managers should consider means to shift terms of thinking from activities to results in order to change the perception of their programs from costs that can be eliminated to the value they contribute to the organization.

The training facilitator through direct feedback from participants and team leaders measured level 1, reaction data, informally. The withdrawal rate also was an indicator of initial reaction. No written instruments or quantification were developed because the training phase was to be offered only once and no decision to provide ongoing training was necessary.

Level 2, learning or confidence, was confirmed by allowing participants to practice several simulations as a group, receiving feedback from the facilitator until they indicated their comfort with the material.

Level 3, application and implementation, was confirmed by reviewing documentation completed by the group that prioritized lists of ideas for specific hazards identified. Some of the most significant involved modifying the assembly line to reduce lifting and awkward positions, replacing small casters with large casters to reduce the rolling resistance of carts, installing turntables to make certain upholstery stations more efficient, installing tables and reducing bundle sizes at the sewing stations to eliminate lifts from the ground and reduce the lift weights and improving seating in the sewing department. The training facilitator also accompanied the team on several plant tours and they visited various production areas with a history of injuries.

Business impact, or level 4 data, was derived exclusively from insurance carrier claim records known as "loss runs." This data is recorded in monetary form and requires no conversion. Care was taken to use "developed" claims data, ensuring that accurate and conservative values were subsequently used in the ROI calculation.

15.2.2 Standards or Guiding Principles

Guiding Principle 1: When higher levels of evaluation are conducted, data must be collected at lower levels.

Guiding Principle 2: When an evaluation is planned at a higher level, the previous level of evaluation does not have to be comprehensive.

In this case, level 4 business impact and level 5 ROI were planned, so less emphasis was placed on quantifying Level 1 reaction data. This was also true because the training phase of the intervention would occur only once and no decision to provide recurring training was anticipated.

Guiding Principle 3: When collecting and analyzing data, use only the most credible sources.

Guiding Principle 4: When analyzing data, choose the most conservative alternative for calculations.

Insurance carrier claim data that had "developed" was used for the ROI calculation. The ROI calculation, therefore, did not rely upon forecasts, estimates or total-cost-of-risk that included indirect loss costs.

Guiding Principle 5: At least one method must be used to isolate the effects of the solution.

The ROI estimate was based on claims performance data rather than upon insurance premiums, which can be influenced by a number of external factors including insurance market cycle or general economic conditions. Basing the evaluation on claims data, as if NFC was self-insured, eliminated the possibility of external factors.

The original plan anticipated a quasi-experimental design, leaving two of four production units untouched as control groups. The initial results were so significant that management insisted on implementation plant-wide. It therefore became necessary to use the alternative isolation technique of trend line analysis to compare expected claims against actual claim experience. Claim data, rather than insurance premium was selected as the basis for the ROI estimate because it would be quite difficult or even impossible to isolate premium costs from the effects of the financial market and larger underwriting cycles. At the time of this writing, workers' compensation markets were under the influence of an extended "soft market," indicating a period of aggressive discounts applied by underwriters to attract business. Underwriters rarely apply discounts automatically for having certain components of safety programs. Likewise, the experience modifier was a poor indicator of the effectiveness of the intervention because it essentially is a three-year floating composite compared against industry peers. It is not designed to forecast future impact of changes. Insurance brokers working on behalf of NFC needed credible evidence of program effectiveness much more quickly than the time needed for the experience modifier adjustment.

Guiding Principle 6: If no improvement data are available for a population or from a specific source, assume that little or no improvement has occurred.

This study did not rely upon surveys or multiple data sources that could have been absent. The level 4 data represented by the insurance company reports included the entire manufacturing facility.

Guiding Principle 7: Estimates of improvements should be adjusted for the potential error of the estimate.

Results estimates were not used as the basis for this study. However, care was taken to use developed loss data. Workers' compensation claims tend to increase in value for some time. Relying upon undeveloped data could cause improvements to appear greater than they actually are.

Guiding Principle 8: Extreme data items and unsupported claims should not be used in ROI calculations.

Claim data is well documented and objectively supported risk managers sometimes include "indirect costs" of claims to estimate their total cost. They commonly apply a 4:1 indirect-to-direct cost ratio to obtain a total-cost-of-risk. Indirect costs were not specifically quantified for this study and were therefore not included in the calculation. Using the "total-cost-of-risk" approach assuming a 4:1 ratio would have resulted in an incredibly high (2,660 percent) ROI, inviting undue skepticism from management.

Guiding Principle 9: Only the first year of benefits should be used in the ROI analysis of short-term solutions.

The entire intervention, including training, design, testing and production line modifications, required almost a full calendar year. Claim data from the following year was therefore used to estimate the impact in keeping with the methodology, however the ultimate ROI may be considerably higher than estimated because the intervention involved physical modifications to the assembly process and ultimately could be significantly higher than reported for this one-time investment.

Guiding Principle 10: Costs of the solutions should be fully loaded for ROI analysis.

Care was taken to capture all of the costs associated with the intervention, including the hourly wage rate of employees who participated on the team, capital improvements and even the costs associated with remodeling a training room to accommodate team meetings. Although the insurance broker provided the assistance of its risk management consultant at no additional fee to NFC, consulting costs were included in the ROI calculation as if NFC had assumed the consulting costs.

Guiding Principle 11: Intangible measures are defined as measures that are purposely not converted to monetary values.

Remarks from the human resources director, supervisors and employees regarding reduced fatigue or more energy were important indicators that the ergonomic improvements were having intended effects even before the claims data was evaluated. The entire purpose of the ergonomic improvements was to reduce the stress, strain and energy demands on the body that ultimately contributed to injury.

Guiding Principle 12: The results from the ROI Methodology must be communicated to all key stakeholders.

Although financial data was not shared with employees, the OSHA recordable injury statistics and a summary of plant improvements were shared with employees who contributed to the ergonomic improvement teams. Financial results were communicated quarterly with the management and ownership groups and the results were used to communicate with underwriters to negotiate better terms for the insurance renewal. The final ROI estimate was presented to the ownership group as part of an annual stewardship meeting in which brokers reviewed insurance program highlights and achievements.

15.2.3 Example Data Collection Instruments

No special instruments were necessary for the project. Simple lists of ideas, direct observations by the facilitator and workers' compensation claims data were all that were necessary. Figure 15.1 shows the data collection plan.

The quarterly review of claims data provided strong indications that the intervention would have significant ultimate results, but a conservative ROI estimate required the claim data to mature. Adjusters make preliminary estimates of the ultimate value of claims when they are initially received, but these estimates must frequently be increased due to unanticipated developments such as medical complications requiring additional treatment over time. In this case, the post-intervention claim data had developed for approximately the same period of time as had base-line claim data.

15.2.4 Isolation Techniques

The preferred isolation technique involving quasi-experimental design could not be used due to management's enthusiasm. The effects of the project were isolated by comparing the expected losses, based on three years of claims data prior to the project, with the actual claim totals in the year after implementation. No other significant changes occurred to the process and the number of employees and production rates remained constant.

Data Collection Plan						
Program/Project: National Furniture Company Ergonomics Improvement Program **Responsibility** : Blackwell/Jo nes/Jernigan						
Level	**Broad Program Objectives**	**Measures/Data**	**Data Collection Method/ Instruments**	**Data Sources**	**Timing**	**Responsibility**
1	Positive reaction to the awareness program and intent to participate	Comments solicited from leaders and members	Interviews with team leaders Team focus groups	Team leaders Team members	Post-session 30 days	Blackwell
2	Team members can identify risk factors and high-risk tasks	Members can recognize ergonomic risk factors and high-risk tasks from photographs	Interactive session using photographs of workers performing tasks with various risk factors	Team members Team Leaders	30 days	Blackwell
3	Teams meeting at least 2x per month Teams identifying high-risk tasks Teams using R3 worksheet Teams generating ideas for testing	Meetings Completed R3 worksheets Ideas	Team meeting records Completed worksheets Idea list	Team records	1,2,3 months	Team leaders HR Director
4	Reduce OSHA lost workday rate Reduce workers' compensation claim rate Reduce total incurred WC claims Reduce forecast experience mod Reduce WC cost % of labor	OSHA lost workday rate Claims / 100 employees Total claims NCCI experience mod WC % of payroll	OSHA 300 Liberty Mutual loss runs AcuComp® analysis HR records	Company records Liberty Mutual	6, 12 months	Blackwell HR director
5	Target ROI ⇨ 15%	**Comments** :				

Figure 15.1 Data Collection Plan.

15.2.5 Data Conversion Techniques

No data conversion was necessary because claims data is reported in monetary value terms.

15.2.6 Cost Summary

NFC's insurance agent and carrier provided loss control as a value-added service for no additional fee. Had NFC been self-insured, it would have incurred expenses for the consulting and training services it received. Guiding principle ten requires fully loaded costs to be used. Consulting fees of $1,500 per day plus travel, lodging and meal per diem will be included in the example as if NFC had purchased "unbundled" loss control service, as is common for self-insured businesses.

1 day	Initial loss run and OSHA record analysis (no travel)	$1,500
1 day	Training preparation day (no travel)	$1,500
1 day	Initial plant tour, orientation and consulting	$1,725
1 day	Training for ergonomic teams plus travel and lodging	$1,725
3 days	Follow up with ergonomic teams plus travel and lodging	$5,175
3 days	Data analysis and report preparation (no travel)	$4,500
	Total consulting, training and evaluation fees	$16,125
	Engineering modifications	$58,438
	Total cost of intervention	$74,563

15.3 Evaluation Results

The project was an almost immediate success. The teams quickly identified specific risk factors associated with specific tasks and generated alternative solutions to reduce the risk. Using worksheets provided by the consultant, alternative approaches were evaluated on paper and then prioritized for presentation to plant management. Some of the ideas were drawn from materials provided to the team and others were developed based on experiences some team members had at other furniture manufacturers, while other ideas were original ideas. Various suggestions were abandoned while others were implemented throughout the plant.

The plant maintenance staff was able in most cases to utilize stock items from inventory to fabricate engineering modifications in-house, thus keeping costs minimal. The most significant and extensive modification involved the design and installation of pipe-rail fabrication lines to replace the series of upholstery "bucks" that previously had been used. This engineering modification greatly reduced the need to manually lift and move the pieces along the assembly process. Minimal effort is now required to move the pieces between stations.

The human resources director maintained time and attendance records for team meetings and the plant maintenance director tracked materials expenditures. Fully loaded costs, including remodeling the room to accommodate the team meetings, are detailed below. The out-of-pocket cost to NFC, including hourly rates for employees to participate in training and team meetings and for materials to fabricate production line changes, totaled $58,438.

15.3.1 Level 1: Reaction

The instructor sought feedback from participants, team leaders, supervisors and the human resources director to estimate the team members' reaction to the training. The employees' reaction was very positive, with some observing that this was the first time management had sought their participation and input. After the initial training, two employees asked not to participate. They were thanked, excused and not replaced on the teams.

15.3.2 Level 2: Learning

The instructor later attended team meetings to observe and estimate the participants' knowledge retention during team meetings. Greater effort could have been made to quantify levels 1 and 2, however the ROI estimate is based entirely on Level 4 claim cost data. This is consistent with guiding principles 1 and 2, which require evaluation at lower levels by allows lower-level evaluation to be less rigorous.

Observations during subsequent follow-up meetings and observations confirmed that the team members comprehended the basic concepts based on their ability to recognize problem areas, generate alternative ideas and use the methodology to estimate and prioritize the best solutions.

15.3.3 Level 3: Application

The objective of the ergonomic team training was to enable participants, who had little or no prior training in ergonomic risk factor recognition, to develop and estimate the effectiveness of assembly line modification ideas that would reduce risk factors contributing to employee injuries. The team recorded results on a simple spreadsheet that confirmed over 30 ideas had been developed and considered. From this list, twelve modifications with the most potential reduction were implemented. The most significant modifications included installation of assembly rails and other modifications to the assembly process to greatly reduce the frequency of lifts. Others included installing tables adjacent to cutting tables to eliminate lifting bundles of upholstery from the floor, reducing the size and weight of upholstery fabric bundles, replacing worn casters to reduce the effort needed to push or pull buggies, limiting storage height to reduce lifting objects over shoulder height, maintaining casters regularly to prevent fabric from accumulating in caster wheels, repairing rough flooring and installing scissor lifts to facilitate pulling lumber for the chop saw operation.

15.3.4 Barriers to Application

The primary barrier to applying the information and skills obtained in the training aspect of the intervention involved scheduling time for participants representing disparate areas of operation to meet together as a team to generate, evaluate and prioritize alternative solutions. Production lines are highly sensitive to absent team members, so adjustments had to be made to accommodate the monthly meetings.

The most significant barrier involved implementing the team-recommended physical modifications to the assembly process. The assembly line process is remarkably choreographed and even minor physical changes required employees to develop new behavioral routines in order to coordinate product movement from one assembly station to the next.

15.3.5 Enablers to Application

Management was highly motivated to improve safety performance results, so it placed a high priority on allowing reasonable time for the teams to meet. Suggestions with the most promise were implemented almost immediately on a trial basis and quickly adopted throughout the production lines when their potential was confirmed.

15.3.6 Level 4: Business Impact

The annual developed claim costs for the three calendar years prior to the intervention, 2006, 2007 and 2008, averaged $662,959, as shown in Figure 15.2. Team training, idea development, prototype testing and implementation consumed much of 2009, although claims count and average costs began to trend downward almost as soon as implementation commenced. The actual developed claims incurred for 2010 totaled $251,351, representing a $411,608 (62 percent) improvement in claim cost performance as compared to expected losses without the intervention.

In order to further validate the financial improvements and in anticipation of other metrics of interest to management, the claim cost per unit of production was calculated and the number of claims was also tracked. Management's challenge to not impede productivity was met. Productivity increased slightly from an annual number of 713,044 before the intervention to 737,955 post-intervention, representing a 3.84 percent increase. Figure 15.3 shows the impressive result.

According to the human resource director, employee turnover of the more stressful jobs was high due to the physical demands, but management did not perceive this as a problem because many qualified applicants were available to immediately fill vacated

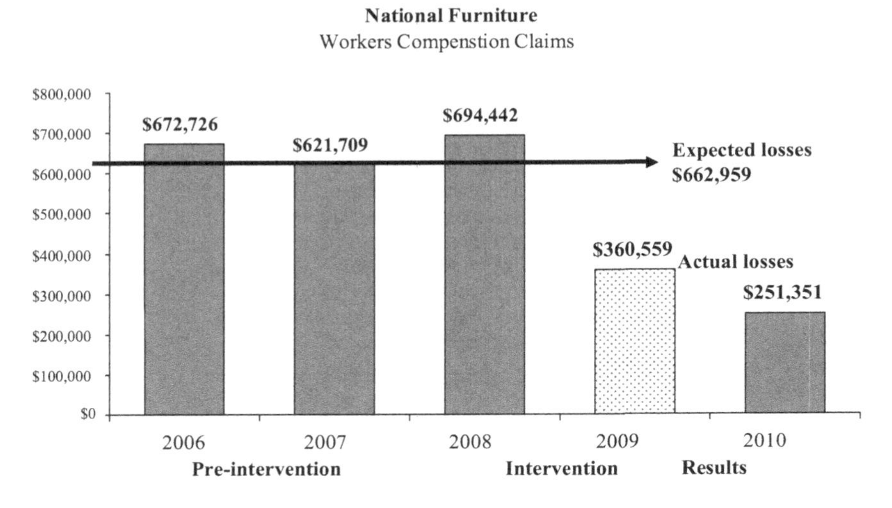

Figure 15.2 Expected losses compared to claims after intervention.

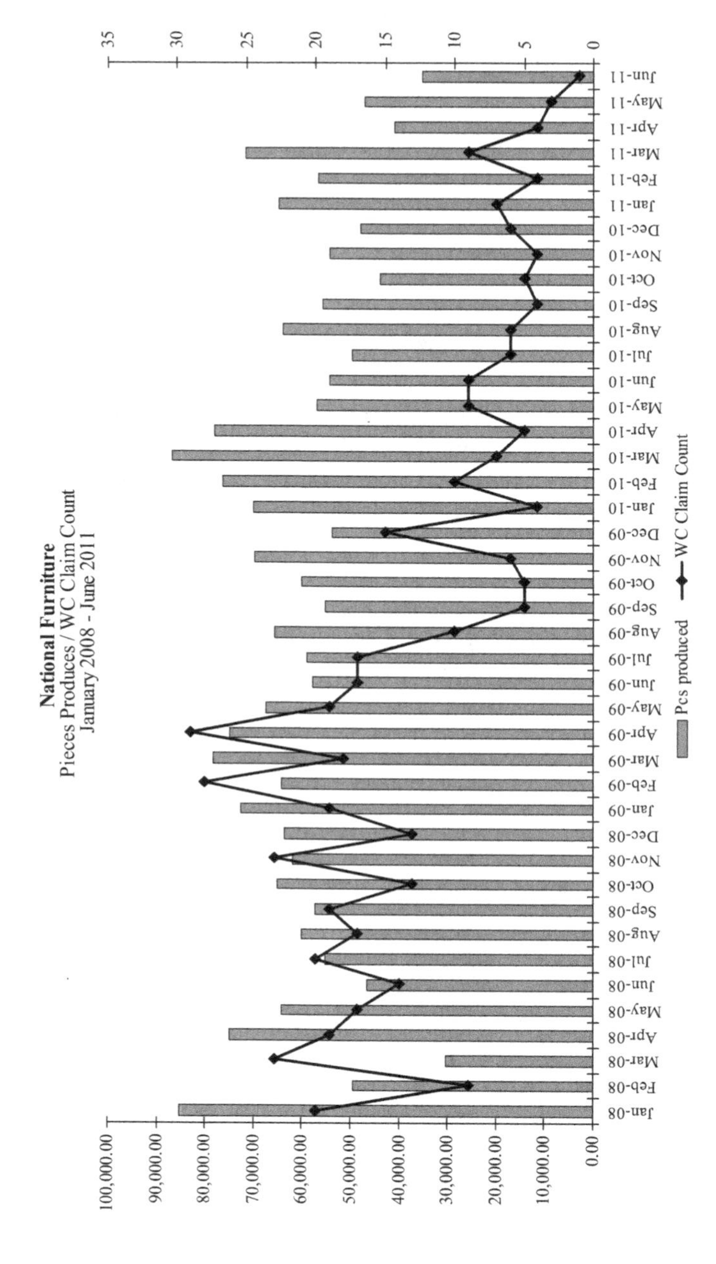

Figure 15.3 Claims Cost Per Unit of Production.

positions. Because management believed turnover costs were extremely minimal due to job market conditions, no attempt was made to measure changes in turnover rates and include those costs in the final ROI calculation.

15.3.6.1 How Results Were Isolated to the Intervention

Guiding principle five requires the effects of the solution to be isolated. NFC is organized into four autonomous production units and it originally was planned to establish control groups to compare units with changes against units that remained unchanged. The impact of the changes was so significant and immediate that management insisted on implementing changes in all production units. This required changing the preferred isolation method from an experimental design to a comparison of expected losses against actual post-intervention losses.

15.3.6.2 How Data Were Converted to Money

Insurance claims data require no conversion to monetary terms.

15.3.6.3 Intangible Benefits

Guiding principle 11 recognizes intangibles, those measures difficult or impossible to quantify or purposefully not converted to money values. During the process of this intervention, many NFC employees had an opportunity to contribute their thoughts, many expressed pride in their innovations.

Several employees remarked that they were less fatigued at the end of the day and had more energy to enjoy their time off after work. Several supervisors and the human resources manager remarked that employees had a "spring in their step" at the end of the day. Since the goal of ergonomic improvements is to reduce the stress placed on the body, this anecdotal observation indicated that the project was on target.

15.3.7 Level 5: ROI

Program Costs are as follows

Engineering modifications	$ 58,438
Consulting, training and evaluation expenses	$ 16,125
Total costs	$ 74,563

Expected annual claim cost w/o intervention	$662,959
Actual claim cost	$251,351
First year post-intervention improvements	$411,608
Net benefit	$337,045

ROI = $337,045 / $74,563 X 100 or 452%

15.4 Communication Strategy

15.4.1 Results Reporting

Accident frequency and claim costs were monitored, compiled quarterly and reported in a series of charts to the chief financial officer, who then reported to the ownership group.

The charts also illustrated production and efficiency levels in order to compare accident costs against production levels and also to ensure that the production line changes did not impede production efficiency. Figures 15.4 and 15.5 provide more detail of the results.

15.4.2 Stakeholder Response

The employees participating on the team provided enthusiastic input and took pride that their ideas were being implemented. This seemed to carry over to the other employees, who either saw or benefited from the improvements. The plant expressed great surprise that such a significant improvement was achieved in a relatively short amount of time, considering that their previous attempts had been unsuccessful. The ownership group was very pleased with the results and has renewed the insurance coverage with the consulting broker for three years, based significantly upon the success of the consulting services.

15.5 Lessons Learned

While a few risk managers hold advanced accounting degrees, the ROI Methodology, while not easy, is not beyond the reach of those with technical, training or operations backgrounds. This framework gives us the ability to speak the language of decision makers without

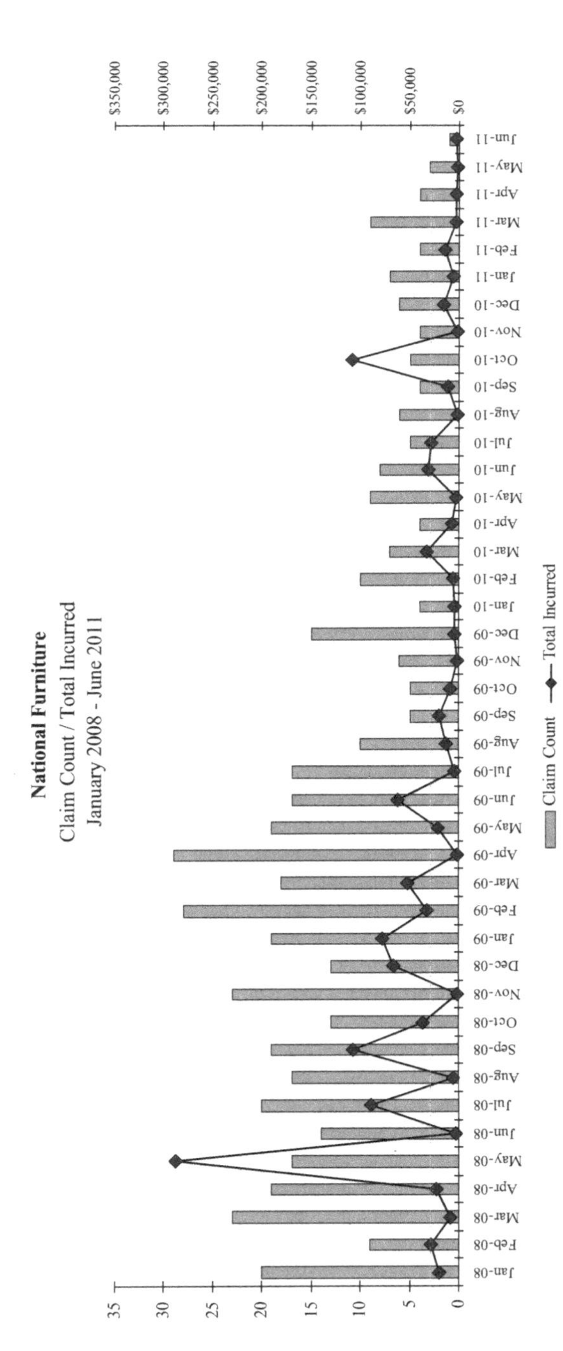

Figure 15.4 Workers' Compensation Claims.

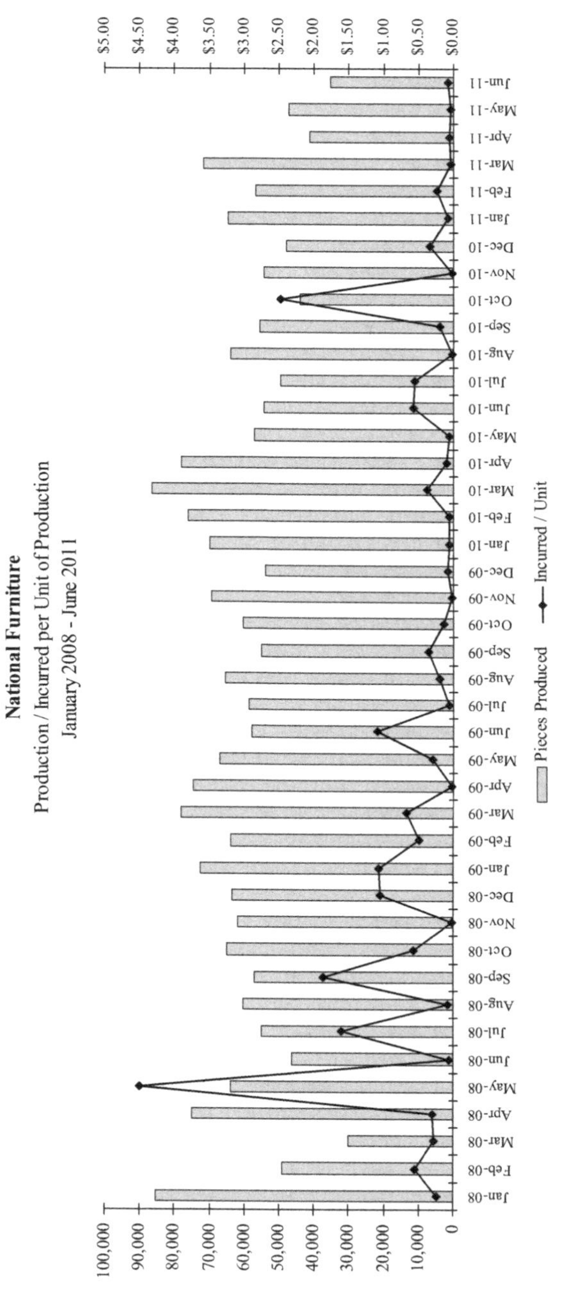

Figure 15.5 Production on Workers' Compensation Claims Per Unit.

attempting to be artificially technical and damage credibility by attempting to be finance experts. The ROI Methodology holds significant potential for risk managers, whether they are employed by insurance agents, carriers, businesses, public sector or nonprofit organizations. It combines critical risk management issues such as prioritization and accountability into a single process that can be integrated into enterprise risk management systems, aligning risk management with other organizational goals. By following a proven, reliable method, we can establish objective credibility for our proposals. This method provides us the ability to monitor program performance and make modifications necessary to make them more effective and cost efficient. Over time, as ROI studies are collected, risk management professionals could accumulate a reliable stockpile of case studies that would be beneficial by allowing us to identify consistently effective interventions and avoid trendy or unsubstantiated methods.

Questions for Discussion

1. Is this case study credible? Explain.
2. Are risk management and safety interventions considered costs-of-doing-business or activities that add value to the enterprise?
3. Which specific risk management or loss prevention programs in your organization exist primarily due to regulatory compliance? Do they also provide a value to the organization? If they were eliminated, would there be any effect on financial performance?
4. What strategies or programs represent potential to provide value if designed and executed more effectively? What would be the financial impact to the enterprise?
5. Would risk management and safety-related employees welcome or feel threatened by augmenting traditional metrics with ROI or other performance-based measures?
6. How could risk management ROI for be used in enterprise planning, budgeting, financial reporting or marketing?
7. What could make this study more credible?

About the Author

George (Sonny) Blackwell joined BancorpSouth Insurance, a regional insurance brokerage, in 2004 as vice president of risk management resources after twenty years of various responsibilities for a major insurance carrier. He helps clients control their cost of risk by defining the major factors that contribute to claims costs and facilitates risk improvement plans. He also advises sales, marketing and underwriting professionals. Sonny earned a Master's of Science in Occupational Safety & Health Education from the University of Southern Mississippi and a Training Professional Certificate from the Phillips Workplace Learning and Performance Institute at the University of Southern Mississippi. He often presents at conferences and workshops and may be contacted at sonnyblackwell@bellsouth.net.

Resources

American National Standard for Occupational Health and Safety Management Systems; ANSI Z-10, (Fairfax, VA, American Industrial Hygiene Association, 2005)

Cascio, Wayne F., *Applied Psychology in Human Resource Management,* (Upper Saddle River, NJ, Prentice Hall, 1998)

Few, Stephen, *Now You See It; Simple visualization techniques for quantitative analysis,* (Oakland, CA, Analytics Press, 2009)

Green, L., Kreuter, M. and Deeds, S., *Health Education Planning: A diagnostic approach* (Mountain View, CA., 1980)

Heinrich, H.W., *Industrial Accident Prevention,* (New York, McGraw-Hill, 1950)

Hubbard, D.W., *The Failure of Risk management: Why it's broken and how to fix it,* (Hoboken, NJ, Wiley, 2009)

Knowles, M., Houlton, E. and Swanson, R., *The Adult Learner,* (Woburn, MA, Butterworth-Heinemann, 1998)

Peterson, Dan, *Safety Management: A human approach,* (Goshen, NY, Aloray, 1988)

Phillips, J.J. and Phillips, P.P., *Measuring for Success: What CEOs really think about learning investments,* (East Peoria, IL, Versa Press, 2010) Phillips, J.J. and Phillips, P.P., *Return on Investment (ROI) Basics,* (Baltimore, Victor Graphics, 2005)

Voluntary Ergonomics Guidelines for the Furniture manufacturing Industry, (High Point, NC, American Furniture Manufacturing Association, 2002)

16

Measuring ROI in Stress Management

Midwest Electric Inc.

Patti P. Phillips

Abstract

This case study begins by describing how the needs for a stress management program were determined and how an organization development solution was evaluated through ROI. The comprehensive approach includes the use of the StressMap® to measure learning, as well as the use of control groups to isolate the effects of the program. A description of how the ROI was measured is included. The specific forms, issues and processes make this a practical case study for organizations interested in a comprehensive, balanced approach to evaluation.

Keywords: Stress management, learning, isolate the effects of a program, needs assessment

16.1 Background

Midwest Electric Inc. (MEI) is a growing electric utility serving several Midwestern states. Since deregulation of the industry, MEI has been on a course of diversification and growth. Through a series of acquisitions, MEI has moved outside its traditional operating areas and into several related businesses.

MEI had been experiencing significant workplace changes as it transformed from a bureaucratic, sluggish organization into a lean, competitive force in the marketplace. These changes placed

Jack Phillips, Patti Phillips, and Al Pulliam, Measuring ROI in Environment, Health, and Safety, (361–386) 2014 © Scrivener Publishing LLC

tremendous pressure on employees to develop multiple skills and perform additional work. Employees, working in teams, had to constantly strive to reduce costs, maintain excellent quality, boost productivity and generate new and efficient ways to supply customers and improve service.

As with many industries in a deregulated environment, MEI detected symptoms of employee stress. The safety and health function in the company suggested that employee stress lowered productivity and reduced employee effectiveness. Stress also was considered a significant employee health risk. Research had shown that high levels of stress were commonplace in many work groups and that organizations were taking steps to help employees and work groups reduce stress in a variety of ways. The vice president of human resources at MEI asked the safety and health department, with the help of the training department, to develop a program to help work groups alleviate stressful situations and deal more productively and effectively with job-induced stress.

16.1.1 Needs Assessment

Because of its size and sophisticated human resource systems, MEI had an extensive database on employee-related measures. MEI prided itself as being one of the leaders in the industry in human resources issues. Needs assessments had been routinely conducted at MEI and the HR vice president was willing to allow sufficient time for an adequate needs assessment before proceeding with the stress management program.

The overall purpose of the needs assessment was to identify the causes of a perceived problem. The needs assessment would accomplish the following objectives:

- Confirm a problem did exist and provide an assessment of the actual impact of this problem
- Uncover potential causes of the problem within the work unit, company and environment
- Provide insight into potential remedies to correct the problem

The sources of data for the needs assessment included company records, external research, team members, team leaders and managers. The assessment began with a review of external research that identified the factors usually related to high stress and the consequences of high stress in work groups. The consequences uncovered specific measures that could be identified at MEI.

This external research led to a review of several key data items in company records, including attitude surveys, medical claims, employee assistance plan (EAP) use, safety and health records and exit interviews The attitude survey data represented the results from the previous year and were reviewed for low scores on the specific questions that could yield stress-related symptoms. Medical claims were analyzed by codes to identify the extent of those related to stress-induced illnesses. EAP data were reviewed to determine the extent to which employees were using provisions and services of the plan perceived to be stress-related. Safety records were reviewed to determine whether specific accidents were stress-related or causes of accidents could be traced to high levels of stress. In each of the above areas, the data were compared with data from the previous year to determine whether stress-related measures were changing. Also, where available, data were compared with expected norms from the external research. Finally, exit interviews for the previous six months were analyzed to determine the extent to which the stress-related situations were factors in an employee's decision to voluntarily leave MEI.

During MEI's needs assessment process, a small sample of employees (10 team members) was interviewed to discuss their work-life situations and to uncover symptoms of stress at work. A small group of managers (five) was interviewed with the same purpose. To provide more detail about this input, a 10 percent sample of employees received a questionnaire to explore the same issues. MEI had 22,550 employees with 18,220 non-supervisory team members.

16.1.2 Summary of Findings

The needs assessment process uncovered several significant findings:

- There was evidence of high levels of stress in work groups, caused by MEI's deregulation, restructuring and job changes—in essence, the change in the nature of work induced high levels of stress in most work groups.

- Stress had led to deterioration in several performance measures, including medical costs, short-term disability, withdrawals (absenteeism, turnover) and job satisfaction.
- Employees often were not fully aware of stress factors and the effect stress had on them and their work.
- Employees had inadequate skills for coping with stress and adjusting to, managing and eliminating highly stressful situations.
- Managers had more insight into the causes of stress but did not have the skills or mechanisms to deal with most stressful situations.

16.1.3 Program Planning and Design

Several inherent factors about work groups and data at MEI influenced the program and its subsequent evaluation. MEI was organized around teams and groups usually were not identical. However, many teams had similar performance measures. The HR database was rich with a variety of measures and with data about employees and work unit factors. Because of the team environment and the important role of the team leader/manager, the program to reduce stress needed to involve the management group in a proactive way. Any efforts to reduce stress needed to shift much of the responsibility to participants and therefore reduce the amount of time off the job. Job pressures in the deregulated environment provided fewer off-the-job opportunities for meeting and development activities.

Although several approaches feasibly could have satisfied this need, four issues surfaced that influenced program design:

- A skills and knowledge deficiency existed, and some type of learning event was necessary.
- Several stress management programs were commercially available, which could prevent developing a new program from scratch.
- Managers needed to be involved in the process to the greatest extent possible.

Departments or work groups of 10 or more people who are committed to improving the satisfaction and effectiveness of their teams will benefit by this more comprehensive approach to stress. The process uses the StressMap® tool as the starting point.

Managers and representative employees will participate in focus groups to identify work satisfiers and distressors and then will collaborate on alleviating systemic sources of stress.

What Group Members Will Learn

- How to identify sources of stress and their personal response to them
- That individuals have the ability to make a difference in their lives
- How to take the first steps to enhance personal health and overall performance
- How to access resources, internally and externally, to help teach personal goals

What the Group/Manager Will Learn

- Group profile of sources of stress and response patterns
- Additional information about sources of both work distress and work satisfaction obtained through focus groups and themes identified when possible
- New stress reduction skills specific to the needs of the group
- Development of recommendations for next steps to improve work satisfaction and productivity

Highlights

- Through completion of a comprehensive self-assessment tool called StressMap®, individuals will be able to immediately score themselves on 21 stress scales dealing with work and home life, as well as learn about their preferred coping styles and the thinking and feeling patterns that impact their ability to manage stress. Anonymous copies of each member's StressMap® will be compiled to create a group score.
- A 3–4 hour StressMap® debriefing session designed to help individuals better interpret their scores will be followed by a four-hour module suited to the needs of the group (such as situation mastery, changing habits, creating climate for agreement). Total of one day.

Precourse Requirements

- Management commitment to the process is essential. Employees will complete the StressMap® tool and submit a confidential copy.

Length and Format

- Lead time of three to four weeks minimum for preparation and communication
- Consultant on-site 1-1/2 days
- Initial follow-up one to two weeks later on-site or by phone to senior management (Subsequent follow-up on impact of the initiative to occur as negotiated with three to four hours of telephone follow-up included).

Figure 16.1 Stress Management for Intact Work Teams.

- Because of the concerns about time away from the job, the actual classroom/formal meeting activities needed to be limited to one or two days.

With this in mind, the program outlined in Figure 16.1 was designed to meet this important need.

16.2 Why ROI?

HR programs usually targeted for a Level 5 ROI evaluation are those perceived to be adding significant value to the company and closely linked to the organizational goals and strategic objectives.

The evaluation is then pursued to confirm the added value. Based on the results of the analysis, these programs may be enhanced, redesigned or eliminated if the results are insufficient. Stress management can be different. If the results are inadequate, the program may not be discontinued but may be altered for future sessions, particularly if behavior changes are not identified in the Level 3 evaluation.

At MEI, the stress management program was chosen for an ROI evaluation for two reasons. First, the HR department was interested in the accountability of all programs, including stress management. Second, positive results would clearly show management that these types of programs, which are preventive in nature, could significantly contribute to the bottom line when implemented and supported by management.

Because the program could have been expensive if applied to the entire company, it was decided to try it on a limited basis to determine its success and then to either adjust the program, discontinue the program or expand the program to other areas in MEI. The evaluation methodology provided the best information to make that decision.

16.2.1 Data Collection Plan

Figure 16.2 shows the data collection plan for the stress management program. Broad objectives were established for Levels 1, 2, 3 and 4 data collection. The data collection plan was comprehensive but necessary to meet all requirements at each of the four levels of data collection. The timing and responsibilities were detailed. For measuring learning, three tools were used. The StressMap® was one measure of learning in the awareness category. Completion of the StressMap® provided insight into stress factors and stress signals. In addition, built into the one-day program was an end-of-course self-assessment to measure learning. Finally, the facilitator had a brief checklist to indicate the extent of learning for the group.

At Level 3 data collection, the completion of the 21-day plan provided some evidence that the participants had changed behavior to reduce stress. A conference call was planned with the facilitator, team manager and the team 21 days after the course. This provided a review of issues and addressed any concerns or barriers to further implementation. A follow-up session was planned with the team, co-facilitated by the manager and facilitator, approximately one to

Program: Stress Management for Intact Groups **Responsibility:** ____________ **Date:** ____________

Level	Broad Program Objective(s)	Data Collection Method	Timing of Data Collection	Responsibilities for Data Collection
I **Reaction, Satisfaction and Planned Actions**	• Positive Reaction	• Standard Questionnaire	• End of 1-Day Course	• Facilitator
	• Suggestions for Improvements			
	• Planned Action	• 21-Day Action Plan	• End of Course	• Facilitator
II **Learning**	• Personal Stress Awareness	• StressMap®	• Prior to Course	• Facilitator
	• Coping Strategies	• Self-Assessment	• End of Course	• Facilitator
	• Stress Reduction Skills	• Facilitator Assessment	• End of Course	• Facilitator
III **Application**	• Change Behavior to Reduce Stress	• Completion of 21-Day Plan	• 21 Days After Course	• No Report
	• Develop Group Action Plan and Communicate to Group	• Conference Call	• 21 Days After Course	• Facilitator
		• Follow-Up Session	• 1–2 Weeks After 1-Day Course	• Facilitator/Manager
	• Access Internal/External Resources	• Review Records	• 6 Months After Course	• Program Coordinator
	• Application of Skills/Knowledge	• Follow-Up Questionnaire	• 6 Months After Course	• External Consultant
IV **Business Impact**	• Reduce Medical Care Costs	• Group Records	• 6 Months After Course	• Program Coordinator
	• Reduce Absenteeism	• Group Records	• 6 Months After Course	• Program Coordinator
	• Reduce Turnover	• Group Records	• 6 Months After Course	• Program Coordinator
	• Increase Productivity	• Group Records	• 6 Months After Course	• Program Coordinator
	• Increase Job Satisfaction	• Follow-Up Questionnaire	• 6 Months After Course	• External Consultant

Figure 16.2 Data Collection Plan.

two weeks after the one-day program, to discuss changes in behavior and to address barriers. To determine the extent to which the participants were using internal or external resources to address stress-related problems, records of those requests were scheduled for review for approximately six months. Finally, a detailed follow-up questionnaire was planned for six months after the program to collect both level 3 and 4 data. This questionnaire was intended to capture sustained behavior changes, indicate barriers to improvement and identify impact measures for both groups and individuals.

Group records were expected to reveal changes in medical costs, absenteeism, turnover and productivity six months after the program. In addition, increased job satisfaction was to be determined from the follow-up questionnaire, which would be administered six months after the program (the same questionnaire described earlier).

16.2.2 ROI Analysis Plan

Figure 16.3 shows the ROI analysis plan. For most data items, the method to isolate the effects of the program would be obtained in a control group arrangement in which the performance of the group involved in the program would be compared with the performance of a carefully matched companion control group. In addition, for most of the data items, trend-line analysis was scheduled for use. Historical data were projected in a trend and compared with the actual data to determine the impact of the program.

The methods of converting data involved a variety of approaches, including tabulating direct costs, using standard values, using external data and securing estimates from a variety of target audiences. The cost categories represented fully loaded costs for the program. Expected intangible benefits from the program were based on the experience of other organizations and other stress reduction programs. The communication target audience consisted of six key groups ranging from corporate and business unit managers to participants and their immediate supervisors.

16.2.3 Management Involvement

Management involvement was a key issue from the beginning and was integrated throughout the design of the program. The manager served as the team leader for the program, although a facilitator provided assistance and conducted a one-day workshop.

Program: Stress Management for Intact Groups **Responsibility:** ________________ **Date:** ________

Data Items (Usually Level 4)	Methods of Isolating the Effects of the Program	Methods of Converting Data	Cost Categories	Intangible Benefits	Other Influences/Issues	Communication Targets
Medical Health Care Costs—Preventable Claims	• Control Group Arrangement • Trendline Analysis	• Direct Costs	• Needs Assessment • Program Development • Program Materials • Participant Salaries/Benefits • Participant Travel (if applicable) • Facilitator • Meeting Facilities (Room, Food, Beverages) • Program Coordinator • Training and Education Overhead • Evaluation Costs	• Improved Communication • Time Savings • Fewer Conflicts • Teamwork • Improvement in Problem Solving	• Match Groups Appropriately • Limit Communications with Control Group • Check for Team-Building Initiatives During Program • Monitor Restructuring Activities During Program • 6 Grtoups Will Be Monitored	• Program Participants • Intact Team/Manager • Senior Manager/ Management in Business Units • Training and Education Staff • Safety and Health Staff • Senior Corporate Management • Prospective Team Leaders
Absenteeism	• Control Group Arrangement • Trendline Analysis	• Supervisor Estimation • Standard Value				
Employee Turnover	• Control Group • Trendline Analysis	• External Study—Cost of Turnover in High Tech Industry • Management Review				
Employee Job Satisfaction	• Control Group Arrangement • Management Estimation	• Management Estimation				
Employee/Group Productivity	• Control Group Arrangement • Trendline Analysis	• Standard Values • Management Estimation				

Figure 16.3 Evaluation Strategy: ROI Analysis.

Figure 16.4 illustrates the tool used for identifying initial problems as the work group began using the stress management program. With this brief questionnaire, the manager identified specific problem areas and provided appropriate

Before you begin the stress reduction program for your team, it is important to capture specific concerns that you have about your work group. Some of these concerns may be stress-related and therefore may be used to help structure specific goals and objectives for your team. For each of the following potential areas of improvement, please check all that apply to your group. Add others if appropriate. Next to the item, provide specific comments to detail your concerns and indicate if you think this concern may be related to excessive stress.

- Employee Turnover. Comments: ______
- Employee Absenteeism. Comments: ______
- Employee Complaints. Comments: ______
- Morale/Job Satisfaction. Comments: ______
- Conflicts with the Team. Comments: ______
- Productivity. Comments: ______
- Quality. Comments: ______
- Customer Satisfaction. Comments: ______
- Customer Service. Comments: ______
- Work Backlog. Comments: ______
- Delays. Comments: ______
- Other Areas. List and Provide Comments: ______

Figure 16.4 Manager Input: Potential Area for Improvement.

comments and details. This exercise allowed program planning to focus on the problems and provided guidance to the facilitator and the team.

Figure 16.5 illustrates manager responsibility and involvement for the process. This handout, provided directly to the managers, details 12 specific areas of responsibility and involvement for the managers. Collectively, initial planning, program design and detailing of responsibilities pushed the manager into a higher-profile position in the program.

16.2.4 Control Group Arrangement

The appropriateness of control groups was reviewed in this setting. If a stress-reduction program was needed, it would be appropriate and ethical to withhold the program for certain groups while the experiment was being conducted. It was concluded that this approach was appropriate because the impact of the planned program was in question. Although it was clear that stress-induced problems existed at MEI, there was no guarantee that this program would correct them. Six control groups were planned. The control

With the team approach, the team manager should:

1. Have a discussion with the facilitator to share reasons for interest in stress reduction and the desired outcome of the program. Gain a greater understanding of the StressMap® and the OD approach. Discuss recent changes in the work group and identify any known stressors. This meeting could be held with the senior manager or the senior management team.
2. Identify any additional work group members for the consultant to call to gather preliminary information.
3. Appoint a project coordinator, preferably an individual with good organizing and influencing skills, who is respected by the work group.
4. Send out a letter with a personal endorsement and signature, inviting the group to participate in the program.
5. Allocate eight hours of work time per employee for completion of StressMap® and attendance at a StressMap® debriefing and customized course.
6. Schedule a focus group after discussing desired group composition with the facilitator. Ideal size is 10 to 22 participants. The manager should not attend.
7. Attend the workshop and ensure that direct reports attend.
8. Participate in the follow-up meeting held after the last workshop, either in person or by conference call. Other participants to include are the HR representative for your area, the Safety and Health representative for your area and your management team. The facilitator will provide feedback about the group issues and make recommendations of actions to take to reduce work stress or increase work satisfaction.
9. Commit to an action plan to reduce workplace distress and/or increase workplace satisfaction after thoughtfully considering feedback.
10. Communicate the action plan to your work group.
11. Schedule and participate in a 21-day follow-up call with the consultant and your work group.
12. Work with your team (managers, HR, safety and health, facilitator) to evaluate the success of the action plan and determine the next steps.

Figure 16.5 Manager Responsibility and Involvement.

group arrangement was diligently pursued because it represented the best approach to isolating the effects of the program, if the groups could be matched.

Several criteria were available for group selection. Figure 16.6 shows the data collection instrument used to identify groups for a control group arrangement. At the first cut, only those groups that had the same measures were considered (that is, at least 75 percent of the measures were common in the group). This action provided an opportunity to compare performance in the six months preceding the program.

Next, only groups in the same function code were used. At MEI, all groups were assigned a code depending on the type of work,

To measure the progress of your team, a brief profile of performance measures for employees and your work group is needed. This information will be helpful to determine the feasibility of using your group in a pilot study to measure the impact of the stress management program. Changes in performance measures will be monitored for six months after the program.

Listed below are several categories of measures for your work group. Check the appropriate category and please indicate the specific measure under the description. In addition, indicate if it is a group measure or an individual measure. If other measures are available in other categories, please include them under "Other."

Key Performance Measures Dept__________________

Performance Category	Measure	Description of Measure	Group Measure	Individual Measure
Productivity	1.		O	O
	2.		O	O
Efficiency	3.		O	O
	4.		O	O
Quality	5.		O	O
	6.		O	O
Response Time	7.		O	O
	8.		O	O
Cost Control/ Budgets	9.		O	O
	10.		O	O
Customer Satisfaction	11.		O	O
	12.		O	O
Absenteeism	13.		O	O
Turnover	14.		O	O
Morale/ Job Satisfaction	15.		O	O
	16.		O	O
Other (please specify)	17.		O	O
	18.		O	O
	19.		O	O
	20.		O	O

Group Characteristics

Average tenure for group ______years
Average job grade for group ______
Number in group _____
Group function code ____________
Average age ______
Average educational level ______

Figure 16.6 Manager Input.

such as finance and accounting or engineering. Therefore, each experimental group had to be in the same code as the matched control group. It was also required that all six groups span at least three different codes.

Two other variables were used in the matching process: group size and tenure. The number of employees in the groups had to be within a 20 percent spread and the average tenure had to be within a two-year range. At MEI, as with many other utilities, there was a high average tenure rate.

Although other variables could have been used to make the match, these five were considered the most influential in the outcome. In summary, the following criteria were used to select the two sets of groups:

- Same measures of performance
- Similar performance in the previous six months
- Same function code
- Similar size
- Similar tenure

The six pairs of groups represented a total of 138 team members and six managers for the experimental groups and 132 team members and six managers for the control groups.

16.3 Program Results

16.3.1 Questionnaire Response

A follow-up questionnaire, Figure 16.7, served as the primary data collection instrument for participants. A similar, slightly modified instrument was used with the managers. In all, 73 percent of the participants returned the questionnaire. This excellent response rate was caused, in part, by a variety of actions taken to ensure an appropriate response rate. Some of the most important actions were the following:

- The team manager distributed the questionnaire and encouraged participants to return it to the external consulting firm. The manager also provided a follow-up reminder.

Check one: 0 Team Member 0 Team Leader/Manager

1. Listed below are the objectives of the stress management program. After reflecting on this program, please indicate the degree of success in meeting the objectives.

OBJECTIVES	Failed	Limited Success	Generally Successful	Completely Successful
PERSONAL				
• Identify sources of stress in work, personal and family worlds				
• Apply coping strategies to manage stressful situations				
• Understand to what degree stress is hampering your health and performance				
• Take steps to enhance personal health and overall performance				
• Access internal and external resources to help reach personal goals				
GROUP				
• Identify sources of stress for group				
• Identify sources of distress and satisfaction				
• Apply skills to manage and reduce stress in work group				
• Develop action plan to improve work group effectiveness				
• Improve effectiveness and efficiency measures for work group				

2. Did you develop and implement a 21-day action plan?
Yes 0 No 0
If yes, please describe the success of the plan. If not, explain why. ____

____

3. Please rate, on a scale of 1–5, the relevance of each of the program elements to your job, with (1) indicating no relevance and (5) indicating very relevant.

_____StressMap® Instrument _____Action Planning
_____Group Discussion _____Program Content

4. Please indicate the degree of success in applying the following skills and behaviors as a result of your participation in the stress management program.

Figure 16. 7 Impact Questionnaire.

Figure 16. 7 *(Cont.)*

	1	2	3	4	5	
	No	Little	Some	Significant	Very Much	No Opportunity To Use Skills
a) Selecting containable behavior for change						
b) Identifying measures of behavior						
c) Taking full responsibility for your actions						
d) Selecting a buddy to help you change behavior						
e) Identifying and removing barriers to changing behavior						
f) Identifying and using enablers to help change behavior						
g) Staying on track with the 21-day action plan						
h) Applying coping strategies to manage stressful situations						
i) Using control effectively						
j) Knowing when to let go						
k) Responding effectively to conflict						
l) Creating a positive climate						
m) Acknowledging a complaint properly						
n) Reframing problems						
o) Using stress talk strategies						

5. List (3) behaviors or skills you have used most as a result of the stress management program.

____

____

6. When did you first use one of the skills from the program?

_____During the program

_____Day(s) after the program (indicate number)

_____Week(s) after the program (indicate number)

7. Indicate the types of relationships in which you have used the skills.

- O Coworkers
- O Manager or supervisor
- O MEI employee in another function
- O Spouse
- O Child
- O Friend
- O Other (list): ______________________________

PERSONAL CHANGES

8. What has changed about your on-the-job behavior as a result of this program? (positive attitude, fewer conflicts, better organized, fewer outbursts of anger, etc.)

____

____

____

(Continued)

Figure 16. 7 *(Cont.)*

9. Recognizing the changes in your own behavior and perceptions, please identify any specific personal accomplishments/improvements that you can link to this program. (time savings, project completion, fewer mistakes, etc.)

____

____

____

10. What specific value in U.S. dollars can be attributed to the above accomplishments/improvements? Although this is a difficult question, try to think of specific ways in which the above improvements can be converted to monetary units. Use one year of data. Along with the monetary value, please indicate the basis of your calculation.

$ ______________

Basis ____

____

____

11. What level of confidence do you place in the above estimations? (0% = No Confidence, 100% = Certainty) __________%

12. Other factors often influence improvements in performance. Please indicate the percent of the above improvement that is related directly to this program. __________%

Please explain. ____

____

____

____

GROUP CHANGES

13. What has changed about your work group as a result of your group's participation in this program? (interactions, cooperation, commitment, problem solving, creativity, etc.)

____

____

____

14. Please identify any specific group accomplishments/improvements that you can link to the program. (project completion, response times, innovative approaches)

____

____

____

15. What specific value in U.S. dollars can be attributed to the above accomplishments/improvements? Although this is a difficult question, try to think of specific ways in which the above improvements can be converted to monetary units. Use one year of values. Along with the monetary value, please indicate the basis of your calculation.

$ ______________

Basis ____

____

____

____

16. What level of confidence do you place in the above estimations? (0% = No Confidence, 100% = Certainty) __________%

Figure 16. 7 *(Cont.)*

17. Other factors often influence improvements in performance. Please indicate the percent of the above improvement that is related directly to this program. ______________%

____

____

____

18. Do you think this program represented a good investment for MEI?

Yes 0 No 0

Please explain. ____

____

____

____

19. What barriers, if any, have you encountered that have prevented you from using skills or knowledge gained in this program? Check all that apply. Please explain, if possible.

0 Not enough time
0 Work environment does not support it
0 Management does not support it
0 Information is not useful (comments)
0 Other ____

____

____

20. Which of the following best describes the actions of your manager during the stress management program?

0 Very little discussion or reference to the program
0 Casual mention of program with few specifics
0 Discussed details of program in terms of content, issues, concerns, etc.
0 Discussed how the program could be applied to work group
0 Set goals for changes/improvements
0 Provided ongoing feedback about the action plan
0 Provided encouragement and support to help change behavior
0 Other (comments) . . ____

____

21. For each of the areas below, indicate the extent to which you think this program has influenced these measures in your work group.

	No Influence	**Some Influence**	**Moderate Influence**	**Significant Influence**	**Very Much Influence**
a) Productivity					
b) Efficiency					
c) Quality					
d) Response Time					
e) Cost Control					
f) Customer Service					
g) Customer Satisfaction					
h) Employee Turnover					
i) Absenteeism					
j) Employee Satisfaction					
k) Healthcare Costs					
l) Safety and Health Costs					

(Continued)

Figure 16. 7 (*Cont.*)

Please cite specific examples or provide more details.

22. What specific suggestions do you have for improving the stress management program? Please specify.
    0 Content
    0 Duration
    0 Presentation
    0 Other

23. Other comments:

- A full explanation of how the evaluation data would be used was provided to participants.
- The questionnaire was reviewed during the follow-up session.
- Two types of incentives were used.
- Participants were promised a copy of the questionnaire results.

16.3.2 Application Data

The application of the program was considered an outstanding success with 92 percent of the participants completing their 21-day action plans. A conference call at the end of the 21 days showed positive feedback and much enthusiasm for the progress made. The follow-up session also demonstrated success because most of the participants had indicated changes in behavior.

The most comprehensive application data came from the six-month questionnaire administered to participants and managers. The following skills and behaviors were reported as achieving significant success:

- Taking full responsibility for one's actions
- Identifying or removing barriers to change behavior
- Applying coping strategies to manage stressful situations

- Responding effectively to conflict
- Creating a positive climate
- Acknowledging a complaint properly

Coworkers were the most frequently cited group in which relationships had improved through use of the skills, with 95 percent indicating application improvement with this group.

16.3.3 Barriers

Information collected throughout the process, including the two follow-up questionnaires, indicated few barriers to implementing the process. The two most frequently listed barriers were the following:

- There is not enough time.
- The work environment does not support the process.

16.3.4 Management Support

Manager support seemed quite effective. The most frequently listed behaviors of managers are listed below:

- Managers set goals for change and improvement.
- Managers discussed how the program could apply to the work group.

16.3.5 Impact Data

The program had significant impact with regard to both perceptions and actual values. On Figure 16.7, the follow-up questionnaire, 90 percent of the participants perceived this program as a good investment for MEI. In addition, participants perceived that this program had significantly influenced several elements:

- Employee satisfaction
- Absenteeism
- Turnover
- Healthcare cost
- Safety and health cost

This assessment appears to support the actual improvement data, outlined below. For each measure below, only the team data were collected and presented. Because managers were not the target of the program, manager performance data were not included. An average of months five and six, instead of the sixth month, was used consistently for the post-program data analysis to eliminate the spike effect.

16.3.5.1 Healthcare Costs

Healthcare costs for employees were categorized by diagnostic code. It was a simple process to track the cost of stress-induced illnesses. Although few differences were shown in the first three months after the program began, by months five and six, an average difference of $120 per employee per month was identified. This apparently was caused by the lack of stress-related incidents and the subsequent medical costs resulting from the stress. It was believed that this amount would be an appropriate improvement to use. The trend-line projection of healthcare costs was inconclusive because of the variability of the medical care costs prior to the program. A consistent trend could not be identified.

16.3.5.2 Absenteeism

There were significant differences in absenteeism in the two groups. The average absenteeism for the control group for months five and six was 4.65 percent. The absenteeism rate for the groups involved in the program was 3.2 percent. Employees worked an average of 220 days. The trend-line analysis appeared to support the absenteeism reduction. Because no other issues were identified that could have influenced absenteeism during this time period, the trend-line analysis provided an accurate estimate of the impact.

16.3.5.3 Turnover

Although turnover at MEI was traditionally low, in the past two years it had increased because of significant changes in the workplace. A turnover reduction was identified using the differences in the control group and experimental group. The control group had an average annual turnover rate of 19.2 percent for months five and six. The experimental group had an average of 14.1 percent for the same two months. As with absenteeism, the trend-line analysis supported the turnover reduction.

16.3.5.4 *Productivity*

Control group differences showed no significant improvement in productivity. Of all the measures collected, the productivity measure was the most difficult to match between the two groups, which may account for the inconclusive results. Also, the trend-line differences showed some slight improvement, but not enough to develop an actual value for productivity changes.

16.3.5.5 *Job Satisfaction*

Because of the timing difference in collecting attitude survey data, complete job satisfaction data were not available. Participants did provide input about the extent to which they felt the program actually influenced job satisfaction. The results were positive, with a significant influence rating for that variable. Because of the subjective nature of job satisfaction and the difficulties with measurement, a value was not assigned to job satisfaction.

16.3.6 Monetary Values

The determination of monetary benefits for the program was developed using the methods outlined in the ROI analysis plan. The medical costs were converted directly. A $120-per-month savings yielded a $198,720 annual benefit. A standard value had routinely been used at MEI to reflect the cost of an absence. This value was 1.25 times the average daily wage rate. For the experimental group, the average wage rate was $123 per day. This yielded an annual improvement value of $67,684. For employee turnover, several turnover cost studies were available, which revealed a value of 85 percent of annual base pay. As expected, senior managers felt this cost of turnover was slightly overstated and preferred to use a value of 70 percent, yielding an annual benefit of $157,553. No values were used for productivity or job satisfaction. The total annual benefit of the stress management program was $423,957. Table 16.1 reflects the total economic benefits of the program.

The medical costs were converted directly. A $120-per-month savings yielded $198,720 annual benefit. Other values are as follows:

Unit Value for an Absence
$123 x 1.25 = $153.75

Table 16.1 Annual Monetary Benefits for 138 Participants.

	Monthly Difference	Unit Value	Annual Improvement Value
Medical Costs	$120	-	$198,720
Absenteeism	1.45%	$153.75	$ 67,684
Turnover	5.1% (annualized)	$22,386	$157,553
Total			$423,957

Unit Value for Turnover
$31,980 x 70% = $22,386

Improvement for Absenteeism
138 employees x 220 workdays x 1.45% x $153.75 = $67,684

Improvement for Turnover
138 employees x 5.1% x $22,386 = $157,553
No values were used for productivity or job satisfaction.

16.3.7 Intangible Benefits

Several intangible benefits were identified in the study and confirmed by actual input from participants and questionnaires. The following benefits were pinpointed:

- Employee satisfaction
- Teamwork
- Improved relationships with family and friends
- Time savings
- Improved image in the company
- Fewer conflicts

No attempt was made to place monetary values on any of the intangibles.

16.3.8 Program Costs

Calculating the costs of the stress management program also followed the categories outlined in the evaluation plan. For needs assessment, all the costs were fully allocated to the six groups.

Although the needs assessment was necessary, the total cost of needs assessment, $16,500, was included. All program development costs were estimated at $95 per participant or $4,800. The program could have possibly been spread through other parts of the organization and then the cost would ultimately have been prorated across all the sessions. However, the costs were low because the materials were readily available for most of the effort and the total development cost was used.

The salaries for the team members averaged $31,980, while the six team managers had average salaries of $49,140. The benefits factor for MEI was 37 percent for both groups. Although the program took a little more than one day of staff time, one day of program was considered sufficient for the cost. The total salary cost was $24,108. The participant travel cost ($38 per participant) was low because the programs were conducted in the area. The facilitator cost, program coordination cost and training and development overhead costs were estimated to be $10,800. The meeting room facilities, food and refreshments averaged $22 per participant, for a total of $3,968. Evaluation costs were $22,320. It was decided that all the evaluation costs would be allocated to these six groups. This determination was extremely conservative because the evaluation costs could be prorated if the program was implemented over other areas.

Table 16.2 details the stress management program costs. These costs were considered fully loaded with no proration, except

Table 16.2 Program Costs.

Cost Category	**Total Cost**
Needs Assessment	$16,500
Program Development	$4,800
Program Materials (144 × $95)	$13,680
Participant Salaries/Benefits	
Based on 1 day	
138 × $123 × 1.37 and 6 × 189 × 1.37	$24,108
Travel and Lodging	
144 x 38	$5,472
Facilitation, Coordination, T&D Overhead	$10,800
Meeting Room, Food and Refreshments $3,168	144 x 22
Evaluation Costs	$22,320
TOTAL	**$100,848**

for needs assessment. Additional time could have been used for participants' off-the-job activities. However, it was concluded one day should be sufficient for the program.

16.3.9 Results: ROI

Based on the given monetary benefits and costs, the return on investment and the benefits/costs ratios are shown below.

$$\text{BCR} = \frac{\$423{,}957}{\$100{,}848} = 4.20$$

$$\text{ROI} = \frac{\$423{,}957 - \$100{,}848}{\$100{,}848} = 320\%$$

Although this number is considered quite large, it is still conservative because of the following assumptions and adjustments:

- Only first-year values were used. The program should actually have second- and third-year benefits.
- Control group differences were used in analysis, which is often the most effective way to isolate the effects of the program. These differences were also confirmed with the trend-line analysis.
- The participants provided additional monetary benefits, detailed on the questionnaires. Although these benefits could have been added to the total numbers, they were not included because only 23 participants of the 144 supplied values for those questions.
- The costs are fully loaded.

When considering these adjustments, the value should represent a realistic value calculation for the actual return on investment.

16.4 Communication Strategies

Because of the importance of sharing the analysis results, a communication strategy was developed. Table 16.3 outlines this strategy. Three separate documents were developed to communicate with the different target groups in a variety of ways.

Table 16.3 Communication Strategies.

Communication Document	Communication Target	Distribution
Complete report with appendices (75 pages)	• Training and Education Staff • Safety and Health Staff • Intact Team Manager	Distributed and discussed in a special meeting
Executive Summary (8 pages)	• Senior Management in the Business Units • Senior Corporate Management	Distributed and discussed in routine meeting
General interest overview and summary without the actual ROI calculation (10 pages)	• Program Participants	Mailed with letter
Brochure highlighting program, objectives and specific results	• Prospective Team Leaders	Included with other program descriptions

16.5 Policy and Practice Implications

Because of the significance of the study and the information, two issues became policy. Whenever programs are considered that involve large groups of employees or a significant investment of funds, a detailed needs assessment should be conducted to ensure the proper program is developed. Also, an ROI study should be conducted for a small group of programs to measure the impact before complete implementation. In essence, this influenced the policy and practice on needs assessment, pilot program evaluation and the number of impact studies developed.

16.6 Questions for Discussion

1. What is the purpose of the needs assessment?
2. What specific sources of data should be used?
3. Critique the data collection plan.

4. What other methods could be used to isolate the effects of the program?
5. Critique the methods to convert data to monetary values.
6. Are the costs fully loaded? Explain.
7. Is the ROI value realistic? Explain.
8. Critique the communication strategy.

17

Measuring ROI in a Safety Incentive Program

National Steel

Jack Phillips

Abstract

This case addresses measuring the effectiveness of an incentive program designed to influence employee behavior and reduce accidents in a manufacturing environment. Although top executives were concerned about the safety and wellbeing of employees, they also wanted to reduce the cost of accidents. The cost of accidents was reaching the point that it was an obstacle to the company becoming a low-cost provider of products in a competitive industry. This case demonstrates that an evaluation of an HR program can be implemented with minimal resources. It also demonstrates that, although some programs may achieve a return on investment, it is sometimes best to communicate the ROI results to a limited audience.

Keywords: Incentive program, cost reduction, accident reduction, audience, success measures

17.1 Background

National Steel is a large manufacturing operation with divisions in the Southeastern, Southwestern and Midwestern United States. Each division has multiple plants. The company also has several foreign plants in operation and two others in the construction phase. The plants produce steel products such as bar stock and plate steel. They also fabricate specialized fasteners used in the

---

Jack Phillips, Patti Phillips, and Al Pulliam, Measuring ROI in Environment, Health, and Safety, (387–396) 2014

commercial building industry. The nature of National Steel's manufacturing business requires that safety always gets top priority with management and the workforce. Concern for employee safety is a significant issue. Additionally, domestic and foreign competition is a major factor in National Steel's strategy to become a low-cost producer.

The company had always been concerned about the human element of a safe work environment, but economic issues became a significant concern, driven by the cost of accidents. The company had long had a group, the Central Safety Committee, in place to continually review safety issues, direct accident investigations and to establish policy and best practices. The committee was made up of a senior line officer (who served as the sponsor), one line manager, two foremen, six members of the work force, the corporate manager of safety and the vice president of HR.

17.1.1 A Performance Problem

The committee had recently informed senior management of National Steel that the Midwestern division was experiencing unacceptable accident frequency rates, accident severity rates and total accident costs. For a two-year period, these costs had been in the $400,000 to $500,000 range, annually—much too high for the Central Safety Committee and management to accept.

17.1.2 Needs Assessment

The safety manager was directed to meet with managers and employees in the three plants of the Midwestern division to seek causes of the problem and to work with the division manager to implement appropriate solutions. A team of HR specialists completed the assignment. The team also analyzed the cost and types of accidents. The team of specialists concluded the following:

- Employee safety habits were inconsistent; employees were not focusing enough attention on safety.
- Employees knew and understood safety guidelines and practices, therefore, training was not an issue.
- A significant number of accidents and accident-related costs involved injuries of a questionable nature.

- Some type of monetary incentive would likely influence employee behavior.
- Peer pressure could possibly be used to help employees focus on safety practices and to avoid the costs of seeing a physician when it was unnecessary.

17.2 The Solution

As a result of the assessment, the team recommended that the division implement a group-based safety incentive plan at the three plants. A monetary incentive had been successful in another division during previous years. The division manager reviewed the details of the plan and even helped craft some of the components. He accepted the recommendations and agreed to sponsor the implementation. The two objectives of the recommended plan are listed below:

1. Reduce the annual accident frequency rate from a level of sixty to a much lower level of approximately twenty or less.
2. Reduce the annual disabling accident frequency rate from a level of 18 to 0.

17.2.1 The Measure of Success

The HR team expressed the need to track certain measures on a continual basis. Once the incentive plan's objectives were established, the team identified the specific data needed to analyze safety performance. The measures identified were the following:

- Number of medical treatment cases
- Number of lost-time accidents
- Number of lost-time days
- Accident costs
- Hours worked
- Incentive costs

The monthly data were collected. Because the data collection system had been in place before the implementation of the safety incentive plan, no additional data collection procedures were needed. The same system was also used by the HR specialists when the Central Safety Committee asked them to review the problem and make recommendations. Additionally, management was interested in a payback for the incentive program. Although accident reduction and severity were major concerns, there was also a need to achieve low-cost provider goals. Management requested to see figures that demonstrated that the benefits from the plan exceeded the costs of implementation. The team members concluded that they could use the same tracking system to determine the return on investment.

17.2.2 The Incentive Plan

The incentive plan consisted of a cash award of $75, after taxes, to each employee in the plant for every six months the plant worked without a medical treatment case. A medical treatment case was defined as an accident that could not be treated by plant first-aid and, therefore, needed the attention of a physician. Each plant had a workforce of about 120 employees. A team effort at each plant was important, because the actions of one employee could impact the safety of another. Peer pressure was necessary to keep employees focused and to remind them to avoid unnecessary physician costs. Therefore, the award was paid at each of the three plants independently. An award was not paid unless the entire plant completed a six-month period without a medical treatment case. When a medical treatment case occurred, a new six-month period began.

17.2.3 Implementation of the Incentive Plan

The plant managers implemented the plan at the beginning of the new year, so results could easily be monitored and compared with performance in previous years. Each plant manager announced the plan to employees and distributed the guidelines for payout. The managers communicated the details of the plan and answered questions during the regular monthly safety meeting at each plant. Thorough communication ensured that each employee clearly understood how the plan functioned and what the group had to accomplish to receive an award.

17.2.4 Cost Monitoring

Two groups of costs were monitored: the total accident costs and the incentive compensation costs. Total accident costs were monitored prior to the safety incentive plan as part of collecting routine safety performance data. The additional costs related directly to incentive compensation were also tabulated. Because the $75 cash was provided after taxes, the cost to the division was approximately $97.50 per employee for each six-month period completed without a medical treatment case. Additional administrative costs were minimal because the data used in analysis were already collected prior to the new plan and because the time required to administer the plan and calculate the award was almost negligible. No additional staff was needed and no overtime for existing staff could be directly attributed to the plan. However, a conservative estimate of $1,600 per year of plan administration costs was used in the tabulation of incentive costs.

17.3 Data Collection and Analysis

In addition to the two-year data history, data were collected during a two-year period after the plan implementation to provide an adequate before-and-after comparison. Medical treatment injuries, lost-time injuries, accident frequency rates and accident costs were all monitored to show the contribution of the safety incentive plan. The data shown in Table 17.1 document the accident costs for the four-year period. The data reveal significant

Table 17.1 Accident Costs and Frequency for All 3 Plants.

	Year 1 Before Plan	**Year 2 Before Plan**	**Year 3 After Plan**	**Year 4 After Plan**
Accident Frequency	61.2	58.8	19.6	18.4
Disabling Frequency	17.4	18.9	5.7	3.8
Medical Treatment Injuries	121	111	19	17
Lost-Time Injuries	23	21	6.8	5.2
Actual Cost of Accidents	$468,360	$578,128	$18,058	$19,343

reductions in accident costs for the two-year period after the implementation of the plan.

When comparing the average of years 3 and 4 (after the incentive plan) with the 1- and 2-year average (before the incentive plan), the accident frequency was reduced by 68 percent, while the disabling accident frequency was reduced by 74 percent. The annual cost of accidents (the average of the two years before and the two years after the incentive plan) dropped from $523,244 to $18,701, producing a significant savings of $504,543.

The objective of the plan to reduce the annual accident frequency rate to less than 20 was met with a post-plan average of 19. The objective of reducing the annual disabling accident frequency rate from 18 to 0 was not achieved. Although the average after two years dropped significantly from 18.15 annually to 4.75 annually, this was still short of the target. Both calculations are shown below.

BEFORE INCENTIVE PLAN	AFTER INCENTIVE PLAN
Accident Frequency 61.2 + 58.8 = 120	19.6 + 18.4 = 38
Annual Average (÷ 2) = 60 ANNUALLY	= 19 ANNUALLY

Annual Improvement. 60 – 19 = 41
% Accident Improvement. 60 . 41 = 68%

BEFORE INCENTIVE PLAN	AFTER INCENTIVE PLAN
Disabling Frequency 17.4 + 18.9 = 36.3	5.7 + 3.8 = 9.5
Annual Average (÷ 2) = 18.15 ANNUALLY	4.75 ANNUALLY

Annual Improvement. 18.15 – 4.75 = 13.4
% Disabling Improvement. 13.4 ÷ 18.15 = 74%

These impressive results demonstrated a positive business impact. The incentive plan resulted in a safer work environment, fewer accidents and fewer disabling accidents. Although these results improved the overall safety program considerably, the issue of cost savings remained unanswered. Did the incentive plan bring greater monetary benefits than the cost incurred to implement and administer it? Table 17.2 details the additional cost issues. There was also the issue of how much the incentive plan influenced improvement in the measures when compared with other actions that may have influenced improvements.

Table 17.2 The Contribution of the Safety Incentive Plan for All 3 Plants.

	Year 1 Before Plan	Year 2 Before Plan	Year 3 After Plan	Year 4 After Plan
Needs Assessment Costs (spread)	—	—	$1,200	$1,200
Plan's Annual Administration and Evaluation Costs	—	—	$1,600	$1,600
Safety Incentive Plan Payout Costs	—	—	$58,013	$80,730
Actual Cost of Accidents	$468,360	$578,128	$18,058	$19,343
Total Cost of Accidents and Prevention	$468,360	$578,128	$78,871	$102,873

Plan implemented beginning in Year 3

The needs assessment cost of $2,400 consisted of capturing the time and travel expenses for the HR Team to conduct 20 interviews with plant operations and management staff and to develop recommendations. The cost to administer the incentive plan was $1,600 annually. The cost of the incentive plan payout must be captured and included in the total cost of accident prevention. Payout is determined by calculating the amount of incentive awards paid to employees in Years 3 and 4. Table 17. 3 provides a breakdown of the payout.

17.4 Data Interpretation and Conclusion

The contribution of the safety incentive plan was determined by adding the accident and administrative costs for years 3 and 4 to the safety incentive plan payout costs and then comparing this total with the accident costs of years 1 and 2. As Table 17.2 shows, the total costs were reduced significantly. Accident costs from years 1 and 2 totaled $1,046,488, for an average of $523,244 annually. Accident and prevention costs for Years 3 and 4 totaled $181,744, for an average of $90,872 annually. This was an annual improvement of $432,372.

The Central Safety Committee discussed the issue of isolating the effects of the safety incentive plan. As a group, committee

Table 17.3 The Safety Incentive Plan Payout.

Plant 1 Payout, Year 3:	115 employees X $97.50 X 1 PAYOUT =	$11,213
Plant 2 Payout, Year 3:	122 employees X $97.50 X 2 PAYOUTS =	$23,790
Plant 3 Payout, Year 3:	118 employees X $97.50 X 2 PAYOUTS =	$23,010
	Total Payout, Year 3	$58,013
Plant 1 Payout, Year 4:	115 employees X $97.50 X 2 PAYOUTS =	$22,425
Plant 2 Payout, Year 4:	122 employees X $97.50 X 2 PAYOUTS =	$23,790
Plant 3 Payout, Year 4:	118 employees X $97.50 X 3 PAYOUTS =	$34,515
	Total Payout, Year	$80,730
	Total Payout, Years 3 and 4	$138,743

members decided the incentive plan should be credited for most of the improvement. They felt that it was the incentive plan that influenced a new safety awareness and caused peer pressure to work. After much debate, they accepted an estimate from the in-house expert, the safety manager, using data from an industry trade group that presented convincing evidence that management's attention to safe work habits had been shown to reduce the cost of accidents by 20 percent. Before these data were presented, the improvement was going to be attributed entirely to the incentive plan because no other factors that could have influenced safety performance during this period had been identified. Also, the safety record at the other two divisions showed no improvement during the same time period.

17.5 Calculating the Return on Investment

To determine the return on investment, the costs and monetary benefits of the incentive plan were needed. The annual cost of incentive payouts (two-year average of $69,372) added to annual administration costs ($1,200 plus $1,600) provided a total incentive plan cost of $72,172. The benefits were calculated starting with the annual monetary benefits of $432,372. Because the Central Safety Committee accepted the suggestion of the safety manager—that management attention played a role in influencing the improvement (20 percent)—then an adjustment had to be made. Therefore, 80 percent of $432,372 resulted in an estimated impact of $345,898. The ROI became:

$$\text{ROI} = \frac{\text{Net Benefits}}{\text{Costs}} = \frac{\$345{,}898 - \$72{,}172}{\$72{,}172} = 3.79 \times 100 = 379\%$$

17.6 Communication of Results

The results of the safety incentive plan were communicated to a variety of target audiences to show the contribution of the plan. First, the division president summarized the results in a monthly report to the chief executive officer of the corporation. The focal point was on the reduction in accident frequency and severity, the reduction in costs, the improvement in safety awareness and the return on investment for the incentive plan.

The HR Department presented the results in its monthly report to all middle and upper division management with the same focus that was presented to the CEO. Return on investment information was reserved for management because it was felt that employees might not understand this issue. Plant safety is important without a positive ROI.

The results were presented to all plant employees through the monthly safety newsletter. This communication focused on the reduction in medical treatment injuries, as well as improvements in lost-time accidents, disabling accidents and accident frequency. It also recognized employees for their accomplishments, as did all the communication.

Finally, the results were communicated to all division employees through the division newsletter. This communication focused on the same issues presented to plant employees. Communications were positive and increased the awareness of the need for the continuation of the incentive plan.

17.7 Questions for Discussion

1. What questions would you have asked during the needs assessment?
2. Should reaction, learning and application data be collected and presented?
3. What are your thoughts about the way the Central Safety Committee decided to isolate the effects of the safety incentive plan?
4. What other alternatives could have been explored to isolate the effects?
5. Would you have sought approval to collect additional data from employees during the follow-up evaluation regarding what caused the improvements? Why would you want additional data? How would you have justified the cost of this additional work?
6. Were there additional costs that should have been included? Should the cost of communication be included?
7. How would you have communicated the results differently?
8. How credible is this study?

18

Measuring ROI in a Job Safety Training Program at a Major Food Retailer

Darry Dugger, PhD

Abstract
A major food retailer holds a competitive edge in the market because of its proprietary production and distribution processes. Yet, the company's largest production plant has developed a consistent problem with missed production schedule standards, which has interrupted its delivery routine and given the plant a 10 percent redelivery rate, twice the company standard. At the same time, the rate of injuries at the plant continued to increase and the plant had unacceptable employee turnover rates. Management decided to implement an aggressive pilot program to improve new employee training in safety procedures, which it is believed would improve all of those measures. In fact, the pilot program not only cut injuries and improved production, the company was estimated to realize an almost $20 benefit for each dollar spent on the training.

Keywords: Job safety, production schedules, turnover, safety, safety training

18.1 Background

The client company is a major food retailer with more than 1,000 specialty retail units in 36 states across the U.S. The company has been in business for approximately eighteen years and is a

Jack Phillips, Patti Phillips, and Al Pulliam, Measuring ROI in Environment, Health, and Safety, (397–412) 2014 © Scrivener Publishing LLC

publicly traded company. A key division is the Dough Production Operations, with the mission to produce and deliver fresh dough product to the retail operating units. Dough Production Operations includes approximately 25 production facilities with approximately 1,000 employees, with integrated production and distribution operations, operating seven days a week, 363 days a year. Dough Production Operations is considered a competitive advantage for the company because of its proprietary production and distribution processes.

Since 2005, approximately 20 percent of production plants have been underperforming against established financial goals and objectives. Included among these facilities is Plant A, which is the largest production plant in the system. Plant A services about 30 percent more retail units than the next largest production facility. Plant A sales and profits are critical contributors to the division's bottom-line results.

Company operation reports and production metrics showed that Plant A consistently has missed production schedule standards, which has interrupted its delivery routine and given the plant a 10 percent redelivery rate. The company standard is 5 percent. Redeliveries require using additional inventory, fuel and other resources as well as increasing production and distribution hours and overtime hours. Addressing this performance problem will improve operation efficiency overall and help the division achieve established financial goals. Plant A's performance problem represents approximately $15,000 cost opportunity per month.

18.1.1 Key Business Issues

A needs analysis determined that one of the key business issues affecting the division's ability to achieve its financial objectives was an increasing trend in workplace injuries. Over the previous two years, there had been a steady climb in workplace injuries. In 2005, the workplace injuries increased 30 percent over 2004; in 2006, they increased 16 percent over 2005. About 60 percent of the increased injuries occurred in 80 percent of the underperforming production facilities. Overall, the injuries cost the division $1.1 million annually.

A cause analysis determined that the increase in injuries was caused by unsafe acts or behavior, which accounted for 68 percent of total injuries in 2006; and unsafe conditions, which accounted for 32 percent of the total injuries that year. As determined by the cause analysis, the unsafe conditions were effects of poor OSHA compliance, inadequate preventive maintenance practices, equipment specifications and inadequate training.

The cause analysis also determined that core employees, those who had been employed for more than one year with the company and worked on the production floor, performed the basic task of the production process satisfactorily. They followed the production schedule without variance. However, employees who had less than six months tenure with the company appeared to have an attitude of, "I just do what they tell me to do," and they demonstrated inconsistent job performance.

18.1.2 Baseline Evaluation

Employees in a 2006 survey gave Plant A a score of 2.7 on a 5-point scale for training hourly employees. The average score for the rest of the production facilities in this category was at or above 4.4 points. Plant A's employee turnover rate that year was 130 percent, compared to the company standard of 40 percent. Human resources staffing reports showed that Plant A lost five employees each period, totaling sixty employees annually leaving the Plant. Data gathered from company staffing reports indicate that 60 percent of employees leave the company due to poor training or unclear job expectations. Based on this analysis, it was projected that thirty-six employees would leave the company for these reasons over the next twelve months, costing the company an estimated $106,200.

As a result of the needs assessment, a root cause analysis determined that employees do not receive adequate consistent safety awareness training during the new hire orientation process. Due to inadequate staffing levels, management circumvents the employee on-boarding and training process to get new employees on the production floor more quickly. While this practice satisfies the immediate need to fill the schedule, in the long term it contributes to increased employee turnover, increased workplace injuries and diminishing operation efficiency.

Before the 2006 needs assessment, several performance interventions had been implemented to address workplace safety in the division. Management tried various safety incentive plans, safety games and a variety of safety communication plans designed to motivate employee behavior to improve workplace safety. However, all the efforts fell short of addressing the root cause problem of the problems. Senior management has become frustrated and increasingly concerned that the problem will worsen and have an even greater impact on the ability of the division to achieve its financial targets and maintain its competitive advantage in the marketplace. They want this problem fixed immediately.

18.1.3 Learning Intervention

Management set the goal of improving and maintaining operations efficiency by reducing worker compensation injuries by 15 percent compared to 2006, reducing employee turnover by 10 percent compared to 2006 and improving staging error rates necessitating redelivers to the company's 5 percent standard. The goals will be achieved by implementing a Job Aid for New Hire Orientation learning intervention. The learning intervention focuses on following a ten-step new-hire onboarding process, under which initial safety awareness training will be conducted for all hourly production employees within the first 90 days of employment.

18.2 Evaluation Methodology

The following program objectives were developed for this project following the Phillips ROI Methodology and are based on a needs assessment analysis. The Phillips ROI Methodology evaluates five different types of data focusing on result rather than inputs.

The Level 1 data measures *Reaction*. The objective for this level is for the participants to experience positive reaction and satisfaction with the knowledge and skills in the presentation. Reaction and satisfaction data is collected using a paper-based survey on which employees can rate six items on a five-point scale. The survey also includes an additional five items designed to gather descriptive data to measure the managers' satisfaction level with the intervention. The targeted audience will complete the survey immediately following the initial training on the intervention and immediately after using the tool in a real-life situation. This level of evaluation is designed to gather data to answer the following questions concerning the effectiveness of the intervention:

- Do managers find the job aid easy to use?
- Is the job aid easily and timely accessible?
- To what extent does the job aid answer managers' questions?
- Does the job aid increase or decrease managers' time in the hiring process?

Measurement for level 1 is to achieve at least 80 percent of the participants rating their overall satisfaction level with the program at satisfied or very satisfied, a 4 or 5 on the Job Aid Evaluation Survey.

The Level 2 data measures *Learning*. The objective for this level of evaluation is for the participants to understand and demonstrate company orientation processes and concepts as well as behavior change gained through the learning process. The learning measurement is reached through structured, on-site observation during the training presentation to assess the knowledge and skill level of the participants. The human resources manager and the learning and development manager will observe each manager using the job aid during an actual hiring process. The observers will use a Structured Observation Checklist tool to capture the behavior of the manager. Each participant is expected to score at least 80 percent overall on the observation evaluation. The key evaluation questions for this level are:

- Do participants understand what they are required to do to execute the 10-step process?
- Are participants confident to apply their newly acquired skills consistently once training is completed?

The Level 3 data measures *Application*. The application objective is for participants to consistently use and follow the 10-step orientation process with every new employee. This level is important to evaluate how successful the participants transferred the training to the real workplace. This level of evaluation is also useful to measure the effectiveness of the Job Aid tool. A Level 3 evaluation will answer the following questions:

- Have participants' behaviors changed to incorporate and demonstrate the knowledge and skills learned in the training to apply at their jobs?
- Is the Job Aid tool a supporting factor in the behavior change?
- If the participants are not applying their skills and knowledge, why not? It is also important to capture the enablers that supported desired behavior.

Measurement for Level 3 is participants performing and following the 10-step process 100 percent of the time during the structured observation evaluation.

The Level 4 data measures *Impact*. Senior management wants to know how the intervention would influence workplace injuries, employee turnover and operations efficiency. Based on the root cause analysis determination of inadequate staffing as a driver of poor training practices, increased workplace injuries and increased staging errors, employee turnover data were collected, analyzed and converted to monetary values for ROI analysis.

The Level 5 *Return on Investment* compares the benefits of a program with the total cost of the program. In this study, an ROI analysis is conducted on employee turnover data and the standard ROI target of 25 percent is established as a target for this program.

18.2.1 Data Collection Plan

The data collection process begins with the review of objectives and measures of success, identification of the appropriate data collection methods and the most credible sources of data and the determination of the timing of data collection. For this study, data were collected using a participant program questionnaire, structured observation checklist and business performance monitoring, as illustrated in Figure 18.1.

The participant's program questionnaire, Figure 18.2, was developed to measure the participants' reaction and satisfaction to the learning intervention. It was delivered at the end of the training presentation with the expectation that each participant complete it.

Level	Broad Objectives	Measures	Data Collection Method/Instruments	Data Sources	Timing	Responsibilities
1	**Reaction and Satisfaction** Positive reaction and satisfaction with the knowledge and skills presented in the presentation	At least 80% of the mangers rate their satisfaction level with the program at satisfied or very satisfied (4 or 5 on a 5-point scale) using the Job Aid Evaluation Survey.	End of program / Job Aid Evaluation Questionnaire in hardcopy format	Plant A manager participants	Immediately at the end of the training program	Regional Learning and Development Manager / Facilitator
2	**Learning** Participants understand and demonstrate company orientation process and concepts. Confidence to use the knowledge and skills	Average score of 80% using the Structured Observation Checklist evaluation Average overall score of 4.0 for the following item on the Job Aid Evaluation Survey: Your confidence that you will use the tool in the future.	During the training program presentation / Structured Observation Checklist	Participants	During mock on-boarding training scenario	Regional Learning and Development Manager / Facilitator
3	**Application/ Implementation** Participants consistently use and follow the 10-step orientation process with every new employee.	Each manager performs and follows a 10-step process 100% of the time during the structured observation evaluation.	End of training program / Structured Observation Checklist	Participants	During orientation training at the workplace	Regional Learning and Development Manager / Facilitator
4	**Business Impact** Reduce worker compensation injuries by providing basic safety awareness training to new employees. Improve operations efficiency by reducing staging errors.	Workplace injuries reduced by 15% compared to 2006 total injuries Employee turnover reduced by 60% Achieve staging error rate of 5% or less	Business performance monitoring Business performance monitoring Business performance monitoring	Monthly Worker Compensation Injury Report for Plant A Monthly Staffing Report for Plant A Production Exception Report for Plant A	60-90 days post program training 30-60 days post program training 2-week post program training (14 shifts)	Evaluation team: Regional Manager, Regional L&D Manager and Regional HR Manager
5	**ROI**	**Target 25%**				

Figure 18.1 Data Collection.

Direction: To continue to assess the value and usability of this tool it is critical to get you feedback as a participant in this project. By drawing a circle, please rate satisfaction level for each statement listed below on a scale of 0 to 5 (0 = did not apply, 1 = very dissatisfied, 5 = very satisfied). Please provide additional feedback to the questions below.	
The overall content included in the tool and relevance to your job.	0 1 2 3 4 5
The usability of the tool to the interview process.	0 1 2 3 4 5
The value of this tool for improving your skills in the recruiting process.	0 1 2 3 4 5
The look and feel of the tool and its application to your job.	0 1 2 3 4 5
The overall helpfulness of the information the tool provided to your job.	0 1 2 3 4 5
Your confidence in using the tool in the future.	0 1 2 3 4 5
1. What did you like most about the tool?	
2. What did you like least about the tool?	
3. What was not helpful regarding the tool?	
4. What would you do to improve the tool?	
5. Is there anything about the tool that you think would discourage its use for this type of intervention? What and why?	

Figure 18.2 Job Aid Hiring Process Evaluation Survey.

The structured observation checklist, Figure 18.3, was designed to measure the participants' knowledge and skill to apply the 10-step process of the learning intervention. To measure change in behavior, each manager completed a survey after using the job aid in an actual interviewing environment.

The L&D manager, using a structured observation tool developed for the study, also observed the participants' performance and skill level with the job aid. The survey and structured observation tools were used as knowledge and skills verification during the training and again at the end of training as a follow-up evaluation. The business performance monitoring process consists of review and tracking essential data to measure level 4 evaluations captured on current operations and human resources reports.

Because of the skills verification component of the learning intervention, it was necessary to collect data at two different intervals during the data collection process for levels 2 and level 3. It

Part 1: Observation Form

The person being observed:			
Is able to execute the steps outline below in conducting an effective new hire orientation session:	Acceptable	Needs Practice	Omitted
Step 1: Introductions, agenda and team building exercise.			
Step 2: Review company benefits.			
Step 3: Review company mission statement.			
Step 4: Review employee handbook.			
Step 5: Complete new hire paperwork.			
Step 6: Safety orientation.			
Step 7: Harassment Awareness.			
Step 8: GMP orientation.			
Step 9: Order uniforms			
Step 10: Overview of training schedule, Adjourn			

Part 2: Need for Follow-Up	
Activities Requiring Coaching	**Potential Performance Improvement Actions**
1.	
2.	
3.	

Figure 18.3 Structured Observation Checklist.

is critical to the success of the program that the participants have demonstrated knowledge and hands-on practical skills in the learning intervention. The timing strategy for data collection for level 4 is critical to identifying trends as well as the immediate impact the learning intervention has on the business impact objective.

18.2.2 Isolating Effects of the Intervention

The intervention selected for this project was implemented as a pilot in Plant A. The participant sample consisted of five managers assigned to Plant A. Because level 3 and level 4 evaluations are a new approach for the organization, it is best to avoid complicating the evaluation process by using techniques such as forecasting or trend line analysis. Senior management was much more comfortable with using participant and/or supervisor estimates through the questionnaire approach because these approaches have been used historically for other ROI evaluations of key company initiative implementations. Therefore, the technique used for this project to isolate the effects of the training intervention is the participant estimation approach.

Managers participating in the program are a credible source to determine the value of the training and transferability of the skills and knowledge to the job. Participants in this study have hands-on experience in the workplace and the issues that contribute to the performance problems affecting Plant A's ability to achieve financial objectives. The participants were included in the development and evaluation process of the training intervention and have a clear understanding of the performance problem and expected outcomes of the training.

18.2.3 Results of Training

The data collected to analyze the results of this intervention suggest that, overall, the participants experienced a high satisfaction level for this learning intervention. The participants' aggregate score is 4.3, which is a satisfaction level between satisfied and very satisfied. The descriptive data findings are consistent with this conclusion. Write-in comments suggest there is overall approval of the learning intervention in terms of content, purpose and usability.

Additionally, the data analysis revealed managers performed the required steps of the learning intervention 74 percent at an acceptable level. The goal of this intervention is for managers to follow a consistent process during onboarding activities. Because the job aid presented a different approach and required managers to adapt to an organized format, key stakeholders considered 74 percent compliance acceptable. A score of 74 percent suggests a significant improvement in manager behavior and application of the on-boarding process.

These findings suggest that if managers continue to follow the Job Aid 10-Step orientation process at least 74 percent of the time when processing new hires over the next year, twenty-seven fewer employees will leave the company, resulting in $76,650 potential cost saving to the company (36 potential terminations x 74 percent = 27 fewer terminations).

18.3 Calculating ROI

The following cost categories were considered when capturing cost associated with this learning intervention: analysis/design/development, delivery and research and evaluation. Figure 18.4 shows the overall cost estimates for each category. Figure 18.5 captures the cost category estimates associated with replacing an hourly employee due to unwanted turnover. Data captured for this project are needed to calculate the benefit-to-cost ratio and the return-on-investment percent of the intervention.

Annual value of improvement ($79,650) and the total cost of the program ($4,004) determine the benefit-to-cost ratio and the return on investment of the program intervention. The following methods are used to determine the benefit-to-cost ratio and return-on-investment calculations.

18.3.1 Benefit-to-Cost Ratio

The formula for calculating the benefit-to-cost ratio is BCR = Program Benefits / Program Cost. Applying this formula, the BCR calculation for this project is:

$$(BCR) = \$79,650 / \$4,004 = \$19.89:1$$

For each dollar invested in this intervention, the company realized a $19.89 financial benefit for the first year after the program implementation.

Category	Cost Estimate
Analysis Cost	$1219
Development Cost	$920
Delivery Cost	$1010
Operations/Maintenance Cost	$180
Evaluation Cost	$675
Total Program Cost	$4004

Figure 18.4 Overall Cost Categories.

	Hours	Avg. $	Total
Advertising			$50
Background			$32
Interview time for Mgr	1	30	$30
Training time 2 wks	80	11	$880
Mgr- 1 wk of pain x train	20	29	$580
Fill time 2 wks OT for prod	80	17	$1,320
New Hire paper & orient	2	29	$58
Total Cost per Turn Over			$2,950

Figure 18.5 Cost Categories: Hourly Employee Turn Over.

18.3.2 ROI

The formula for calculating ROI is ROI (%) = Net Program Benefits / Program Cost x 100. Using this formula, the ROI for this intervention is:

$$(ROI) = 79{,}650 - 4{,}004 \ / \ 4{,}004 \times 100 = 1889\%$$

This intervention generated a 1,900 percent return on the company investment. A 74 percent reduction in employee turnover resulted in a more stable, efficiently trained workforce.

Figure 18.6 Illustrates training process benefits for the primary measures.

18.4 Barriers and Enablers

The participants provided the following feedback regarding their perspective barriers and enablers of the job aid training intervention.

18.4.1 Enablers

- The job aid is very easy to use and follow.
- The job aid is very helpful in keeping the onboarding process on track and consistent.

	Pre-trng. 6/Month Average	Post-trng. 6 Month Average	Pre/post Difference	Participant's Estimate of Impact	Unit Value	Annual Impact of Training Estimate
W. C. Injuries	2.7	.33	2.37	1.9 (2.37 x 80%)	$8,000	$233,280
Employee Turnover	60%	.15%	45%	.45 (45% x 100%)	$2,950	$79,650
Staging Errors	10%	7.0%	3.0%	2.4 (3.0% x 80%)	$1,500 each redelivery	$2,160

Calculations

Workers Comp:
Change in number of injuries in one year = 36 injuries x 80% = 29.16
Savings = $8,000 x 29.16 = $233,280
Turnover:
Change in numbers leaving in one year = 60 terminations x .45 = 27
Savings = $2,950 x .27 = $79,650
Staging Errors:
Reduce staging errors by 3% = .15% fewer errors x 80% = .12
Savings = $1500 x .12 = $2,160

Figure 18.6 Annual Monetary Values for Primary Measures.

- Stakeholders support the use of the job aid because it provides a systematic approach to onboarding new employees.
- Implementing the job aid resulted in an impressive benefit-to-cost ratio.
- The intervention resulted in an exceptional ROI.

18.4.2 Barriers

- Some managers were concerned that the job aid provided too much structure and lacked the ability to allow managers to deviate when the situation dictated.
- Stakeholders were concerned about the time commitment to implement the job aid across the system in all production facilities and about it being used consistently.

18.5 Communication Plan

Senior management was a main focus for communication during and after the training process, partly because their buy-in was critical in securing approval for the project to continue in the remaining production facilities. It was also critical to gain support of the regional manager and Plant A management team participants in the project. Without their support and feedback, it would have been difficult to measure and evaluate the overall success of the program, as well as identify areas that may need to be addressed or modified before proceeding with a division rollout.

The communication delivery format was a written report containing an overview of the details of the project as well as specifics of how the ROI forecasting was generated. The meeting place was a conference room with access to video conferencing and conferencing telecommunications. Key stakeholders invited to attend were sent a copy of the final report by email before the meeting. The purpose of the presentation was to communicate the status of the strategic plan project, communicate the findings of the study and garner feedback and support for recommendations and next steps.

The communication strategy is outlined in Table 18.1.

Table 18.1 Job Aid Intervention Communication Plan.

Communication Document	Target Audience	Communication Method	Communication Timing	Location	Meeting Facilitator
ROI impact report with Appendices.	VP/Ops (sponsor) Regional Manager HR staff	Distributed hard-copy report. PPT presentation.	Scheduled People meeting prior to next leadership meeting.	Corporate office	National L&D Mgr Dir/HR
Executive Summary Report	Div /SVP HR/SVP Operations leadership team	Distribute hard-copy and electronic report.	Special meeting Next leadership meeting	TBD / ASAP Corporate office	National L&D Mgr Dir/HR
Top-line review including level 1-5 results.	Plant A mgrs (participants) Division plant managers	PPT presentation. Hardcopy report.	Recap meeting Next manager conference call after leadership meeting.	Plant A Conf. call	National L&D Mgr Regional Mgr.

18.6 Conclusion

Based on the results of the data analysis, the job aid intervention was successful in achieving established goals for all four levels of evaluation. Additionally, the benefit-to-cost ratio suggests that each dollar invested in the intervention equaled $19.89 in financial benefit for the company; and the intervention generated a ROI of approximately 1,900 percent. The ROI analysis also revealed other primary measures—number of injuries, employee turnover and staging errors—showed a positive trend as a result of the training program. These factors should be considered when determining next steps of this project and the option of implementing the learning intervention division-wide.

18.7 About the Author

Dr. Darry B. Dugger has practiced in the field of human resources and training and performance improvement for more than 25 years. During this time, he held leadership-level positions with several major retail and restaurant companies and is considered an expert in strategic human resources and organizational development. Dugger earned his Ph.D. in education with specialization in training and performance improvement from Capella University. He also holds a Master of Arts in Human Resources Development from Webster University in St. Louis, Missouri, and has a professional certification from the ROI Institute. He is an associate faculty member at the University of Phoenix, where he teaches graduate and undergraduate courses in management and human resources.

19

Measuring ROI in a Work-at-Home Program

Family Mutual Health and Life Insurance Company (FMI)

Patti P. Phillips

Abstract

This case study shows the power of a work-at-home project designed to ease the environmental problems of traffic and congestion caused by the long daily work commute of more than 300 employees. The project cut average daily commute times from one hour and forty-four minutes to 15 minutes, saving each participating employee the cost of 490 gallons of fuel per year and keeping an estimated total 1,478 tons of carbon emissions out of the air. Employees taking part in the project also reported significant intangible benefits, including reduced stress and absenteeism and increased job satisfaction and engagement. The company not only got a boost in its image as an environment-friendly concern, it saved money through increased productivity, lower office expenses and less employee turnover. From an environmental perspective, the study shows how an important project can have significant impact by lowering carbon emissions. It represents a win-win project for participants, their initially reluctant managers and the organization. Perhaps the greatest winner is the environment. While this type of project may not be suitable for every organization, this is an example of how such a project can be implemented for many organizations.

Keywords: Work at home, productivity, emissions, environmentally friendly, turnover

Jack Phillips, Patti Phillips, and Al Pulliam, Measuring ROI in Environment, Health, and Safety, (413–442) 2014

19.1 FMI: PART A

19.1.1 Background

Family Mutual Health and Life Insurance Company (FMI) has enjoyed a rich history of serving families throughout North America for almost 80 years. Their focus has been on health and life insurance products and they are regarded as a very innovative and low-cost health insurance provider. The executives are proud of their cost control efforts and the low prices they can offer. Company advertisements regularly highlight their low-cost approach, quality of service and ethical practices.

FMI has grown significantly in recent years due to increased healthcare concerns in North America, particularly in the USA. Rising healthcare cost has forced the company to raise premiums several times in recent years, while still maintaining a cost advantage over other suppliers.

19.1.1.1 The Challenge

Lars Rienhold, CEO, is proud of the accomplishments of FMI and is perhaps its biggest fan. A man of considerable and contagious personality, he is continually trying to offer affordable health and life insurance policies, provide excellent customer service and be a responsible citizen. As part of this effort, Lars wanted to ensure that FMI was doing all it could to help the environment. While FMI's carbon footprint is relatively low compared to manufacturing companies, its headquarters was located in a congested area. Lars became concerned about helping the environment in as many ways as possible. During a recent trip to Calgary, Canada, he saw a television report about a local company that had implemented a work-at-home program. The report presented the actual amount of carbon emissions that had been prevented with this project. Lars thought that FMI should be able to implement a similar program, including the possibility of employees working from home. He brought this idea to Melissa Lufkin, executive vice president of human resources. The message was short. "I want us to make a difference. I think this is the way to do it." Although her team already had examined the work-at-home issue, Melissa agreed to explore the possibility in a more formal way.

19.1.1.2 Exploring the Situation

Melissa began her investigation by discussing the issue with the operations chief. Although there was some resistance, John Speegle, executive vice president of operations, was interested in exploring the idea. John was concerned about the lack of a productivity increase in the past three years with the largest segment of employees, the claims processors and the claims examiners. There were 950 employees involved in processing or examining claims submitted by customers or healthcare providers. Claims examiners reviewed claims that were disputed or when an audit sparked a review of a particular claim. The number of claims processors and examiners had grown to the point that the office space they occupied in Building 2 was overflowing. This impeded productivity, not to mention made it an uncomfortable environment in which to work. Given the company's continued growth, it was likely that a new building space or perhaps a new facility was needed to manage the growth.

John concluded, "I'm interested in the possibility of employees working from home if it can be managed properly. Let's explore the possibility if all parties are in agreement to pursue it." John was interested in lowering the real estate cost of new office space, which averaged about $17,000 per person per year and improving productivity, which was at a rate of 33.2 claims processed and 20.7 claims examined per day.

Melissa discussed the issue with Linda Green, the vice president of claims, to identify her concerns about processors and examiners working at home. Although this issue had been discussed in previous meetings and many people had said that these jobs could be easily managed remotely, Melissa had never received direct communication on the topic. Linda was supportive but raised several concerns. "Some of our managers want to keep tabs on what is going on and they feel like they have to be there to resolve issues and problems - and they want to see that everyone is working and busy. I am afraid it is a matter of control, which they may have a hard time giving up if people work remotely." Melissa realized that it would take some extra effort with these managers, who would have to view this initiative as necessary and feasible in their world. Linda added, "I realize that the right approach might make their jobs easier, but right now they may not be at that point."

Next, Melissa met with the IT department and discussed how they could equip workstations at home with the latest technology. She found a supportive audience in Tim Holleman, senior vice president and chief information officer, who thought that employees could be setup with adequate security and technology to work at home in the same manner as they were working onsite. Tim added, "They can have full access to all databases and they could be using high-speed processes. It would cost FMI a substantial amount the first year, but may not represent a very significant cost in the long run."

Melissa later discussed potential issues with the legal department. Margaret Metcalf, chief legal officer, was cautious, as expected and said several legal issues would have to be addressed from a liability perspective. She asked about other companies that were pursuing this route and Melissa agreed to furnish examples and make contacts with them to discover what problems they had encountered.

Melissa then contacted Anne Burson, executive vice president of sales and marketing, to uncover any customer service issues that might arise. Anne was in favor of the move as long as customer service would not suffer. She remarked, "The claims examiners are in direct contact with the customers and I want to make sure that acceptable customer service is maintained. Also, many of the processors have to make routine direct contact with healthcare suppliers, as well as patients, and we want to maintain these contacts at an acceptable level. Other than these concerns, I can see that this would probably help morale and might even improve our service. Let's give it a try."

Finally, Melissa met with her Chief Financial Officer, Rodrick Harper, to discuss the project and the plan to measure its success. Melissa previously had talked with Rod about measuring success and he expressed some desire to show the value of major human resources initiatives. Melissa was eager to show the value of HR programs and had challenged her staff to measure success, even using ROI. Rod volunteered a member of his team to work with Melissa on these types of projects. When Melissa discussed the project with him, including the measurement plans and a financial ROI, Rod's interest really piqued. He said, "Let's make sure this is very credible analysis and that it is very conservative. Frankly, I think we want to be involved when you discuss ROI. I think it's proper that we use a standard approach to analysis and we would like to be involved in this every step of the way, if you don't mind." Melissa was pleased with the support, but somewhat anxious about working with the Finance and Accounting Team to evaluate a program that she ultimately would own. However, she felt the project was necessary and would be advanced by the very good relationship she and her team had with the Finance and Accounting Group.

Melissa and her staff explored the attitudes of the employees to determine how they would perceive a work-at-home program. She was not sure how many would take advantage of the opportunity, but she was certain most would be interested. The staff conservatively estimated that at least a third would opt to participate in the program. For many in this group, working at home would be a huge motivator and would probably make a difference in retaining them at FMI. From that perspective, the staff suggested that it be explored. Melissa cautioned, "They may have issues at home that they want to address, but we must be able to get eight hours of work out of them. They cannot discontinue daycare, trying to manage childcare and work as well. If they have an elderly or disabled person at home, this cannot be a way for them to deal with both situations. We must have full productivity, and that is essential."

With this positive reaction (and a few concerns), Melissa and her team decided to undertake this substantial project. After much discussion, the group decided to engage a consulting group, Workforce Solutions International (WSI), to manage the project. WSI had considerable experience in implementing alternative work systems, particularly work-at-home programs. They knew what questions to ask and what situations were going to occur and more importantly, they were able to anticipate the problems that could derail the project.

19.1.2 The Analysis and Initial Alignment

After some discussion, the group asked WSI for a proposal. Included in the RFP to WSI was a forecasting component for the project. Essentially, WSI was asked to bid on analyzing the need for the project to determine its feasibility, forecast its value, design the appropriate program and implement and monitor the success of the program. Success would be measured at the ROI level. Armed with this information, WSI was prepared to begin work on the proposal.

19.1.2.1 The Consultants

Deborah Rousseau was selected by WSI as the lead consultant for this project. Deborah had previous experience with flexible work systems, had managed many successful projects and was an outstanding consultant. Deborah believed in showing the value of their work and she guided the proposal process toward an agreement to deliver the four components:

1. Clarifying that the solution is needed and connecting it to the appropriate business measures
2. Forecasting the impact and ROI of the project
3. Implementing the program with claims processors and examiners
4. Showing the value of the project using the ROI Methodology.

With this focus on results, Deborah knew that she had to skillfully present the best proposal and the most focused implementation possible. There was no room for error. WSI was obligated to deliver the value desired by the clients.

To make the proposal meaningful, Deborah asked the client if they could forecast ROI after they verified the solution. The client agreed. In essence, the proposal was developed in two parts. The first proposal validated the solution and provided a forecasted ROI. The forecast would be developed and approved by the client before the program would be implemented. The second proposal focused on implementation and an impact study with ROI. This seemed reasonable because the analysis required to develop the forecast was part of the analysis that would verify the proper solution to drive the business measures. WSI proposed $31,000 for the first proposal ($21,000 for the initial analysis and assessment and $10,000 for forecasting ROI), which included a briefing to senior executives.

19.1.2.2 The Analysis

When the first contract was awarded, Deborah began meeting with appropriate individuals from the HR Team, including representatives from employee relations, learning and development, recruiting, compensation and HR planning. She examined records, conducted employee focus groups and conducted a survey of a small, selected group of employees to understand their desire, need and intentions to work at home if the option was available. In this survey, employees were asked about benefits from this type of arrangement. The focus groups and the survey revealed that this solution should drive business measures.

Part of this analysis involved the examination of other case studies of work-at-home projects, to understand the payoffs of those projects and the barriers to success. This analysis focused on potential improvements in productivity, reduction in absenteeism, improvements in healthcare costs and a reduction in real estate costs. The potential effect on healthcare costs was weak and it was removed as a potential impact measure that could be influenced by this new arrangement. By using the employee feedback, analysis of other studies and examination of internal records, Deborah and the HR Team agreed that this solution could drive important business measures.

19.1.2.3 Alignment

Deborah's next task was to identify specific business measures. Her key input for this task was provided by the executive vice president of operations (EVP), who thought that this program could reduce real estate and productivity costs. It was obvious that the real estate costs could be improved, unless the cost of maintaining an at-home office proved to be excessive or the same office space for each participant continued to be maintained at FMI. Deborah worked with the EVP to set clear objectives for office space and productivity. After some discussion, the EVP suggested that processors and examiners could both process an average of at least one more claim than they had been producing and the office expense should be dramatically reduced, in the neighborhood of 20 percent for the first year. When discussing the actual ROI of the study, the EVP was reluctant to set an objective. However, when Deborah suggested that the ROI should be more than an investment in a building, for example, the EVP agreed to set a goal. Given that FMI would average about 15 percent for capital expenditure investments, Deborah suggested that an ROI of about 25 percent would be appropriate and the EVP agreed.

The vice president of claims confirmed the objectives regarding productivity and real estate costs with Deborah and then focused on turnover reduction. The annual turnover rate at the time was 22.3 percent and they felt that an improvement rate of at least 5 percent to 17.3 percent should be achievable if the project was successful. They also reviewed the absenteeism rate and thought it could drastically improve from a current level of 7.3 percent to a new level of 4.0 percent. Deborah addressed the critical issues regarding implementation during this discussion.

Deborah met with Ginger Terry, environmental coordinator in the procurement function, to collect data about carbon emissions from automobiles and set a goal to show the actual reduction in carbon emissions that could be realized by eliminating the office commute. Ginger had compiled data about the commute time of employees for these two groups and the average time was estimated to be one hour and forty-four minutes each day. In a work-at-home arrangement, this time could be reduced to about 15 minutes, assuming a visit to the office every seven days. Deborah realized that the environmental benefits would not add to the ROI in monetary terms, but would be a substantial intangible for the citizens of this city and of the country as a whole.

Finally, she met with the CFO, chief legal counsel, chief information officer and the HR Team. The principal focus of the meeting with this group was to review the tentative objectives for additional refinement and concurrence.

Deborah's meeting with the HR team generated some important information, in terms of what employees and managers must learn to make the process effective and successful. Several questions surfaced about working without distractions, such as childcare issues, other people in the residence or elder parent care. Associates also would have to adjust their working habits from an office to a home environment. They would need to adopt the discipline and structure necessary to be effective, by following consistent rules, regulations and working hours. It was also noted that managers must be able to effectively provide coaching and counseling along the way and be there for associates to address particular issues.

Deborah also explored the issues of perception and desired reaction with the executives. The executives expressed their belief that the process was needed and ultimately would be motivational and rewarding for participants. With this data, Deborah began to develop the objectives that would lead to the ROI forecast.

19.1.3 Questions for Discussion

1. Critique the way in which the analysis has been conducted.
2. Are there additional questions or issues you would explore? If so, what are they?
3. Write the objectives for all five levels.
4. Complete the V-model showing the connection between the upfront assessment, objectives and evaluation at five levels.

19.2 FMI: PART B

19.2.1 Objectives and Alignment

From the discussions, subsequent analyses and potential solutions, Deborah, Melissa and the HR Team could develop all of the objectives at different levels. The objectives are developed from the needs at different levels.

19.2.1.1 Objectives

Table 19.1 shows the objectives for the project by different levels, ranging from reaction to ROI. Deborah secured agreement on the objectives from those stakeholders involved.

19.2.1.2 Alignment Model

The alignment model is shown in Figure 19.1. It shows the connection between the upfront needs assessment, the objectives and the evaluation. Deborah found it helpful to construct this model to clearly determine whether any pieces were missing. She worked through the analysis in order beginning with Level 5. The project's value became obvious early in discussions. She then clearly explored the business needs with different stakeholders. Job performance needs were revealed in concerns voiced by the senior vice president of claims. The learning needs evolved from that conversation and the reaction needs were consequently developed from these discussions. Previous studies in which Deborah and her firm had been involved dictated some of the learning and reaction needs, e.g., projects sometimes fail because people do not fully understand the rules, do not understand the work process or have the incorrect perception of the process. The objectives came directly from the needs assessment

Table 19.1 Detailed Objectives

After implementing this project:

Reaction

- Employees should see the work-at-home project as satisfying, important, rewarding and motivational.
- Managers must see this project as necessary, appropriate and important to FMI.

Learning

- Employees must know the realities of working at home, the conditions, roles and regulations.
- Employees must have the discipline and tenacity to work at home.
- Managers must be able to explain company policy and regulations for working at home.
- Managers must be able manage remotely.

Application

- Managers should conduct a meeting with direct reports to discuss policy, expected behavior and next steps.
- At least 30 percent of eligible employees should volunteer for at-home assignments within one month.
- At-home offices are built and should be properly equipped.
- Work-at-home employees should work effectively at home.
- The at-home workplace should be free from distractions and conflicting demands.
- Managers should properly administer the company's policy.
- Managers should manage the remote employees effectively.

Impact

For those involved in the program:

- Commute time should be reduced to an average of 15 minutes per day.
- Office expense per person should reduce by 20 percent in six months.
- Productivity should increase by 5 percent in six months.
- Employee turnover should reduce to 12 percent in six months.
- Unplanned absenteeism should be reduced.
- Employee stress should be reduced.
- Carbon emissions should be reduced.
- The company's image as a green company should be enhanced.
- Employee engagement should improve.

ROI

- The company should achieve a 25 percent return on investment.

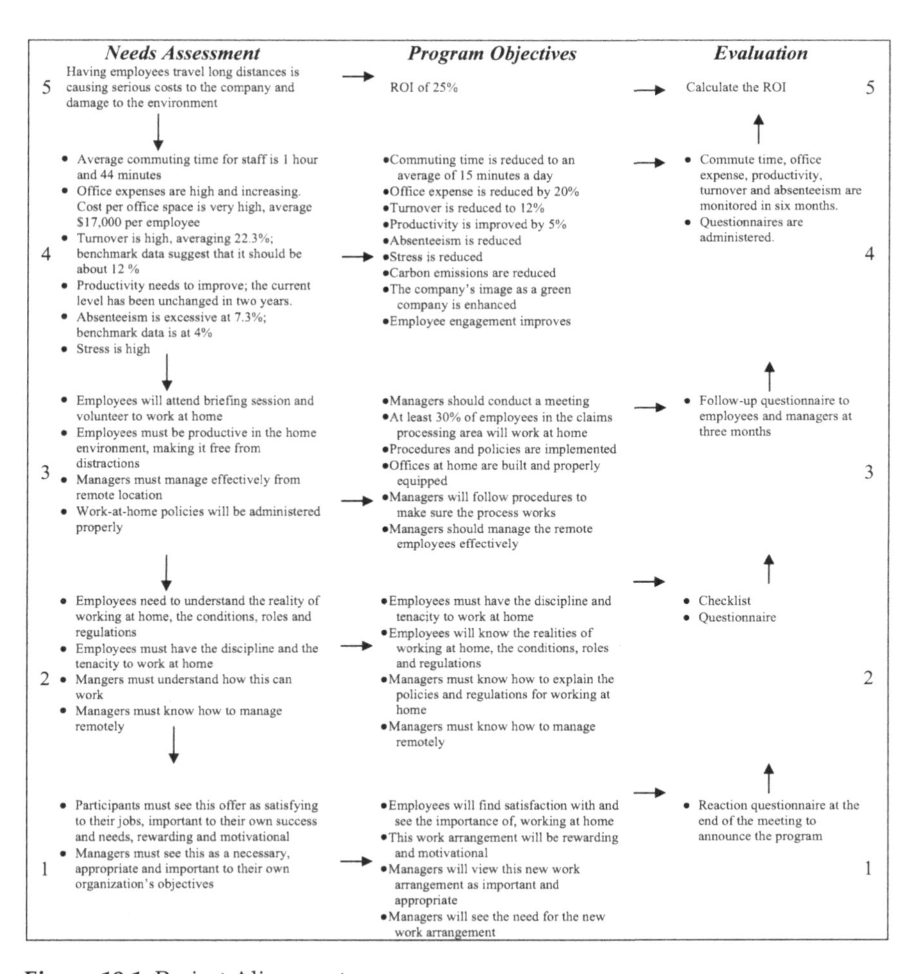

Figure 19.1 Project Alignment.

and were specifically developed based on each need. At this point, evaluation was tentative in terms of how the data would be collected. More detail on the evaluation side would be provided as the project unfolded, but the V-model provides the alignment necessary at the different levels of needs assessment, objectives and evaluation.

19.2.2 Questions for Discussion

1. What is the value of having objectives at all five levels? Please discuss.
2. Is it helpful or necessary to develop a V-model on a program that is destined for implementation? Please explain.

19.3 FMI: PART C

19.3.1 ROI Forecast Process

With a clear understanding of the solution and the connection to the business impact measures, a forecast was now possible. Although Deborah could have forecast reaction, learning and application, she limited her forecast to impact and ROI. This essentially is what was requested in the RFP, with most of the emphasis on the ROI number itself. Deborah developed the forecast, following the assumptions from the various involved stakeholders.

19.3.1.1 Impact Forecast

The first important input to the forecast was the expected number of employees who would participate. This was a voluntary program for which both the advantages and disadvantages were clearly laid out for employees, along with conditions and regulations. Not everyone would be willing to go down this path. As stated earlier, Deborah and the HR Team thought that about one-third of employees would sign up for this program. One-third of 950 is 317, so the forecast is based on 317 participating employees. Based on the percentage makeup of the two groups, this translates into 237 and 80, respectively, for processors and examiners.

Figure 19.2 shows the development of the monetary forecast, following estimated improvement in business measures. The estimated business impact was obtained directly from the chief of operations and the vice president of claims. The monetary value of a claim also was obtained by these stakeholders, estimated to be $10 cost for processing a claim and $12 for review of a claim. The office expenses were estimated to be $17,000 and the cost of a turnover taken directly from a similar study (where the cost of turnover was pegged as a percent of annual pay) was provided at $25,400. With this in mind, the calculations are listed below:

19.3.1.2 Estimated Costs and Forecast of the Project

The costs of the project were estimated to be about $1 million. This estimate is the total cost including the amount of the initial analysis to determine whether this was the proper solution and the development of that solution. The majority of the charges

Anticipated Participation

- Target Group: 950
- Predicted Enrollment: 1/3
- 950 x 33 1/3% = 317
- Allocation: 237 processors
  80 examiners

Estimated Impact

- Productivity: 1 additional claim processed
  1 additional claim examined
- Office expenses: 20% reduction
  $17,000 x 20%=$3,400
- Turnover reduction: 22.3% to 12%=10.3% improvement

Converting Productivity to Money

- Value of one claim=$10.00
- Value of one disputed claim=$12.00
- Daily improvement=1 claim per day
- Daily improvement=1 disputed claim per day
- Annual value=237 x 220 work days x 1 x 10.00= $521,400
- Annual value=80 x 220 days x 1 x 12.00=$211,200

Office Expense Reduction

- Office expenses in company office: $17,000 annually per person
- Office expenses at home office: $13,600 first year per person
- Net improvement: $3,400, first year
- Total annual value=317 x 3,400=$1,077,800

Converting Turnover Reduction to Money

- Value of one turnover statistic = $25,400
- Annual improvement related to program=10.3%
- Turnover prevention: 317 x 10.3% = 33 turnovers prevented
- Annual value=$25,400 x 33= $838,200

Figure 19.2 Forecast of Monetary Benefits.

were in the IT support and maintenance, administrative and coordination categories. When the monetary benefits are combined with the cost, the ROI forecast is developed, as shown in Figure 19.3.

19.3.1.3 Presentation of Results

Although this number is quite impressive, Deborah cautioned the team not to make the decision solely on the ROI calculation. In her presentation to the senior executives, Deborah stressed that there were significant intangibles, first and foremost being

Total Forecasted Monetary Benefits

Benefits = $521,400 Processor Productivity
211,200 Examiner Productivity
1,077,800 Office Costs
838,200 Turnover Reduction
$2,648,600

Costs = $1,000,000

$$BCR = \frac{\$2,648,600}{1,000,000} = 2.65$$

$$ROI = \frac{\$2,648,600 - 1,000,000}{1,000,000} \times 100 = 165\%$$

Figure 19.3 Forecasted ROI.

the contribution to the environment, which is not included in this calculation. Other factors such as job satisfaction, job engagement, stress reduction and image were huge intangibles that should be directly influenced from this. However, because these projects need to be based on good business decisions, the ROI forecast is credible and conservative and based on only one year of value. Much more value will be realized after the first year, because most of the office setup expenses will occur in the first year.

Deborah also cautioned that for these results to materialize, the program would have to be implemented with a focus on results and the objectives set for the program. One by one, she presented each of the objectives and stressed that they would be communicated to all the stakeholders, including the employees. This would ensure that everyone would clearly grasp his or her responsibilities to make the program successful. While it should deliver a significant ROI, most of the emphasis should be placed on the intangibles.

19.3.2 Questions for Discussion

1. What value does a forecast add to the situation? Is it needed in today's climate?
2. How helpful would it be to include reaction, learning, application and intangibles in the forecast?
3. What prevents credible forecasts from being an option pursued by consultants and clients?

19.4 FMI: PART D

19.4.1 The Solution

The details of the solution were developed with proper input. For this program to be successful the design had to be acceptable and the execution must be flawless.

19.4.1.1 Design

The design of the program followed the traditional work-at-home model, in which employees work a full 40-hour week in a home office designated for this work. The office was equipped with the appropriate interconnectivity to the company, databases and functions, much like an office in one of FMI's buildings. The pertinent ground rules for this arrangement included the following:

1. The office must be free of distractions. For example, it is recommended that a television not be located in the room.
2. Employees must work on a set schedule, if they are required to have direct contact with customers, which most are. Employees must log on at the time they begin their work and log off when they have completed work for the day.
3. The system has mechanisms for monitoring the work being accomplished. Each activity can easily be tracked to provide a user performance profile. In essence, the system determines if a person is working and records the results.
4. The home office must be designed for efficiency, good health and safety.
5. Employees were urged to take short breaks and re-energize as necessary and to always take a lunch break. The total amount of expected actual work time was 40 hours.
6. Employees were required to negotiate expectations and agreements with the family and significant others.
7. When employees must take time off for personal errands, visits to the doctor or other breaks, this time will be subtracted from their time worked. The employees will be required to make up that time during the week.

8. Employees must stay in touch with the office and periodically make contact with the immediate manager.
9. Employees must sign a work-at-home pledge and attend a session on "working at home."
10. Because there was an initial investment in equipment, computers and connections, employees were required to sign a two-year commitment to continue to work for FMI, with certain conditions. If they were to leave the company before the end of two years, they would be required to pay back the setup charges, estimated to be about $5,000.

The principal stakeholders agreed on the design. It was reviewed by a group of employees in a focus group and then modified to produce the final set of regulations.

19.4.1.2 Execution

With the design was finalized, the program was launched via communications to the target group of 950 employees. Employees received memos explaining the program and were asked to attend briefing sessions during formal working hours to discuss the work-at-home arrangement. In all, 21 employee meetings were held for the 950 employees and managers held meetings with their respective teams to discuss the advantages and disadvantages of the process. Employees were given three weeks to make a decision and to enroll in the program.

19.4.2 Questions for Discussion

1. Critique the design of the work-at-home arrangement
2. Discuss the implementation and execution
3. What precautions must be taken in an experiment involving only one segment of the company?

19.5 FMI: PART E

19.5.1 ROI Planning

The next logical phase of the process was to plan for the ROI Study. This involved completing the data collection plan and the ROI analysis plan. This phase emerged from the objectives and the input that went into the V-model.

19.5.1.1 Data Collection Plan

The starting point for the data collection plan, shown as Figure 19.4, are the objectives listed in Table 19.1. The measures are further defined along with those objectives. Methods of data collection are identified, sources of data are pinpointed and the timing for data collection is determined. The plan concludes with the responsibilities for collecting the data. The data collection is comprehensive and primarily focused on interviews, questionnaires and monitoring the data in the system.

19.5.1.2 ROI Analysis Plan

Figure 19.5 shows the ROI analysis plan, which details the analysis for the impact study. This document begins with the impact measures planned for analysis. The first column lists each of the data items, followed by the method of isolating the effects of the program on the data and the method of converting data to money. The intangibles anticipated from the project are listed after the particular cost categories. The individuals or groups targeted for the results are then identified, along with any influences that might make a difference in this evaluation.

When completed, the data collection plan and the ROI analysis plan provide a road map to conduct the study. The various stakeholders approved these documents, allowing the work to begin.

19.5.1.3 Data Collection and Integration

Data collection followed the data collection plan using interviews and questionnaires. The interviews were few in number but they did provide an opportunity to explore the issues that were included on detailed follow-up questionnaires. All participating employees and their managers received the questionnaires. In total, 342 questionnaires were distributed to employees and 45 managers. Figure 19.6 shows the data integration plan and how the data was collected and integrated to form the results.

19.5.2 Questions for Discussion

1. Critique the date collection plan
2. Critique the ROI Analysis Plan
3. How helpful is the data integration figure?
4. What improvements would you recommend for data collection and analysis?

Evaluation Purpose: Measure Success of Program

Program: FMI Work-at-Home Project | **Responsibility:** HR/Consultants | **Date:** March 30

Level	Broad Program Objectives(s)	Measures	Data Collection Method/ Instruments	Data Sources	Timing
	Reaction				
1	• Employees should see the work-at-home project as satisfying, important, rewarding and motivational	• Rating scale (4 out of 5)	• Questionnaires	• Participants	• 30 days
	• Managers must see this project as necessary, appropriate and important to FMI.		• Interviews	• Managers	• 30 days
	Learning				
2	• Employees must know the realities of working at home, the conditions, roles and regulations	• Rating scale (4 out of 5)	• Questionnaires	• Employees	• 30 days
	• Employees must have the discipline and tenacity to work at home		• Interviews		
	• Managers must be able to explain company policy and regulations for working at home			• Managers	• 30 days
	• Managers must be able to manage remotely				
	Application				
3	• Managers should conduct a meeting with direct reports to discuss policy, expected behavior and next steps	• Checklist	• Data Monitoring	• Company Records	• 30 days
	• At least 30 percent of eligible employees should volunteer for at-home assignments within one month	• Sign Up	• Data Monitoring	• Company Records	• 30 days
	• At-home offices are built and properly equipped		• Questionnaires		
	• The home workplace should be free from distractions and conflicting demands			• Participants	• 90 days
	• Managers will properly administer the company's policy	• Rating Scale (4 out of 5)		• Managers	• 90 days
	• Managers should effectively manage the remote employees				
	Impact				
4	• Commute time should be reduced to an average of 15 minutes per day	• Direct Costs	• Business Performance Monitoring	• Company Records	• 6 months
	• Office expense per person should reduce by 20 percent in six months	• Claims per day			
	• Productivity should increase by 5 percent in six months	• Voluntary turnover			• 6 months
	• Employee turnover should reduce to 12 percent in six months	• Rating scale (4 out of 5)			
	• Unplanned absences should be reduced				• 6 months
	• Stress should be reduced	• Rating scale (4 out of 5)	• Survey	• Participants	• 90 days
	• Carbon emissions should be reduced			• Managers	• 90 days
	• The company's image as a green company should be enhanced				
	• Employee engagement should improve				
	ROI				
5	Achieve a 25% return on investment				

Figure 19.4 Data Collection Plan.

Program: FMI Work-at-Home Project **Responsibility:** HR/Consultants **Date:**

Data Items (Usually Level 4)	Methods for Isolating the Effects of the Program/ Process	Methods of Converting Data to Monetary Values	Cost Categories	Intangible Benefits	Communication Targets for Final Report	Other Influential Issues during Application
• Office expenses	• Control group • Expert estimates	• Standard value based on costs	• Initial analysis and assessment • Forecasting Impact and ROI • Solution development • IT support and maintenance • Administration and coordination • Materials • Facilities and refreshments • Salaries plus benefits for employee and manager meetings • Evaluation and reporting	• Reduced commuting time • Reduced carbon emissions • Reduced fuel consumption • Reduced sick leave • Reduced absenteeism • Improved job engagement • Improved community image • Improved image as environmental friendly company • Enhanced corporate social responsibility • Improved job satisfaction • Reduced stress • Improved recruiting image	• Participants • Managers • HR team • Executive group • Consultants • External groups	• Must observe marketing and economic forces • Search for barriers/ obstacles for progress
• Productivity	• Control Group • Participant estimates	• Standard values				
• Turnover	• Control Group • Participant Estimates	• External studies				

Figure 19.5 ROI Analysis Plan.

Method	Level 1 Reaction	Level 2 Learning	Level 3 Application	Barriers / Enablers	Level 4 Impact	Costs
Initial Participant Questionnaire	X	X	X		X	
Initial Manager Questionnaire	X	X	X		X	
Participant Interviews	X	X	X		X	
Follow-up Questionnaire: Participants	X		X	X	X	
Follow-up Questionnaire: Managers	X		X	X		
Company Records			X		X	X

Figure 19.6 Data Collection Methods and Integration.

19.6 FMI: PART F

19.6.1 Results: Reaction and Learning

The data were collected following the data collection plan. The levels of data present the results.

19.6.1.1 Reaction Data

Reaction data was collected early in the program and focused on both reactions from the employees involved in the project and their managers. Although open verbal and informal positive reactions were detected early in the process, four particular measures were collected on the questionnaire directly from the employees.

1. The satisfaction with the new work arrangement
2. The importance of this approach to their success
3. The rewarding effect of this opportunity for them
4. The motivation effect of this arrangement. The company anticipated this new work arrangement would provide a more motivated employee who would produce more.

These four measures scored high numbers and Table 19.2 shows the results. The reaction from the employee perspective averaged 4.4 on a 5-point scale.

From the managers' perspective, it was important to understand how managers perceived this new work arrangement. Although there were many aspects of this issue, the objectives focused on

Table 19.2 Reaction Data.

- From Participating Employees
- Rating of 4.6 out of 5 on satisfaction with new work arrangement
- Rating of 4.7 out of 5 on importance of new work arrangement to their success
- Rating of 4.2 out of 5 on the rewarding effect of the new work arrangement
- Rating of 4.1 out of 5 on motivational effect of new work arrangement

From Managers

- Rating of 4.2 out of a 5 on importance of the work alternative
- Rating of 4.1 out of 5 on appropriateness of the work alternative
- Rating of 4.3 out of 5 on the need for the work alternative

three points: how managers perceived this program to be necessary, appropriate and important to the company. Managers must see the necessity of this program in today's work climate when considering problems encountered with commuting and the environment, as well as the desire for the flexibility of working from home. Table 19.2 shows the managers' reactions, which exceeded expectations. The ratings averaged 4.2 out of 5 on the three items. In summary, the reaction surpassed the expectations of the implementation team.

19.6.2 Learning

Although this project is not a classic learning solution, where significant skills and knowledge must be developed for the program to be successful, there is still a learning component. Employees must understand their roles and responsibilities and managers must understand the policies of working at home. They also must have the ability to explain the policies and successfully address any performance issues that can develop in the unique environment of a remote workforce. The managers and employees provided self-assessment input on their questionnaires, which typically show the learning from the two groups. The managers were given an opportunity during meetings to practice the performance discussions so they would be able to address the issues effectively. The facilitator of the meeting was required to confirm that each manager could successfully address those issues and that every manager involved in the program properly demonstrated their ability to do so. As Table 19.3 shows, the self-assessment ratings exceeded the expectations on five measures from employees, averaging 4.3 out of 5. Managers averaged 4.1 out of 5 on two measures. The confidence to explain the policy was 3.9 out of 5, just short of the goal of 4.0. Still, there was confidence that learning had occurred so that the program could be properly implemented. Also, each manager successfully demonstrated four types of performance discussions through role playing.

19.6.3 Results: Application

19.6.3.1 Application Data

These types of programs can easily go astray if employees are not following the policies properly and the managers are not managing the process appropriately. Consequently, application and tracking the implementation of the process became a very important data set. Table 19.4 shows the key items monitored which are directly connected to objectives. In all, 93 percent of managers conducted meetings with employees to discuss the work-at-home arrangement. Although 100 percent would usually be expected, a few managers either did not have direct employees or had no employees who were interested in working at home. The possibility remained that some managers did not conduct the meetings when they should have. Because of this, a complete briefing involving all employees

Table 19.3 Learning Data.

From Employees
• Rating of 4.0 of 5 on the discipline and tenacity to work
• Rating of 4.1 of 5 on the tenacity to work at home
• Rating of 4.3 of 5 on roles and responsibilities
• Rating of 4.3 of 5 on conditions and regulations
• Rating of 4.2 of 5 on the realities of working at home
From Managers
• Rating of 4.2 of 5 on key elements of the policy for working at home
• Rating of 3.9 of 5 on the confidence to explain policy
• Successful skill practice demonstration on performance discussions – all checked

Table 19.4 Application Data.

- Ninety-three percent of managers conducted meetings with employees to discuss working at home.
- Thirty-six percent of eligible employees volunteered for at-home work assignments (342 participants).
- In total, 340 home offices were built and equipped properly (two employees changed their minds before establishing an office).
- Work-at-home employees rated 4.3 out of 5 on working effectively at home.
- Ninety-five percent of employees reported the workplace was free of distractions and conflicting demands.
- Managers rated 4.1 of 5 on administering policy properly
- Managers rated 3.8 of 5 on managing remote employees effectively

covered most of the issues that the managers were exploring in the meeting. The meeting with the managers represented reinforcement and showed their connection to the project.

After all the briefings and information sharing, 36 percent of eligible employees volunteered for work-at-home assignments, representing 342 participants. This participation was better than expected and left the project team pleased. In the follow-up data, the participants rated 4.3 out of 5 on working effectively at home; in addition, 95 percent of the employees reported that the workplace was free of distractions and conflict. The managers rated themselves 4.1 out of 5 on properly administering policy. However, some had difficulty when it came to the rating of managing employees remotely, which was 3.8 out of 5.

19.6.3.2 Barriers and Enablers

With the recognition that many issues could derail the success of this program, the barriers and enablers were captured. Table 19.5 shows the barriers and enablers to success and as expected, there were some classic barriers; however, the barriers did not prove to be very significant. The greatest obstacle was the lack of manager support, with 18 percent of participants indicating this as a concern. Following closely behind was lack of necessary support from staff that would normally support them in their office work. Additionally, 15 percent indicated communication breakdown, while 11 percent thought this program would limit their career progression. A few felt that they would be left out

Table 19.5 Barriers and Enablers to Success.

Barriers	**Percent Indicated**
Managers support is lacking	18%
Lack of support staff	16%
Communication breakdowns	13%
Career progression is limited	11%
Left out of decisions	9%
IT support is lacking	7%
Lack of social interactions	5%
Enablers	**Percent Indicated**
Personal cost savings	89%
Flexibility to schedule	71%
Convenience of work	71%
Work life balance	64%
I have all the tools	54%
Support of manager	31%
Support of staff	14%

of decision-making and that IT support would be lacking, while some indicated that they were concerned about the lack of social interaction.

Regarding the enablers, the number one enabler on this list is the personal cost savings. Many employees signed up for this arrangement as a way to save costs by not paying to commute. Next was the flexibility of having some adjustments in their work schedule and taking time for personal activities to be made up. After that was the convenience of working in a home setting and improved work-life balance. Most said they had all the tools to make it work. Only 31 percent said the support of the managers helped to make it more successful. Finally, 14 percent said staff support helped them. These barriers and enablers provided an opportunity for process improvement.

19.6.4 Results: Impact

19.6.4.1 Isolating the Effects of the Program

As the impact data were collected, the key question was how much of this improvement actually was connected to this specific project. While several methods were considered, a classic approach was used. The work-at-home group was considered an experimental group and compared to a matched group that would serve as a control group. The comparison group was matched with the experimental group on job category, length of service with the company, gender, age and marital and family status. With these multiple variables it was difficult to get a perfect match, but the team felt that there was a very good comparison between the two groups. As a backup, expert estimates were used.

19.6.4.2 Impact

The impact data were monitored and included three measures: productivity, office expense and turnover. The team decided not to value this program on absenteeism and instead left it as an intangible. Although absenteeism is probably connected to the program, the HR team thought it would be best to avoid absenteeism as a measurable objective. If there is too much focus on this measure, some employees may decide to work while sick. Table 19.6 shows

Table 19.6 Impact Data.

Business Performance	Work-at-Home Group	Comparison Group	Change	Number of Participants
Daily Claims Processed	35.4	33.2	2.2	234
Daily Claims Examined	22.6	20.7	1.9	77
Office Expense Per Person	$12,500	$17,000	$4,500	311
Annualized Turnover* (*Processors and Examiners)	9.1%	22.3%	13.2%	311

the impact data of both the experimental group and comparison group six months after the project began. The differences are significant, representing distinct improvements in the three measures and exceeding the objectives of the project. Having the data identified and isolated to the project, the analysis moves to the next step, converting data to money.

19.6.4.3 Converting Data to Money

Table 11.7 shows how each of the data sets was converted to monetary value. As the table explains, the method for converting the productivity improvement to value was using standard values. The value previously was developed by a group of experts and analysts

Table 19.7 Converting Data to Money.

Productivity improvement

- Cost (value) of processing one claim = $10.00
- Cost (value) of examining one disputed claim = $12.00
- Daily improvement = 2.2 claims per day
- Daily improvement = 1.9 disputed claims per day
- Annual value = 234 × 220 work days × 2.2 × 10.00 = $1,132,560
- Annual value = 77 × 220 days × 1.9 × 12.00 = $ 386,232

Office Expense Reduction

- Office expenses in company office: per person $17,000 annually
- Office expenses at home office: per person $12,500 first year; $3,600 2nd year
- Net improvement: $4,500, first year
- Total annual value = 311 × 4500 = $1,399,500

Turnover Reduction

- Value of one turnover statistic = $25,400
- Annual improvement related to program=41 turnovers prevented, first year
- Annual value = $25,400 × 41 = $1,041,400

Total Annual Benefits

- Productivity – processing one claim $1,132,560
- Productivity – examining one disputed claim $386,232
- Office expense reduction – $1,399,500
- Turnover reduction – $1,041,400

Total $3,959,692

in finance and accounting. The number was rounded to $10 for claims processing and $12 for claims examination. The calculation shows the annual cost saving.

Office expenses were rounded numbers taken directly from the procurement function. The projected cost for at-home employees was $13,600. However, the actual cost was rounded to $12,500 for the first year. This was compared to the annual cost to maintain the office for all employees, $17,000 per person. The first-year value included cost of a computer, desk and other items that would certainly be there as long as that person works at home. The net improvement was $4,500. The second-year value shows a cost reduction of $3,600. To be conservative, only the first-year value was used in the comparison.

For turnover reduction, several turnover cost studies were performed on jobs in the insurance industry, using the ERIC database. The cost ranged from 90 percent to 110 percent, which seemed consistent and credible to the project team. The 90 percent figure was used and when multiplied by the average salary, yielded $25,400. In all, 41 turnovers were prevented in the first year based on six months of experience. The total annual benefits for the program equal $3,959,692.

19.6.5 Costs

The costs of the entire project as developed and monitored were estimated and are shown in Table 19.8. These costs include the initial analysis to determine whether this was the proper solution, the

Table 19.8 Project Costs.

• Initial Analysis and Assessment	$ 21,000
• Forecasting Impact and ROI	$ 10,000
• Solution Development	35,800
• IT Support and Maintenance	238,000
• Administration and Coordination	213,000
• Materials (400 @ $50)	20,000
• Facilities and Refreshments – 21 meetings	12,600
• Salaries Plus Benefits for Employee and Manager Meetings	418,280
• Evaluation, Monitoring and Reporting	23,000
Total First Year Costs	**$991,680**

ROI forecast and the actual development of the solution. Most of the charges are for IT support and maintenance, administration and coordination categories.

19.7 Questions for Discussion

1. Calculate the benefit cost ratio and ROI for this project
2. Interpret these two calculations and what they mean
3. Are the results of this study credible? Explain.

19.8 FMI: PART G

19.8.1 Results: ROI and Intangibles

19.8.1.1 ROI Calculations

The ROI is calculated when the costs are totaled and the monetary benefits are tallied. Table 19.9 shows the calculation of the benefit cost ratio and the ROI. The ROI calculation at 299 percent greatly exceeds the initial objective of 25 percent. However, important results are not included in the calculation. The intangibles are critical to this study.

19.8.1.2 Intangible Benefits

Table 19.10 shows a list of the intangible benefits connected to the project. A list of expected intangibles is included on the participants' and managers' questionnaires. To compile these results, at least 10 percent of respondents had to indicate a 3, 4 or 5 on a 5-point

Table 19.9 ROI Calculations.

$$\text{BCR} = \frac{\text{Consulting Monetary Benefits}}{\text{Consulting Costs}} = \frac{\$3{,}959{,}692}{\$991{,}680} = 3.99$$

$$\text{ROI} = \frac{\text{Net Consulting Benefits}}{\text{Consulting Costs}} = \frac{\$3{,}959{,}692 - \$991{,}680}{\$991{,}680} \times 299\%$$

Table 19.10 Intangible Benefits.

- Reduced commuting time
- Reduced carbon emissions
- Reduced fuel consumption
- Reduced sick leave
- Reduced absenteeism
- Improved job engagement
- Improved community image
- Improved image as an environmental-friendly company
- Enhanced corporate social responsibility
- Improved job satisfaction
- Reduced stress
- Improved recruiting image

scale where 3 is moderate influence, 4 is significant influence and 5 is very significant influence. These are powerful intangibles, including those connected to the environment. Participants and managers can clearly see the connection. These may be the most important data sets in the minds of some executives, because it is the intangible image of helping the environment that often drives these types of projects. When these data sets are combined with the very high ROI, it is easy to see the tremendous payoff of this program.

19.8.1.3 Fuel Savings

Because the individuals involved in this program have eliminated their commute, with the exception of an occasional required visit to the office, the fuel savings were significant. The average daily commute time reduced from 104 minutes to 15 minutes. When considering the average speed (30 mph), the average miles per gallon of gasoline (20 mpg) and the cost of fuel ($3 per gallon), a savings of $1,470 per year is realized in fuel costs alone.

19.8.1.4 Carbon Emissions

From the perspective of the top executive, the principal motivating factor of this program was to reduce carbon emissions. With reduced fuel consumption, carbon emissions were consequently reduced. A total of 490 gallons of fuel per person was saved, for a total of 152,390 gallons each year. This translates into 1,478 tons of carbon emissions.

19.8.1.5 *What Makes the Results Credible*

It is important to understand what makes this data credible. It is impressive in terms of the numbers, but some specific things make it very credible.

1. The impact data improvements were taken directly from the records and are not estimated.
2. The effects of the program were isolated from other factors using a comparison group.
3. Several impact measures were not converted to money, although they have significant value.
4. All the costs were included, including the very heavy start-up cost.
5. Only the first-year values are used in the analysis. This program will have a lasting effect as long as each individual is employed with FMI.
6. When estimates were used, they were taken from the most credible sources.
7. All of the data sets and methods have buy-in from the appropriate operating executives and key managers.

There is no reason to dispute the results presented in this process.

19.9 Questions for Discussion

1. Which audiences should receive the results of this study?
2. What specific methods should be used to communicate results?
3. What specific improvements could be made to this program going forward?

Index

Also of Interest

Check out these other titles from Scrivener Publishing

Biofuels Production, Edited by Vikash Babu, Ashish Thapliyal, and Girijesh Kumar Patel, ISBN 9781118634509. The most comprehensive and up-to-date treatment of all the possible aspects for biofuels production from biomass or waste material available. *NOW AVAILABLE!*

Biogas Production, Edited by Ackmez Mudhoo, ISBN 9781118062852. This volume covers the most cutting-edge pretreatment processes being used and studied today for the production of biogas during anaerobic digestion processes using different feedstocks, in the most efficient and economical methods possible. *NOW AVAILABLE!*

Bioremediation and Sustainability: Research and Applications, Edited by Romeela Mohee and Ackmez Mudhoo, ISBN 9781118062845. Bioremediation and Sustainability is an up-to-date and comprehensive treatment of research and applications for some of the most important low-cost, "green," emerging technologies in chemical and environmental engineering. *NOW AVAILABLE!*

Sustainable Energy Pricing, by Gary Zatzman, ISBN 9780470901632. In this controversial new volume, the author explores a new science of energy pricing and how it can be done in a way that is sustainable for the world's economy and environment. *NOW AVAILABLE!*

Green Chemistry and Environmental Remediation, Edited by Rashmi Sanghi and Vandana Singh, ISBN 9780470943083. Presents high quality research papers as well as in depth review articles on the new emerging green face of multidimensional environmental chemistry. *NOW AVAILABLE!*

Energy Storage: A New Approach, by Ralph Zito, ISBN 9780470625910. Exploring the potential of reversible concentrations cells, the author of this groundbreaking volume reveals new technologies to solve the global crisis of energy storage. *NOW AVAILABLE!*

Bioremediation of Petroleum and Petroleum Products, by James Speight and Karuna Arjoon, ISBN 9780470938492. With petroleum-related spills, explosions, and health issues in the headlines almost every day, the issue of remediation of petroleum and petroleum products is taking on increasing importance, for the survival of our environment, our planet, and our future. This book is the first of its kind to explore this difficult issue from an engineering and scientific point of view and offer solutions and reasonable courses of action. *NOW AVAILABLE!*